Materials Challenges
Inorganic Photovoltaic Solar Energy

RSC Energy and Environment Series

Series Editors:
Laurence Peter, *University of Bath, UK*
Heinz Frei, *Lawrence Berkeley National Laboratory, USA*
Roberto Rinaldi, *Max Planck Institute for Coal Research, Germany*
Tim S. Zhao, *The Hong Kong University of Science and Technology, Hong Kong*

How to obtain future titles on publication:
A standing order plan is available for this series. A standing order will bring delivery of each new volume immediately on publication.

For further information please contact:
Book Sales Department, Royal Society of Chemistry, Thomas Graham House, Science Park, Milton Road, Cambridge, CB4 0WF, UK
Telephone: +44 (0)1223 420066, Fax: +44 (0)1223 420247
Email: booksales@rsc.org
Visit our website at www.rsc.org/books

Materials Challenges
Inorganic Photovoltaic Solar Energy

Edited by

Stuart J C Irvine
Centre for Solar Energy Research, OpTIC Centre, Glyndwr University,
St Asaph, UK
Email: s.irvine@glyndwr.ac.uk

THE QUEEN'S AWARDS
FOR ENTERPRISE:
INTERNATIONAL TRADE
2013

RSC Energy and Environment Series No. 12

Print ISBN: 978-1-84973-187-4
PDF eISBN: 978-1-84973-346-5
ISSN: 2044-0774

A catalogue record for this book is available from the British Library

Published by The Royal Society of Chemistry,
Thomas Graham House, Science Park, Milton Road,
Cambridge CB4 0WF, UK

Registered Charity Number 207890

For further information see our web site at www.rsc.org

Printed and bound by CPI Group (UK) Ltd, Croydon, CR0 4YY

Preface

This book provides an up-to-date account of exciting developments in thin film inorganic photovoltaic (PV) solar energy. For many years the thin film PV market was led by amorphous silicon and showed the potential for thin film products on glass substrates. This has grown rapidly over the past 10 years with new thin film PV materials going to large volume manufacture such as the polycrystalline thin film PV materials, cadmium telluride and copper indium diselenide. Amorphous silicon has also undergone a transformation with more stable and more efficient multi-junction cells.

The book is very timely because thin film PV is established in the market for large-scale solar energy production but is still small, and arguably in its infancy, compared with the predominant crystalline silicon PV module products. This has generated a wealth of research over the past 10 years to find solutions to challenges such as achieving higher conversion efficiency, greater long-term stability and reduction in manufacturing costs. The latter turns out to be important for materials research as some of the materials currently used in thin film PV might become in short supply in the future and are subject to commodity price fluctuations. For this reason the introduction, Chapter 1, includes an overview of techno-economic considerations that provide a context for the materials challenges covered in this book. The contents give an up-to-date summary of the latest research but also examine some of the fundamental considerations that underpin the technology. The fundamental considerations in thin film PV are covered in Chapter 2 and each of the remaining chapters develops different aspects of these under-pinning considerations. The chapters on absorber materials (Chapters 3, 5, 6 and 7) cover materials systems from thin film silicon through to multi-junction III–V devices. There are also chapters common to assisting all the thin film PV materials systems meeting the challenges such as transparent conducting oxides, where in Chapter 4 we have an account of the pioneering

RSC Energy and Environment Series No. 12
Materials Challenges: Inorganic Photovoltaic Solar Energy
Edited by Stuart J C Irvine
© The Royal Society of Chemistry 2015
Published by the Royal Society of Chemistry, www.rsc.org

work carried out at the US National Renewable Energy Laboratory (NREL) in Colorado, while in Chapters 8 and 9, the topics of light capture and photon management are covered. This makes a truly exciting combination of material for anyone who is studying or has interest in the application of thin film PV.

Each chapter provides a coherent and authoritative account of the topics covered. Together they provide a stimulating view of new possibilities in thin film PV to meet the challenges of increasing the adoption of PV solar energy and reducing our carbon emissions from the use of fossil fuels. These challenges include the need for both high beginning-of-life efficiency and stable long-term performance. This directly impacts the levelised cost of electricity and the carbon impact. In the long term, with the expected growth in thin film PV production, we must look to improving sustainability through the choice of abundant materials and minimising the use of materials through greater efficiency of utilisation both within the device and in manufacturing.

The idea for this book came from the excellent research collaboration in the UK on thin film PV under the Research Councils UK Energy Programme, PV SUPERGEN. This project ran for a total of eight years from 2004 to 2012 and brought together many inspiring ideas that could potentially transform thin film PV production. As with all research projects there is always so much more that could be done and new materials to explore, but the legacy of this research are the research teams that have built on these early successes. Many of the chapters of this book are authored by members of the PV SUPERGEN collaboration and I extend my gratitude to all of the 50 plus researchers who have contributed to this research. The thin film PV research community is the richer and stronger from this formative collaboration and I hope this book will be an inspiration to all those who are interested in the topic.

Stuart J. C. Irvine
Editor
Centre for Solar Energy Research, OpTIC Centre,
Glyndwr University, St Asaph, North Wales, UK

Contents

RSC Energy and Environment Series No. 12
Materials Challenges: Inorganic Photovoltaic Solar Energy
Edited by Stuart J C Irvine
© The Royal Society of Chemistry 2015
Published by the Royal Society of Chemistry, www.rsc.org

CHAPTER 1

Introduction and Techno-economic Background

STUART J. C. IRVINE[*a] AND CHIARA CANDELISE[b]

[a]Centre for Solar Energy Research, OpTIC, Glyndŵr University, St Asaph Business Park, St Asaph LL17 0JD, UK; [b]Imperial Centre for Energy Policy and Technology (ICEPT), Imperial College London, 14 Princes Gardens, London SW7 1NA, UK
*E-mail: s.irvine@glyndwr.ac.uk

1.1 Potential for PV Energy Generation as Part of a Renewable Energy Mix

Climate change became one of the major drivers for changing the balance of energy generation and supply in the latter part of the 20th century and the beginning of the 21st century. The increase in carbon dioxide (CO_2) concentration in the atmosphere over the past century to a figure approaching 400 parts per million (ppm) is taking it closer to the historical 450 ppm concentration where there was virtually no ice on the planet. The Intergovernmental Panel on Climate Change (IPCC) has set a maximum increase in global temperature of 2 °C which Hansen et al.[1] argue can only be achieved if atmospheric CO_2 falls to 350 ppm to avoid irreversible loss of the ice sheet. Meinshausen et al.[2] put a figure on cumulative CO_2 emissions into the atmosphere of 1000 Gt between 2000 and 2050 would yield a 25% probability of exceeding the 2 °C threshold in global warming.

The world electricity supply is heavily dependent on coal, gas and oil, accounting for 62% of the total for Organisation for Economic Co-operation and Development (OECD) countries in the period January to April 2012,

RSC Energy and Environment Series No. 12
Materials Challenges: Inorganic Photovoltaic Solar Energy
Edited by Stuart J C Irvine
© The Royal Society of Chemistry 2015
Published by the Royal Society of Chemistry, www.rsc.org

according to the International Energy Agency (IEA).[3] In the same period the balance was made up of 19% nuclear, 14% hydro and a mere 5% for other renewable energy such as wind, solar and geothermal. However, this small contribution from renewable energy has been increasing and was up by 1% on the same period the previous year. Vries *et al.*[4] have analysed the potential mix of wind, solar and biomass (WSB) to 2050 and concluded that this could be achieved at an energy cost of 10 US cents per kWh of energy, displacing fossil fuel electricity generation.

Although the annual growth of the photovoltaic (PV) sector has been in the range of 30 to 40% over the past 20 years, it is still at an early stage of potential development both in terms of capacity and price. A number of different scenarios exist to predict the future renewable energy mix that will displace combustion of fossil fuels.[5] For example, the World Business Council for Sustainable Development (WBCSD) predicts 50% electricity generated from renewable energy sources by 2050 with 15% generated by solar PV.[6] Other scenarios give a range for renewable energy generation from 31% from the IEA to a number of studies predicting 50%, including the German Advisory Council on Climate Change, Greenpeace and Shell's sustainable development plan. All the scenarios consider PV solar energy to be a significant part of the energy mix though the extent of penetration into the energy mix changes according to the different scenarios. Looking beyond 2050 the proportion of renewable energy and in particular PV solar energy will continue to grow and the German Solar Industry Association predicts that the proportion of PV solar electricity generation will increase to over 50% of the mix by the end of the century. In a separate study by Fthenakis *et al.*[7] which looked at the potential for combined PV and concentrator solar power (CSP) in the USA, it was predicted that all the electrical energy could be produced from the Sun combined with compressed air energy storage.

In 2011 over 25 GW of PV was installed worldwide, taking the cumulated PV installations to over 50 GW. Most of these installations are based on crystalline silicon (c-Si), but the share of thin film PV has grown over the past decade and currently stands at between 10 and 15%.

In terms of climate change there is a carbon cost in manufacture based on the dependence of electricity used in PV module manufacture on fossil fuel sources. The melting of silicon to form the c-Si requires a temperature of over 1400 °C. In contrast, thin film PV uses processing temperatures below 600 °C and therefore will require less energy. Pehnt[8] carried out a lifecycle analysis of c-Si PV module manufacture. With the current German energy mix, where there is 566 g of CO_2 per kWh of electricity, this leads to an emission of 100 g of CO_2 equivalent per kWh of electricity generated over the lifetime of the PV module. As the proportion of non-fossil fuel energy sources in the energy mix increases this could be halved to 50 g of CO_2 equivalent per kWh. This compares with around 10 g of CO_2 equivalent per kWh for onshore wind and 1.5 MW hydropower. Other factors that will reduce this carbon emission are the efficiency of solar energy conversion and processing temperatures. Thin film PV is currently less efficient than c-Si, roughly 10% compared with 15%

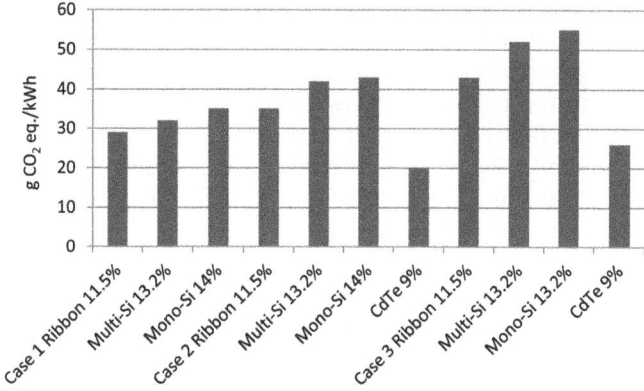

Figure 1.1 Lifecycle emissions (g CO_2 equivalent per kWh) from different types of silicon modules compared with thin film CdTe for: Case 1 – the current geographically specific production of Si; Case 2 – emissions for upstream electricity used in production in Europe; and Case 3 – emissions for equivalent production in the USA (after Fthenakis *et al.*[7]).

but the process energy per square metre is less, which leads to an overall reduction in CO_2 emission. Fthenakis *et al.*[7] estimated that less than 20 g CO_2 equivalent per kWh was emitted for 9% efficient cadmium telluride modules.

Figure 1.1 illustrates that cadmium telluride (CdTe) thin film PV is very competitive in terms of environmental emissions compared with other technologies. Recent improvements in efficiency in thin film CdTe modules to more than 12% would reduce carbon emissions to less than 15 g CO_2 equivalent per kWh. From these estimates it is clear that the adoption of thin film PV modules will make an impact on reducing carbon emissions in PV module manufacture.

In this book we examine the materials challenges for inorganic thin film PV that will influence both the environmental impact and the economic payback, as discussed later in this chapter.

1.2 Historical Development of Thin Film PV

Observation of the photovoltaic effect goes back to Becquerel, first published in 1839.[9] However, practical devices were only realised with the development of high purity silicon for semiconductor devices and the first demonstration of a silicon PV cell at Bell Labs in 1954.[10] The initial devices had a conversion efficiency of 6%, but this rapidly improved and established silicon solar cells as a source of power for the early satellites. By 1980 c-Si cells had reached 16% AM1.5 (air mass) efficiency and have continued to improve to the present day with record efficiency of 25%.[11]

Although the global PV market is dominated by c-Si, there has been rapid growth of other PV cell and module technologies that offer a very wide choice of PV materials, each with its potential advantages and disadvantages compared with the silicon benchmark. For c-Si, the development of cast multi-crystalline silicon has provided a cheaper alternative to single crystal silicon with only a small penalty in loss of efficiency. By 2004 multi-crystalline silicon single cell record efficiency had exceeded 20%.[12]

The earliest thin film PV dates back to the late 1960s with the emergence of amorphous silicon (a-Si) on glass substrates as a much cheaper and lower energy option than the crystalline silicon cells.[13,14] The first a-Si solar cell reports date back to 1976[15] but have never achieved the efficiency of the c-Si counterparts. This is discussed in more detail regarding the fundamental properties in Chapter 2 and in further detail in Chapter 3. In many respects, the history of a-Si has established a cheaper alternative to c-Si and led the way for the emergence of other thin film materials. The techno-economic trade-off between cost of manufacture and module efficiency is discussed later in this chapter and provides a context for the remainder of the book. Single junction a-Si cells have now achieved over 10% stabilised efficiency[16] and over 12% when combined in a tandem cell with a micro-crystalline junction.[17] Early applications of a-Si solar cells were seen in consumer products where the low cost and monolithic integration were important but longer term stability was not as important as for larger scale power applications. Monolithic integration of a-Si onto glass has enabled a range of architectural applications to be explored that would have been difficult or impossible to achieve with c-Si.

The origins of CdTe PV cells goes back to a $Cu_2Te/CdTe$ cell reported in 1976 by Cusano.[18] This sparked a rapid increase in the possibilities for compound semiconductor thin film PV that included, around the same time, the first interest in the Chalcopyrite structure of copper indium diselenide (CIS) with the work of Wagner *et al.*[19] on single crystal material and Kazmerski *et al.*[20] on thin film PV. Further developments on CIS thin film PV led to alloying with gallium to form CIGS where indium can be substituted with gallium to change the bandgap of the absorber. For both CIGS and CdTe cells the absorber is p-type and the preferred n-type heterojunction material has become cadmium sulphide. Although early commercialisation of thin film PV was with a-Si, the more complex polycrystalline chalcogenides have shown the potential to achieve higher module efficiencies and good long-term stability. Large-scale thin film module manufacture of CdTe and CIGS modules has been demonstrated by First Solar and Solar Frontier, respectively. The significance of manufacturing volume in the cost of module manufacture is discussed in Section 1.4 along with the significance of continual improvements of module conversion efficiency.

The past 20 years have seen the development of many alternative materials and designs for PV solar cells suitable for a range of different applications. The drive for low cost materials and low temperature processing has generated a huge amount of research in dye sensitised solar cells (DSC) and organic photovoltaics. The DSC owes its origins to photoelectrochemical cells and the origins of this go back to Becquerel.[9] The principal of these

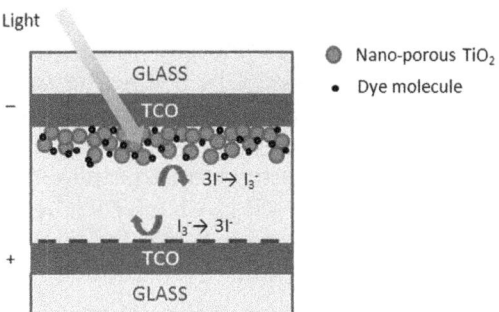

Figure 1.2 Schematic of a dye sensitised solar cell where an incident photon will excite an electron in an absorbing dye molecule which is attached to the surface of a TiO$_2$ particle that can then pass into the transparent conducting oxide (TCO). The charge neutrality of the dye is restored through the iodide redox process at the back contact.

cells[21] is the release of an electron from a suitable dye such as ruthenium organometallics into the conduction band of anatase TiO$_2$. The dye is regenerated from the counter electrode with an iodide redox couple and is shown schematically in Figure 1.2. The highest efficiency achieved for laboratory cells is achieved 11%.[22] Commercial exploitation of this highly manufacturable process has been demonstrated by G24i Power amongst others. In the case of G24i Power the DSC is deposited onto a plastic sheet in a roll to roll process. These cells perform particularly well under low light intensity and are being marketed for consumer products.

A further development of organic semiconductors as an alternative to their inorganic counterparts was the formation of polymer blends that can separate the excitons formed when light is absorbed in organic semiconductors.[23] This is essentially a room temperature process and represents a very low carbon footprint. The excitons are strongly bound so separation of the electron–hole pair is not as easy as with inorganic semiconductors and can only occur at the interfaces. This is made more difficult by the short excitonic diffusion length which is around 10 nm. Hence the success of polymer blends which create a large interface area between the electron donor and electron acceptor polymers to improve collection efficiency. These cells have shown rapid progress over the past 10 years with only 3% efficient cells in 2002 rising to 10% in 2012.[11] These cells are still very much in the research phase, but are likely to be part of the PV materials mix in future generations of PV device applications.

The highest efficiency PV devices are based on III–V epitaxial materials. The highest efficiency single junction cell is not c-Si, as one might expect, but GaAs with a world record efficiency of 28.3%,[11] not too far off the Shockley–Queisser limit described in Chapter 2. These high efficiency cells require very high quality epitaxial layers, which are achieved through lattice matching of each of the layers in the structure. By lattice matching to Ge substrates, it is possible to produce a triple junction of lattice matched layers of InGaP and InGaAs to cover the blue and red parts of the spectrum, respectively, with the infrared covered by a junction formed with the Ge substrate.[24] These triple junction cells have been developed as high performance solar cells for space but there is now increasing interest for use with concentrators (lenses or reflectors) for terrestrial power generation.[25] The III–V cells are covered in greater detail in Chapter 8. The efficiency of these cells has been increasing rapidly in the past 10 years going from 35% in 2002 to 41.6% in 2009[26] for around 500 suns concentration. In December 2012, Solar Junction beat its own previous world record of 43.5% to achieve a National Renewable Energy Laboratory (NREL) verified efficiency of 44% at 947 suns. This shows the same kind of rapid increase over the past decade as the newer organic PV at the other end of the efficiency scale and reflects the advanced semiconductor engineering of epitaxial III–V semiconductors. The challenge for thin film PV moving forward is to find a similar surge in performance. This book explores new developments that could lead to such rapid progress for thin film amorphous and polycrystalline PV.

1.3 The Role of Inorganic Thin Film PV in the Mix of PV Technologies

There is now a wide variety of PV materials and devices that appear to be competing for the same space of low cost per Watt peak (W_p). Conibeer[27] has described the generations of PV devices according to their potential to reduce the cost per W_p. This is reproduced in Figure 1.3 showing the expected trend lines for the three generations of PV materials. The first generation is the currently dominant crystalline silicon cells which are relatively high cost but also high efficiency single junction cells. The second generation is the thin film inorganic monolithic modules that are much cheaper to produce per square metre but are also lower efficiency than the crystalline silicon cells. Both first and second generations are single junction cells which have a theoretical upper limit as explained in Chapter 2. The third generation seeks to breach this single junction limit through either multi-junction cells or other devices that can capture the energy from hot electrons. The cost per square metre is similar to the second generation but achieves a lower cost per W_p through gaining higher efficiency.

Organic PV (OPV) cells are also sometimes referred to as third generation. These can be single or multi-junction but offer the potential for even lower cost per square metre than thin film inorganic PV, largely because of lower deposition and processing temperatures. The limits to OPV efficiency are not understood in the same way as for inorganic PV so how far OPV can go in terms of low cost and high efficiency is not known at this point.[28]

Inorganic thin film PV bridges between second and third generation PV. The cost per square metre is less than for crystalline silicon but has the potential to achieve the same single junction efficiency. There is also the

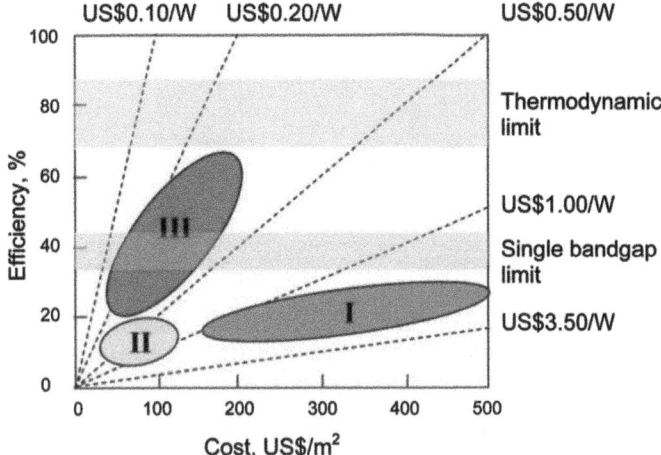

Figure 1.3 Schematic showing the relationship between cell efficiency and cost for the three generations of solar cells, with the dotted lines showing the trajectories for different cost per W_p (after Conibeer[27]).

potential for inorganic thin film PV to incorporate third generation PV concepts to achieve higher efficiency. In this respect the way has already been shown with III–V triple junction cells, but these are high quality single crystal and to truly achieve the potential of third generation PV this would have to be achieved with the lower cost fabrication of inorganic thin films.

Inorganic thin film PV has had to compete with a more mature and larger scale c-Si PV industry, but this scale is now being achieved by some of the thin film PV manufacturers and the significance of this is discussed in the next section. There are other aspects of inorganic thin film PV that can make this class of solar materials attractive over crystalline silicon. The absorber materials are direct bandgap semiconductors so requiring far less material to absorb the available solar radiation than for c-Si. This property is described in Chapter 2. Another advantage is the temperature coefficient of efficiency whereby the efficiency of all PV devices will decrease as the modules get hotter, but the coefficient for thin film PV is approximately half that of c-Si so will have advantages when being operated in hotter climates. Finally the monolithic integration of cells in a thin film module gives greater flexibility over the appearance of the module, making it look more uniform and giving it aesthetic advantages over c-Si modules for building integrated applications.[29]

An example of the potential for thin film PV in architectural design is shown in Figure 1.4, which shows the 85 kW$_p$ CIS array on the technology centre of the OpTIC building in St Asaph, North Wales. The curved façade

Figure 1.4   The 85 kW$_p$ thin film PV façade at the OpTIC building in St Asaph, North
Wales, generating 70 MWh of electricity per year. The modules are Shell
ST36 CIS and the total array area is 1176 m^2.

effect is actually created from a series of flat panels and strings are aligned
with panels of similar elevation to avoid losses as the elevation of the Sun
changes during the day. This façade has been in operation now for 10 years
and the CIS thin film panels continue to perform well with similar output to
their initial performance.

1.4   Costs of Photovoltaics and Recent PV Industry Developments

The cost of PV systems, defined as an integrated assembly of PV modules and
other components by convention called Balance of System (BOS), has been
steadily decreasing over time. There are different metrics for the costs of PV,
which can be measured in terms of:

1. Factory gate costs of individual PV system components (PV module
   and BOS components), *i.e.* the cost of producing them ($ per W$_p$). Cost
   trends and dynamics at component and, in particular, module level
   are global as they can be manufactured and traded in the global
   market and their technological development is affected by worldwide
   R&D efforts. It is important to differentiate production costs from the
   price charged to the final end customer, which is the production cost
   plus the company's mark-up (price–cost margin). Indeed, PV module
   prices are also affected by market forces such as demand–supply
   dynamics and levels of market competition, as further discussed later
   in this section.
2. Cost of investing in a PV system, *i.e.* CAPEX ($ per W$_p$ installed) made
   of PV modules cost and BOS costs. The latter refers to all PV system
   components and cost elements other than the modules, thus

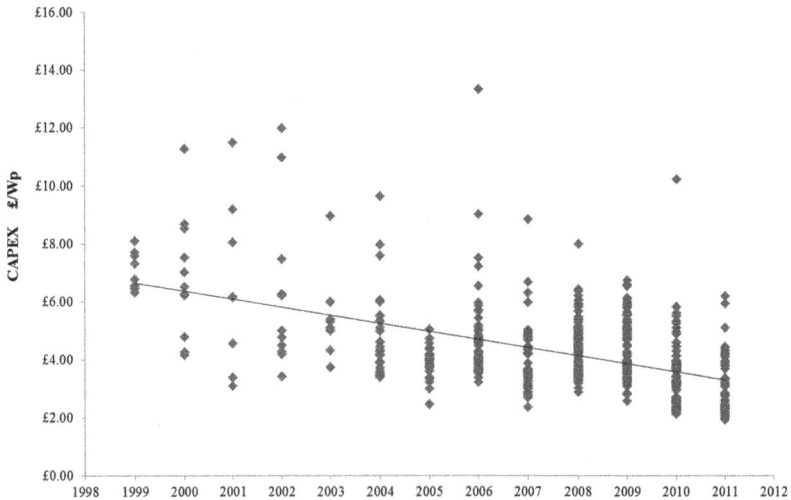

Figure 1.5 PV system price across European countries. Diamonds represent PV system price installed in several European countries over the last decade (Germany, Italy, Spain, Netherlands, Belgium, Austria, Greece, France and the UK). Data are converted to 2011 British pounds, accounting for currency exchange rates and inflation. Source: ref. 33.

including technical components such as inverter, mounting structures, cables and wiring, battery (for off-grid systems), metering (for grid-connected applications) as well as installation, design and commissioning costs. PV system CAPEX vary across market segment (and system size), system types and countries (Figure 1.5). They do not scale linearly with system size and thus tend to be higher in residential markets compared with medium size commercial systems and large utility scale systems. They also differ across countries (as affected by national market size and implementation conditions) and across PV system types (with, for example, building integrated PV systems being more expensive than standard rooftop applications). Despite this variability, PV system CAPEX has been decreasing over time across segments and countries, and is expected to further reduce (see also discussion below). Historically this has resulted from a combination of progressive reductions in module costs (discussed below) and BOS costs.[30–33]

3. Generation cost, usually calculated as the levelised cost of electricity (LCOE) ($ per kWh). LCOE is generally defined as the discounted lifetime PV system CAPEX divided by the discounted lifetime generation of the PV system. It is thus a function of initial capital cost (CAPEX), lifetime of the system, operational and maintenance costs, discount factor and location of the plant, which defines the lifetime generation of the system. LCOE varies considerably according to the type of PV system assumed (being a function of PV system CAPEX) and is very location and

country specific (as PV electricity generation is strongly dependent on climatic conditions and irradiation levels). In 2011, LCOE was estimated to range between 0.25 $ per kWh and 0.65 $ per kWh in the USA and to average around 0.203 € per kWh in Europe.[34]

The following discussion focuses on the costs of the PV module, being the major contributor to the total PV system costs (ranging from 35 to 55% depending on the PV system type and application[35,36]) and being at the centre of most of the available cost reduction and PV roadmapping literature.[37–40] Indeed, analysis and evolution of PV module production costs over time can also help to shed light on future research and technology development priorities as well as in defining successful policy support to emerging technologies and energy technology strategies.[33]

PV module prices have experienced sustained reductions over time. The price dropped from about 90 $ per W_p in the 1970s to about 5 $ per W_p in the early 2000s.[41,42] The impressive PV market growth of the last decade (worldwide cumulative installed capacity increased from 1.4 GW in 2000 to over 67 GW in 2011[43]), coupled with continuous industrial and R&D developments over time, allowed further price reductions.[33,37–40,44] In particular, PV module prices have dropped dramatically in the last couple of years, falling by about 45% between mid-2010 and March 2012 (Figure 1.6).[41]

Such impressive historical reductions reflect the development and deployment of c-Si technologies, which still account for the majority of the PV market (about 87% in 2011[45]). Indeed, as estimated by the experience curves literature, c-Si module technologies have shown an historical learning rate of the order of 20% (ranging from 18 to 22% depending on studies and reference dataset used[33]). In other words, this means a reduction of about 20% in c-Si module prices for every doubling of production capacity (see also Figure 1.7).

The recent dramatic drop in c-Si prices is due to a combination of the following factors (see ref. 33 and 46 for more details):

1. There has been a reduction in the production costs of c-Si PV driven mainly by a combination of: technological development (in particular increase in production cell efficiency) and optimisation of production processes; reduction in the cost of input materials, in particular a drop in silicon feedstock prices due to market oversupply resulting from production capacity expansion triggered by the polysilicon bottleneck experienced by the PV industry in the mid-2000s (see also below); and a massive increase in the scale of production, driven in particular by the fast expansion of the Chinese PV industry production capacity (by 2010 China accounted for 57% of worldwide capacity ramping up from only 8% in 2005[46]).

2. A sustained oversupply situation in the global PV market (annual PV production capacity has been higher than worldwide annual installation, *e.g.* in 2012 it was 50 GW compared with 29 GW in 2010[47]) which, combined with a slowdown in PV demand in key European

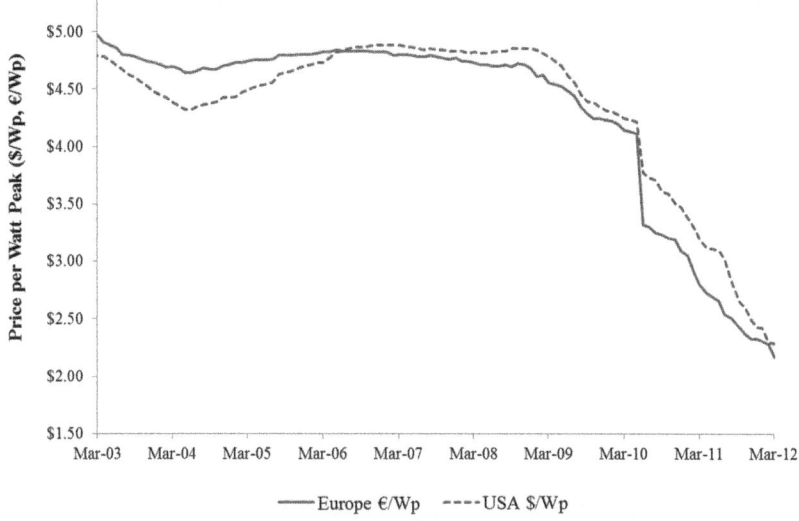

Figure 1.6 PV module retail price index, 2003–2012. The figures presented are average retail prices in Europe and the USA based on a monthly online survey. They encompass a wide range of module prices, varying according to the module technology (with thin film modules generally cheaper than c-Si), the module model and manufacturer, its quality, as well as the country in which the product is purchased. For example, in March 2012 average retail module prices were 2.29 \$ per W_p in the USA and 2.17 € per W_p in Europe, respectively, but the lowest retail price for a c-Si solar module was 1.1 \$ per W_p (0.81 € per W_p) and the lowest thin film module price was 0.84 \$ per W_p (0.62 € per W_p). Source: ref. 41.

countries (such as Germany and Italy, major drivers of PV market growth of the last decade years), has put strong downward pressure on c-Si module prices, reducing manufacturers' margins considerably and triggering worldwide industry consolidation, with several companies have gone out of business since late 2011 to date (both along the c-Si supply chain and among thin film PV manufacturers).

Crystalline silicon PV module prices vary depending on the cost structure of the manufacturer, module quality and efficiency as well as country of production and market features. Nonetheless, average c-Si module prices are reported to be 0.77 € per W_p from Germany and at 0.56 € per W_p from China in June 2013 (which compares with an average market price in Europe of 1.95 € per W_p in March 2010).[48] Production costs of c-Si also vary according to manufacturer and its cost structure, ranging between 1.03 and 0.60 € per W_p.[49,50]

Such recent developments in c-Si costs and prices were largely unexpected and not predicted by the PV cost reduction literature available.[33] The same literature had placed considerable hope in the potential of thin film technologies to deliver higher cost reductions than c-Si[37,38,51-54] for the following reasons.

- They use semiconductor materials which are better absorbers of light than c-Si, allowing much lower material thickness thus reducing costs.
- Their unit of production is more flexible and not constrained by the wafer dimensions, thus allowing larger unit of production (at least as large as a conveniently handled sheet of glass might be). This reduces manufacturing costs allowing large scale continuous production and diversity of use. Roll-to-roll deposition on stainless steel for a-Si technologies are already in production. Flexible substrates, such as stainless foils and polymer films, are even more suitable for roll-to-roll deposition.[55]
- They have the potential for lower energy use in the production process and product recovery, therefore showing a lower energy payback period than c-Si technologies.[51,55,56]

Such potential can be harnessed provided the expected increases in cell and module efficiency and large-scale production capacity are achieved.[37,57] Indeed, significant investment went into inorganic thin film PV in mid-2000s when the PV industry experienced the silicon feedstock bottleneck. This caused an increase in feedstock prices and consequent inversion of the historically negatively sloped experience curve for c-Si module prices (see Figures 1.6 and 1.7). Since then, the production capacity of thin film PV facilities has increased from few MWs to approaching 1 GW, and turnkey production lines with high cost reduction potential have been developed. One company in particular, First Solar, managed to increase its production capacity of CdTe modules from 20 MW in 2005 to above 1 GW by the late

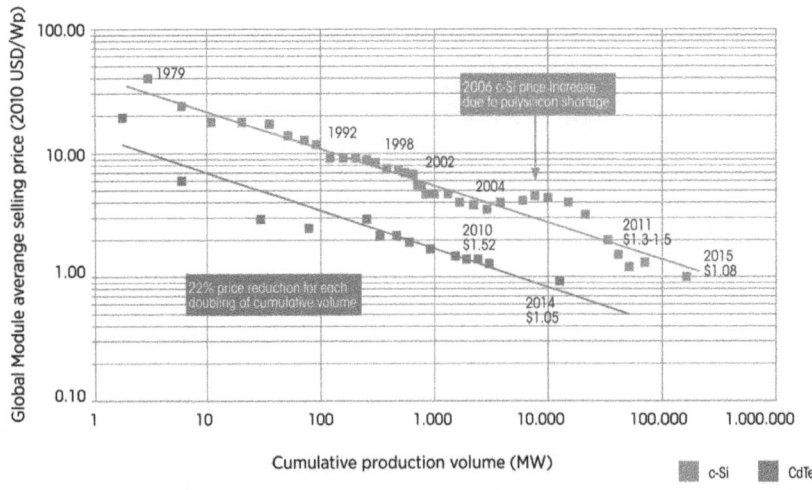

Figure 1.7   The historical PV module price experience curve, 1979–2011. Data points after 2011 are estimated. Source: ref. 50.

2000s.[58,59] It was the first manufacturer to reduce unit production costs below the 1 \$ per W_p threshold in 2009[60] (see also Figure 1.7). Similarly, Solar Frontier, a CIS manufacturer approached a production capacity of 1 GW by 2011.[61] Thin film PV modules are currently the cheapest in the market with average market prices ranging from 0.39 € per W_p for a-Si to 0.57 € per W_p for CdTe in June 2013.[48]

However, despite the considerable technological developments introduced in this chapter and further discussed in Chapters 3, 5 and 6, flat plate thin film PV have not yet achieved module efficiencies comparable with c-Si technologies. This limits their ability to compete in the worldwide market against the incumbent and more mature c-Si, as lower module efficiency makes them less suitable for area constrained applications (such as most rooftop applications) and implies higher area related BOS costs (such as cabling and mounting structures) which partially offset their lower module prices when PV system CAPEX are considered. Indeed several thin film companies have gone out of business during the market consolidation recently experienced by the worldwide PV industry. Among these are the more innovation driven companies (*e.g.* Solyndra, United Solar, Soltecture, Odersun), targeting novel and niche applications for thin film PV technologies such as on flexible substrates or semi-transparent modules suitable for, for example, building integrated solutions or rooftop applications with weight-bearing limitations.

1.5 Role of Materials Cost and Efficiency in Cost of Thin Film PV

The importance of increasing production scale for the reduction of the unit costs of thin film PV technologies is pointed out in the previous section. However, when production scale increases then the input materials cost share also increases (as shown for example for CdTe and CIGS in Figures 1.8

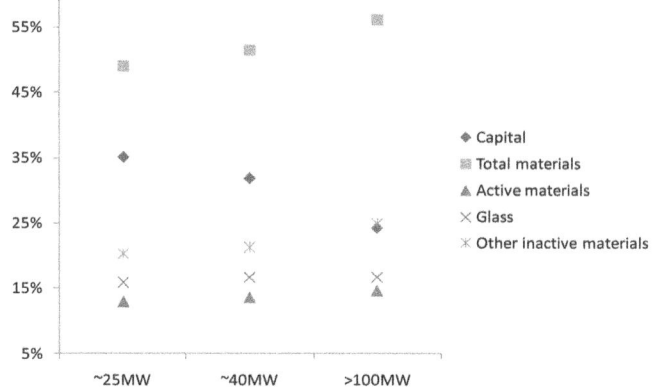

Figure 1.8 CdTe module production cost share for increasing scale of production. Source: ref. 73.

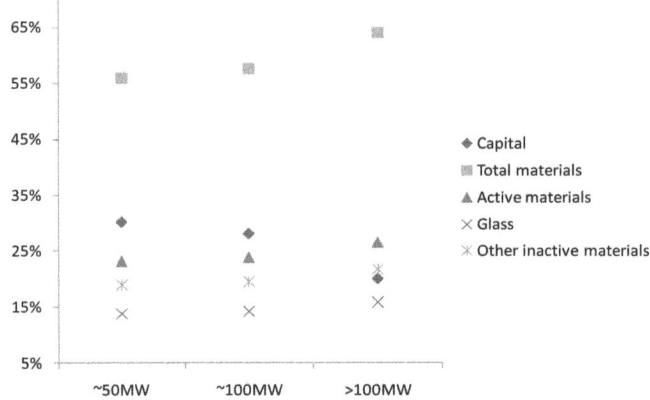

Figure 1.9   CIGS module production cost share for increasing scale of production.
Source: ref. 73.

and 1.9), indicating the need to optimise materials use in order to achieve further cost reductions. An efficient use of input materials is particularly relevant for CdTe and CIGS technologies, as concerns over scarcity (and consequent high price) of key active materials, tellurium (Te) and indium (In), have been highlighted as potential barriers to future market expansion and cost reductions of those thin film PV technologies relying on them. Indeed, due to increased demand for both materials over the past decade, Te and In prices have experienced increased volatility since mid-2000s, causing unprecedented price highs. Driven by exponential demand coming from the liquid crystal display (LCD) screen industry, the price of indium increased dramatically from 85 $ per kg at the beginning of 2003 to 830 $ per kg by the end of 2004,[62] then rose again above 750 $ per kg in 2011[63] after years of volatility. Similarly, the tellurium price experienced an increase of more than 300% in 2005, reaching an annual average price of around 230 $ per kg (from about 35 $ per kg in 2004), due to supply shortfall caused by demand increases from China and solar cell manufacturers outside China.[64] Tellurium prices have been fluctuating since, reaching values above 200 $ per kg in 2010[65] and above 400 $ per kg in 2011 (the average price in late 2011 was around 300 $ per kg).[66]

In the past decade several contributions in the literature have attempted to estimate the potential of thin film PV technologies using scarce materials such as Te and In to expand their production capacity in the future and to contribute to the global warming mitigation challenge. A review of such literature points out the differences in methods and assumptions taken by the various contributors as well as the uncertainties behind critical materials resource assessment and currently available figures for Te and In production and reserves.[67] In particular, the potential to expand production of In and Te is unclear, since data are poor and reporting has been reduced. Resource data are largely absent and the economics of production is complicated by the fact that Te and In are mostly extracted as byproducts of other primary metals, *i.e.* copper and zinc. Moreover, future In and Te demand coming from thin film PV is also subject to uncertainties as materials usage in a PV cell can vary and demand will also depend on future expansion of the PV sector as well as the relative share of thin film PV technologies in the overall PV technology mix. Such uncertainties translate into a wide range of estimates of the impact of Te and In supply constraints on potential future expansion of CdTe and CIGS.[67] The literature review highlighted how, in order to reach conclusive answers on such potential constraint, more analysis and research is needed to better estimate future availability of In and Te. Future availability scenarios should take into account the temporal and economic nature of In and Te extraction and recovery, and their future demand and production expansion should be better understood. Indeed, the relevant literature has been increasingly trying to address such questions.[68–71]

Thus, whether CdTe and CIGS technologies can be considered sustainable solutions in the long term under scenarios of low availability of In and Te materials and very high PV market growth, or rather medium term stepping stones for other PV technologies to come is still an open question. The answer is linked to In and Te future production expansion, to PV market developments and to the future mix of PV technologies available to satisfy PV demand growth. This includes the development of thin film PV technologies based on alternative more abundant materials. Among those, kesterite-based thin film devices such $Cu_2ZnSn(S,Se)_4$ (CZTSSe) and Cu_2ZnSnS_4 (CZTS) solar cells, where indium and gallium are replaced by the readily available materials zinc (Zn) and tin (Sn). This is discussed in detail in Chapter 6.[72]

Nevertheless, recent evidence on In and Te availability and the comparison of estimated maximum annual production achievable by CdTe and CIGS against forecasts of future PV market size seem to indicate that the availability of In and Te is unlikely to constrain CdTe and CIGS technologies *per se* in their ability to scale up production and to supply a significant proportion of future PV market growth.[67,70,71] However, a possible cause of concern for CdTe and CIGS is the impact of an escalation of In and Te price on their production costs and their cost-competitiveness in the wider PV market.[68–70,73–75]

Recent contributions have assessed the impact on CdTe and CIGS production costs of increases in In and Te prices. The absolute maximum In and Te prices which keep CdTe and CIGS technologies cost competitive is debated in the literature, as the calculations vary over time and across models due to high variability of materials prices themselves as well as of the production cost structure assumed (other production cost drivers such as, for example, the prices of other materials, cost of capital and production capacity utilisation also change over time). However, the estimates available clearly show how active materials account for an increasing production cost share of CdTe and CIGS when In and Te prices increase (see, for example,

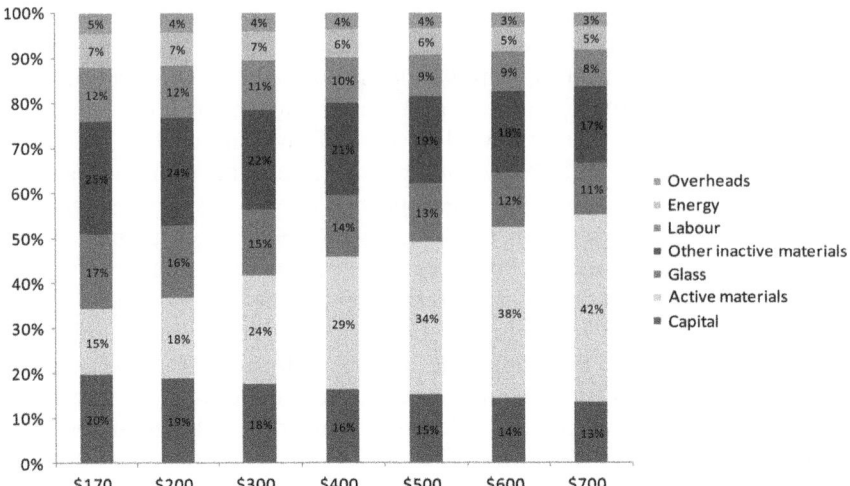

Figure 1.10   CdTe cost breakdown for increasing technical grade tellurium price
(from 170 $ per kg to 700 $ per kg). Note changes over baseline case
scenario of 0.75 $ per W_p production cost, 170 $ per kg tellurium
price, 11.6% efficiency and large-scale production. Source: ref. 73.

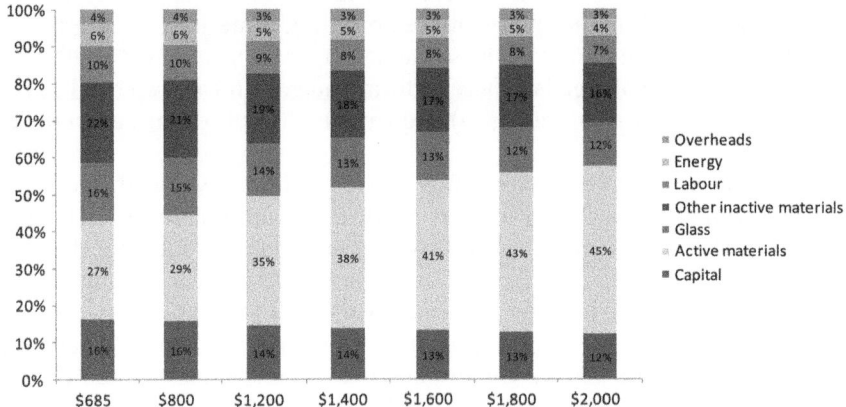

Figure 1.11   CIGS-planar cost breakdown for increasing technical grade indium
price (from 685 $ per kg to 2000 $ per kg). Note changes over
baseline case scenario of 0.97 $ per W_p production cost, 685 $ per
kg indium price, 11% efficiency and large-scale production. Source:
ref. 73.

Figures 1.10 and 1.11) with consequent impact on unit production costs (as
shown in Figure 1.12 for CdTe—in this analysis 0.75 $ per W_p has been
assumed as the baseline unit production cost).[73] However, this impact can be
eased by reducing the intensity of the cell's materials. This can be achieved
by reducing the thickness of the active layer, by improving material

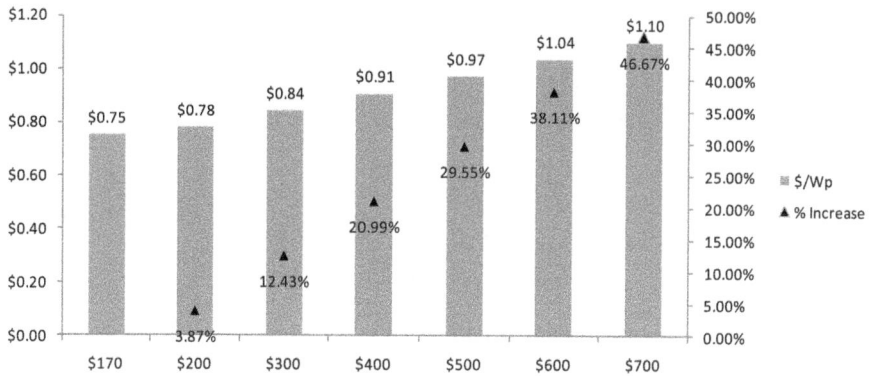

Figure 1.12 CdTe module production cost ($ per W_p) for increasing technical grade tellurium price (from 170 $ per kg to 700 $ per kg). Note percentage production cost increase over baseline case scenario cost of 0.75 $ per W_p, under tellurium price of 170 $ per kg. Source: ref. 73.

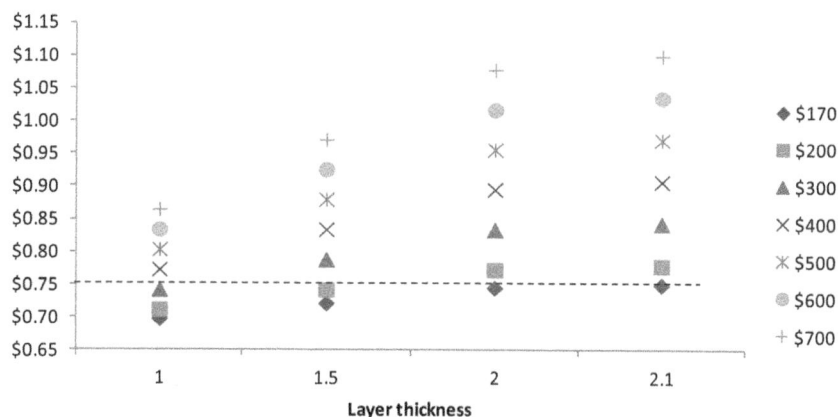

Figure 1.13 CdTe production cost ($ per W_p) for varying layer thickness (from 1 to 2 μm) and tellurium price (from 170 $ per kg to 700 $ per kg). Source: ref. 73.

utilisation in the production process (improving both the effectiveness of deposition processes and the recovery and recycling of the material) and by increasing a cell's efficiency. For example, Figures 1.13–1.15 show how changes in these parameters can help ease the impact of increasing Te price on CdTe production costs (again, against a baseline production costs of

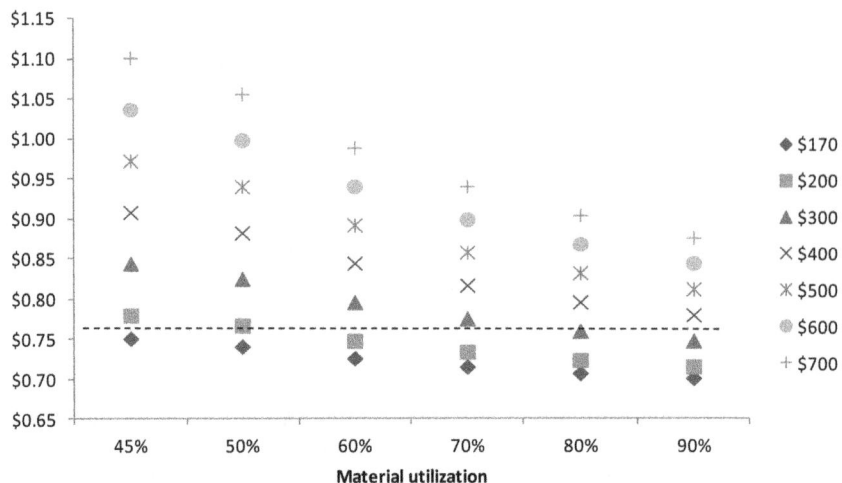

Figure 1.14 CdTe production cost ($ per W_p) for varying material utilization (from 45% to 90%) and tellurium price (from 170 $ per kg to 700 $ per kg). Source: ref. 73.

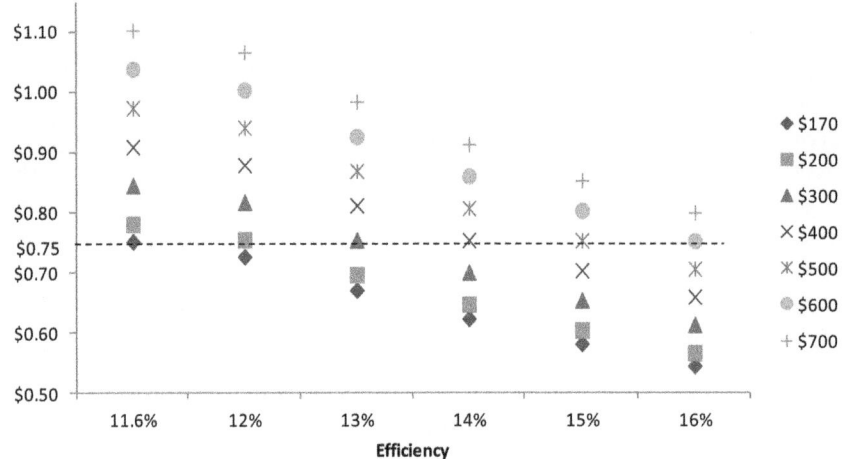

Figure 1.15 CdTe production cost ($ per W_p) for varying module efficiency (up to 16%) and tellurium price (from $170 per kg to $700 per kg). Note changes over baseline case scenario of production cost 0.75 $ per W_p, 170 $ per kg tellurium price, 11.6% efficiency and large-scale production. Source: ref. 73.

0.75 $ per W_p). Similarly, recent contributions show that thin film technologies have significant room to absorb potential critical materials price increases and expand their total potentially available supply base if improvements in their net material intensity are achieved.[75,76] These are covered in Chapters 5 and 6.

1.6 Future Prospects for Cost Reduction and Thin Film PV

Estimates for future cost reductions in PV systems vary according to the source, market segment and country of reference, but they all indicate further decline over time. The European Photovoltaic Industry Association (EPIA) predicts European PV system prices will fall by 36–51% over the next 10 years (see Figure 1.6) and the PV generation cost (LCOE) to decline by around 20% by 2020.[34] Recent US estimates see the utility scale PV system price to decrease to between 1.71 and 1.91 $ per W_p and that for residential systems to 2.29 $ per W_p.[77] The average annual reduction of BOS costs has been estimated to be in the range of 8–9.5%.[78] Grid parity is expected to be achieved in southern Europe (and in high electricity price countries such as Italy) by the end of 2013 and spread across all Europe by 2020.[34] Similarly, PV LCOE is estimated to compete with residential electricity prices in a wide range of US regions by 2020 and grid parity to be achieved in high-cost regions by 2015.[50]

In terms of PV modules both c-Si and thin film technologies are expected to experience further cost reductions. Crystalline silicon module production costs are expected to decrease thanks to increased efficiency and scale of manufacturing as well as process optimisation and reductions in input materials costs. Average production costs figures for c-Si are estimated to be in the 0.85 to 0.50 $ per W_p range.[49,50] Similarly, production costs of thin film PV technologies are also expected to go down further as shown in Table 1.1.

As also introduced in previous sections, increasing scale is crucial for PV technologies to reach lower unit production costs, particularly for thin film PV. This is, for example, clearly shown in Figure 1.16, where CIGS cost projections are presented for different levels of scale.

Increase in cell efficiency is also an important driver for future cost reductions, as it translates into a higher output per area, thus reducing the specific material consumption. As a rule of thumb, an increase in efficiency of 1% is able to reduce costs per W_p by 5–7%.[37] This challenge is particularly relevant to thin film PV, which needs to bridge the gap with c-Si module efficiency to improve its competitiveness in the flat plate modules global market. This would expand the range of applications that thin film PV would become suitable for (including rooftops) and reduce the BOS costs. As a rule of thumb it is estimated that a 1% increase in efficiency reduces BOS cost by between 0.07 and 0.1 $ per W_p.[79] Cell and module efficiency

Table 1.1 Cost reduction potential of thin film PV ($ per W_p)[a]

	2010	2015
Single junction a-Si	0.99	0.55
Tandem junction Si	1.32	0.58
CIGS – co evaporation	1.31	0.63
CIGS – sputtering	1.31	0.69
CdTe	0.73	0.49

[a]Source: ref. 50.

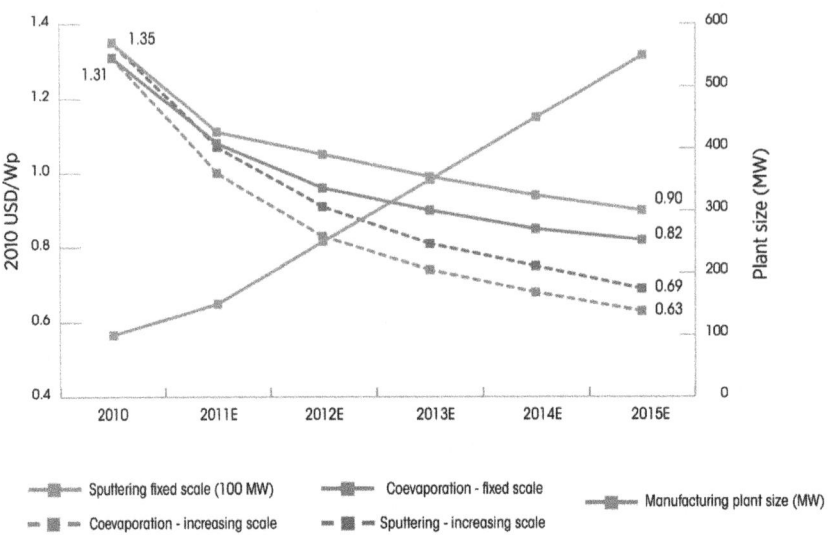

Figure 1.16 Projections of CIGS production costs as function of manufacturing scale, 2010–2015. Source: ref. 50.

increases are expected for both c-Si and thin film PV, with CIGS showing high potential to achieve efficiency levels comparable with c-Si, as discussed in Chapter 6.

Overall, despite the significant technological improvements, production capacity investments and cost reductions that have been achieved, thin film PV has not yet fully harnessed its potential to capture a relevant PV market share. Challenges still lie ahead for thin film PV to improve its competitiveness with respect to c-Si for flat plate applications. Moreover, the potential of thin film PV to deliver innovative devices and systems (such as flexible modules or PV glass), and thus product differentiation, still needs to be fully harnessed. It remains unclear what will be the future prospects for such niche (and more costly, at least initially) PV products, in particular given the consolidation the global PV industry is currently experiencing. However, niche applications such as building integrated photovoltaics are promising emerging market segments where thin film PV is likely to be better placed to deliver.

1.7   Outline of Book and Context of Topics in Terms of Techno-economic Background

This chapter sets the scene for thin film PV in a world dominated still by crystalline silicon PV modules and many new emerging PV technologies. The challenge for any of these PV technologies is to have the potential for:

- high efficiency;
- low materials costs;

- low production costs;
- low embodied energy and low equivalent greenhouse gas emissions;
- good long-term stability.

These are the challenges that form the basis for this book and are addressed for each of the materials systems considered for thin film PV production. In Chapter 2 the fundamentals of thin film PV are discussed and their relevance for achieving high absorption of solar radiation, high efficiency and thin film structures explained. Chapter 3 covers the range of thin film silicon PV cells and expands on the themes of absorption of solar radiation, improving cell efficiency and the challenges of achieving stable performance over a period of many years.

Transparent conducting oxides (TCOs) are common to all thin film PV device structures and in some respects are the most important part of the device, performing the dual role of high transparency and high electrical conductivity. Without this the PV device efficiency would be severely limited. Chapter 4 on this topic is set in the context of the developments at NREL in the USA, which has been a leader in the science and technology of TCOs, and provides insights to the choices that need to be made in selecting and developing TCO thin films for different PV structures.

The leading commercial thin film PV material, CdTe, is described in detail in Chapter 5. This chapter covers some of the historical developments in CdTe solar cells and goes on to look at some of the advanced techniques that can give a greater flexibility in performance and PV module design.

Chapter 6 looks at alternative chalcogenide materials such as the kesterites which replace the high cost In and Ga with lower cost Sn and Zn. In this chapter some techniques are described on how to explore a large number of different phases in discovering new thin film absorber materials.

III–V solar cells, the topic of Chapter 7, are not strictly thin film PV devices though they are fabricated from very high quality thin film materials and represent the pinnacle of achievement not only for compound semiconductors but also for any solar cell material. Recent progress has seen multi-junction cells achieve conversion efficiency under solar concentration of over 40% and this chapter explores the path to achieving over 50% conversion of solar energy. There is a lot to learn in improving PV cell efficiency in the lower cost thin film PV technologies from these very high performance cells.

Chapter 8 makes a comprehensive analysis of the role of light capture in thin film PV and is a key part of the materials challenges as this affects the whole thin film PV structure and the surfaces. The microstructuring and passivation of surfaces is discussed before the chapter moves on to more advanced surface structures involving nano-materials and nano-structures.

The theme of light capture is continued in Chapter 9 where the topic of photon management is covered. Again, the objective is to capture more of the solar radiation into the thin film structure than would be achieved in a conventional thin film PV device. The approaches covered here include fluorescent materials to achieve photon energy conversion either from short wavelengths to longer wavelengths or from the infrared to the visible. The objective is to capture the solar radiation at wavelengths not normally captured by the thin film PV device. Thus a wider range of the solar spectrum can be captured leading to higher efficiency, one of the challenges set in the techno-economic analysis in this chapter.

This book brings together some of the exciting innovations in thin film PV materials and presents challenges for future generations of high efficiency, low cost and highly stable PV module technology. Much of the background to this book and the contributing authors come from the UK research programme on thin film PV, PV21, which has established new approaches and fertile areas of research to enable thin film PV to make a major contribution to future global renewable energy demand.

References

1. J. Hansen, M. Sato, P. Kharecha, D. Beerling, R. Berner, V. Masson-Delmotte, M. Pagani, M. Raymo, D. L. Royer and J. C. Zachos, *Open Atmos. Sci. J.,* 2008, **2**, 217–231.
2. M. Meinshausen, N. Meinshausen, W. Hare, S. C. B. Raper, K. Frieler, R. Knutti, D. J. Frame and M. R. Allen, *Nature,* 2009, **458**, 1158–1163.
3. IEA, *IEA World Energy Outlook 2012*, International Energy Agency, Paris, 2012, available from http://www.worldenergyoutlook.org/publications/weo-2012/.
4. B. J. M. de Vries, D. P. van Vuuren and M. M. Hoogwijk, *Energy Policy,* 2007, **35**, 2590–2610.
5. E. Martinot, *Renewables 2007 Global Status Report*, REN21 Renewable Energy Policy Network, Paris, 2007, available from http://www.ren21.net/Resources/Publications/REN21Publications.aspx.
6. WBCSD, *Pathways to 2050 Energy & Climate Change*, World Business Council for Sustainable Development, Geneva, 2005.
7. V. Fthenakis, J. E. Mason and K. Zweibel, *Energy Policy,* 2009, **37**, 387–399.
8. M. Pehnt, *Renewable Energy,* 2006, **31**, 55–71.
9. A. E. Becquerel, *C. R. Acad. Sci.,* 1839, **9**, 561.
10. D. M. Chapin, C. S. Fuller and G. L. Pearson, *J. Appl. Phys.,* 1954, **25**, 676.
11. M. A. Green, K. Emery, Y. Hishikawa, W. Warta and E. D. Dunlop, *Prog. Photovoltaics,* 2012, **20**, 12–20.
12. O. Schultz, S. W. Glunz and G. P. Willeke, *Prog. Photovoltaics,* 2004, **12**, 553–558.
13. R. C. Chittick, J. H. Alexander and H. F. Sterling, *J. Electrochem. Soc.,* 1969, **116**, 77.
14. W. E. Spear and P. G. LeComber, *J. Non-Cryst. Solids,* 1972, **810**, 727.
15. D. Carlson and C. Wronski, *Appl. Phys. Lett.,* 1976, **28**, 671.

16. S. Benagli, D. Borrello, E. Vallat-Sauvain, J. Meier, U. Kroll, J. Hötzel, J. Spitznagel, J. Steinhauser, L. Castens and Y. Djeridane, High-efficiency amorphous silicon devices on LPCVD-ZNO TCO prepared in industrial KAI-M R&D reactor, presented at 24th European Photovoltaic Solar Energy Conference, Hamburg, September 2009.

17. A. Banerjee, T. Su, D. Beglau, G. Pietka, F. Liu, G. DeMaggio, S. Almutawalli, B. Yan, G. Yue, J. Yang and S. Guha, High efficiency, multi-junction nc-Si:H based solar cells at high deposition rate, presented at 37th IEEE Photovoltaic Specialists Conference, Seattle, June 2011.

18. D. A. Cusano, *Solid State Electron,* 1963, **6**, 217.

19. S. Wagner, J. L. Shay, P. P. Migliorato and H. M. Kasper, *Appl. Phys. Lett.,* 1974, **25**, 434.

20. L. L. Kazmerski, F. R. White and G. K. Morgan, *Appl. Phys. Lett.,* 1976, **46**, 268.

21. M. Gratzel, *Nature,* 2001, **414**, 338–344.

22. N. Koide, R. Yamanaka and H. Katayam, Recent advances of dye-sensitized solar cells and integrated modules at SHARP, *MRS Proc.,* 2009, **1211**, 1211-R12-02.

23. J. Nelson, *Curr. Opin. Solid State Mater. Sci.,* 2002, **6**, 87–95.

24. W. Guter, J. Schone, S. P. Philipps, M. Steiner, G. Siefer, A. Wekkeli, E. Welser, E. Oliva, A. W. Bett and F. Dimroth, *Appl. Phys. Lett.,* 2009, **94**, 223504.

25. R. R. King, D. C. Law, C. M. Fetzer, R. A. Sherif, K. M. Edmondson, S. Kurtz, G. S. Kinsey, H. L. Cotal, D. D. Krut, J. H. Ermer and N. H. Karam, presented at 20th European Photovoltaic Solar Energy Conference, Barcelona, Spain, June 2005.

26. R. R. King, A. Boca, W. Hong, X.-Q. Liu, D. Bhusari, D. Larrabee, K. M. Edmondson, D. C. Law, C. M. Fetzer, S. Mesropian and N. H. Karam, Band-gap-engineered architectures for high-efficiency multijunction concentrator solar cells, presented at 24th European Photovoltaic Solar Energy Conference and Exhibition, Hamburg, Germany, 21–25 September, 2009.

27. G. Conibeer, *Mater. Today,* 2007, **10**, 42–50.

28. P. K. Nayak, G. Garcia-Belmonte, A. Kahn, J. Bisquert and D. Cahen, *Energy Environ. Sci.,* 2012, **5**, 6022–6039.

29. J.-H. Yoon, J. Song and S.-J. Lee, *Sol. Energy,* 2011, **85**, 723–733.

30. G. Barbose, N. Darghouth and R. Wiser, *Tracking the Sun V. An Historical Summary of the Installed Price of Photovoltaics in the United States from 1998 to 2011*, Report No. LBNL-5919E, Lawrence Berkeley National Laboratory, Berkeley, CA, 2012.

31. G. Barbose, N. Darghouth, R. Wiser and J. Seel, *Tracking the Sun IV. An Historical Summary of the Installed Cost of Photovoltaics in the United States from 1998 to 2010*, Report No. LBNL-5047E, Lawrence Berkeley National Laboratory, Berkeley, CA, 2011.

32. R. Wiser, G. Barbose and C. Peterman, *Tracking the Sun. The Installed Cost of Photovoltaics in the U.S. from 1998–2007*, Lawrence Berkeley National Laboratory, Berkeley, CA, 2009.
33. C. Candelise, M. Winskel and R. Gross, *Renewable Sustainable Energy Rev.,* 2013, **26**, 96–107.
34. EPIA, *Solar Photovoltaics Competing in the Energy Sector: On the Road of Competitiveness*, European Photovoltaic Energy Association, Brussels, 2011.
35. Ernst & Young, *The UK 50 kW to 5 MW Solar PV Market*, Ernst & Young, London, 2011.
36. Parsons Brinckerhoff, *Solar PV Cost Update*, Department of Energy and Climate Change, London, 2012.
37. EU PV Technology Platform, *A Strategic Research Agenda for Photovoltaic Solar Energy Technology*, Report prepared by Working Group 3 'Science, Technology and Applications' of the EU PV Technology Platform, Office for Official Publications of the European Communities, Luxemburg, 2007.
38. IEA, *Technology Roadmap. Solar Photovoltaic Energy*, International Energy Agency, Paris, 2010.
39. G. F. Nemet, *Energy Policy,* 2006, **34**, 3218–3232.
40. W. van Sark, G. F. Nemet, G. J. Schaeffer and E. A. Alsema in *Technological Learning in the Energy Sector: Lessons for Policy, Industry and Science*, ed. M. Junginger, Edward Elgar, Cheltenham, 2010, ch. 8.
41. Solarbuzz, Solarbuzz module price survey [online], available at http://solarbuzz.com/facts-and-figures/retail-price-environment/module-prices [accessed January 2013].
42. PV module price data obtained directly from correspondence with Paul Maycock, 2011.
43. EPIA, *EPIA Market Report 2011*, European Photovoltaic Industry Association, Brussels, 2012.
44. A. Jäger-Waldau, *PV Status Report 2012*, European Commission, Joint Research Centre, Institute for Energy and Transport, EUR 25749 EN, Publications Office of the European Union, Luxembourg, 2012.
45. Photon International, *Market survey. Cell and module production 2011.* Photon International, the photovoltaic magazine, 2012. March 2012, pp. 132–161.
46. N. Marigo and C. Candelise, *Economia e Politica Industriale,* 2013, **40**, 5–41.
47. Q-Cell, *Q-Cell Extraordinary General Meeting*, Q-Cell PowerPoint presentation, 9 March, 2012, Leipzig, 2012.
48. PvXchange, PV price index [online], available at http://www.pvXchange.com [accessed August 2013].
49. P. Fath, A. Ramakrishman, N. Rosch and W. Herbst, Cost reduction of crystalline silicon solar modules by advanced manufacturing, presented at 27th European Photovoltaic Solar Energy Conference, Frankfurt, 24–27 September, 2012.

50. IRENA, *Solar Photovoltaics*, IRENA Working Paper, Renewable Energy Technologies: Cost Analysis Series, Volume 1: Power Sector, Issue 4/5, International Renewable Energy Agency, Bonn, 2012.
51. S. Hegedus, *Prog. Photovoltaics*, 2006, **14**, 393–411.
52. T. Surek, *Solar Electricity: Progress and Challenges*, presentation, National Renewable Energy Laboratory, Golden, CO, 2004.
53. K. Zweibel, *The Terawatt Challenge for Thin-film PV*, NREL Technical Report, NREL/TP-520-383502005, National Renewable Energy Laboratory, Golden, CO, 2005.
54. K. Zweibel, *Sol. Energy Mater. Sol. Cells*, 2000, **63**, 375–386.
55. K. L. Chopra, P. D. Paulson and V. Dutta, *Prog. Photovoltaics*, 2004, **12**, 69–92.
56. L. L. Kazmerski, *J. Electron Spectrosc. Relat. Phenom.*, 2006, **150**(2–3), 105–135.
57. NREL, *National Solar Technology Roadmap: Wafer-Silicon PV*, Management Report NREL/MP-520-41733, National Renewable Energy Laboratory, Golden, CO, 2007.
58. First Solar website, www.firstsolar.com [accessed July 2013].
59. M. A. Green, *Prog. Photovoltaics*, 2011, **19**(4), 498–500.
60. First Solar, First Solar Overview: First Solar Datasheet Q3 2011, March 2012, available at http://www.firstsolar.com/Downloads/pdf/FastFacts_PHX_NA.pdf [January 2013].
61. J. Plastow, *Solar Frontier and CIS*, presented at 3rd Thin Film Solar Summit Europe, Berlin, 3–4 March, 2011.
62. J. F. Carlin, Indium, in *US Geological Survey Minerals Yearbook 2004*, US Geological Survey, Reston, VA, 2004, pp. 36.1–36.4, available at http://minerals.usgs.gov/minerals/pubs/commodity/indium/index.html#mcs.
63. MetalPrice, Indium 99.99% price [online], www.metalprice.com, 2012.
64. M. W. George, Selenium and Tellurium, in *US Geological Survey 2005 Minerals Yearbook*, US Geological Survey, Reston, VA, 2005, pp. 65.1–65.10, available at http://minerals.usgs.gov/minerals/pubs/commodity/selenium/index.html#mcs.
65. USGS, *Mineral Commodity Summaries 2010*, US Geological Survey, Reston, VA, 2010, available at http://minerals.usgs.gov/minerals/pubs/mcs/.
66. MetalPrice, Tellurium 99.99% price. www.metalprice.com, 2012.
67. C. Candelise, J. F. Speirs and R. J. K. Gross, *Renewable Sustainable Energy Rev.*, 2011, **15**(9), 4972–4981.
68. V. Fthenakis, *Renewable Sustainable Energy Rev.*, 2009, **13**(9), 2746–2750.
69. M. A. Green, *Prog. Photovoltaics*, 2009, **17**(5), 347–359.
70. M. A. Green, *Sol. Energy Mater. Sol. Cells*, 2013, **119**, 256–260.
71. Y. Houari, J. Speirs, C. Candelise and R. Gross, *Prog. Photovoltaics*, 2014, **22**, 129–147.
72. D. B. Mitzi, O. Gunawan, T. K. Todorov, K. Wang and S. Guha, *Sol. Energy Mater. Sol. Cells*, 2011, **95**, 1421–1436.
73. C. Candelise, M. Winskel and R. Gross, *Prog. Photovoltaics*, 2012, **20**, 816–831.

74. C. Wadia, A. P. Alivisatos and D. M. Kammen, *Environ. Sci. Technol.*, 2009, **43**, 2072–2077.

75. M. Woodhouse, A. Goodrich, R. Margolis, T. James, R. Dhere, T. Gessert, T. Barnes, R. Eggert and D. Albin, *Sol. Energy Mater. Sol. Cells*, 2013, **115**, 199–212.

76. M. Woodhouse, A. Goodrich, R. Margolis, T. L. James, M. Lokanc and R. Eggert, *IEEE J. Photovoltaics*, 2013, **3**(2), 833–837.

77. A. Goodrich, T. James and M. Woodhouse, *Residential, Commercial, and Utility-Scale Photovoltaic (PV) System Prices in the United States: Current Drivers and Cost-Reduction Opportunities*, NREL Technical Report NREL/TP-6A20-53347, National Renewable Energy Laboratory, Golden, CO, 2012.

78. S. Ringbeck and J. Sutterlueti, *Prog. Photovoltaics*, 2013, **21**, 1411–1428.

79. T. Surek, *The Race to Grid Parity: Crystalline Silicon vs Thin Films*, Surek PV Consulting, Denver, CO, 2010.

CHAPTER 2

Fundamentals of Thin Film PV Cells

STUART J. C. IRVINE* AND VINCENT BARRIOZ

Centre for Solar Energy Research, OpTIC, Glyndŵr University,
St Asaph Business Park, St Asaph LL17 OJD, UK
*E-mail: s.irvine@glyndwr.ac.uk

2.1 Introduction

This chapter discusses some of the fundamental properties of photovoltaic (PV) materials and the relationship to conversion of solar radiation into electricity. A common denominator to all PV solar cells is the figure of merit, 'conversion efficiency'. This is the ratio of the electrical output power to the available power from the incident solar radiation. Determining the incident solar radiation is not trivial and under outdoor conditions both the spectrum and intensity of solar radiation will change with time of day, latitude and weather conditions. Thus, standard irradiation conditions have been established for the purpose of efficiency measurement and this are discussed in Section 2.1.1. The chapter goes on to consider the fundamental limits to conversion efficiency for a single junction cell and then how higher efficiencies might be achieved. The relationship between the PV device and the constituent materials is considered in some detail from the solar absorption characteristics through to the effect of electrical defects in practical PV devices.

RSC Energy and Environment Series No. 12
Materials Challenges: Inorganic Photovoltaic Solar Energy
Edited by Stuart J C Irvine
© The Royal Society of Chemistry 2015
Published by the Royal Society of Chemistry, www.rsc.org

2.1.1   The Sun and Solar Energy

Everyone on Earth has a 'local' primary source of energy about 8 minutes away, at the speed of light, from wherever one may be situated. This primary source of energy comes from the Sun and is produced from a nuclear fusion reaction, generating a massive luminosity of 4×10^{20} MW. The Sun is safely situated 1.5×10^{8} km away from Earth. In terms of potential from solar energy, while the world net electricity consumed in 2010 was 20.2 trillion kWh,[1] the solar energy falling onto the Earth's surface over the same period would be around 50 000 times greater.[2] At its surface, the Sun has a black body temperature of 6000 K following Planck's Law for radiated power density. The spectral emission of solar radiation, or more commonly known as spectral irradiance, occurs for an electromagnetic radiation range from the ultraviolet through the visible to the infrared (up to a wavelength 4 µm into the mid-infrared range) as shown in Figure 2.1.[3] Integration over the full solar irradiance spectrum gives a power density equal to 1366 W m^{-2}. This power density is true for extraterrestrial solar irradiance spectra at the zenith, where the photons have not yet entered the Earth's atmosphere.

An important notion in solar energy is the solar irradiance in relation with the air mass (AM) reference relating to the optical path where the light, radiated from the sun, has to travel to the Earth's surface. AM0 and AM1 refer to the light incident at the zenith above the atmosphere and at sea level, respectively. For different angles (θ) from zenith, the air mass can be calculated:

$$AM = \frac{1}{\cos \theta} \tag{2.1}$$

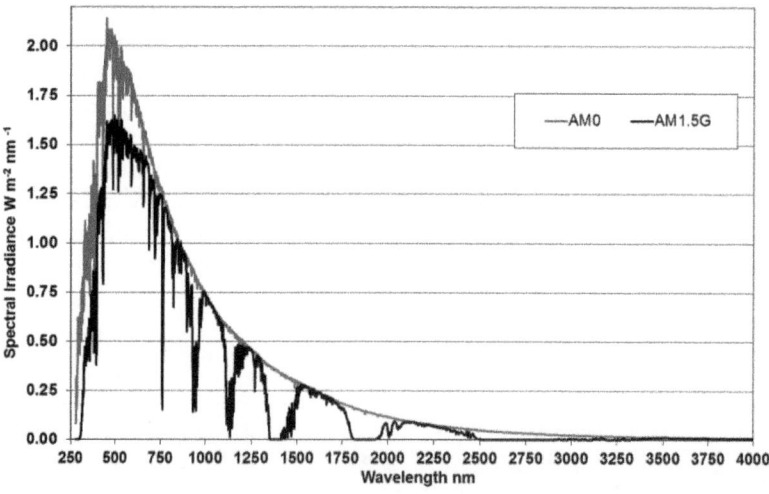

Figure 2.1   Solar irradiance at AM0 (in space) and AM1.5G (at Earth's surface).[3]

Figure 2.1 shows the most commonly used solar irradiance for space and terrestrial applications. For AM1.5G, $\theta_{AM1.5G} = 48.2°$ and G refers to the global solar irradiance containing both direct and diffuse light and equate to a power density of 1000 W m^{-2}. It must be pointed out that, in this standard measurement, it is assumed that the south facing measurement plane is tilted by 37° towards the Equator to average out the effect of seasons from the tilt in the Earth's axis. The $\sim$30% attenuation in power density, seen in Figure 2.1 for AM1.5G, is wavelength dependent and caused by: (1) Rayleigh scattering by small particles in the atmosphere; and (2) absorption bands due to photons being absorbed by various gas molecules in the Earth's atmosphere, mainly water but also carbon dioxide and oxygen.

2.1.2 History of Exploiting Solar Electricity

The photovoltaic effect, which is the basic principle of a solar cell, was discovered in 1839 by A. E. Becquerel,[4] a French physicist who worked on the interaction between solar radiation and materials in electrolyte solutions to produce an electrical current. However, the first inorganic solar cell, made from a selenium wafer and a deposited gold thin film, was reported with a conversion efficiency of less than 1%, by the American inventor C. E. Fritts.[5] The silicon PV solar cell was developed at Bell Laboratories in 1953 by Chapin *et al.*,[6] with conversion efficiency reaching 6%. These silicon solar cells were then used to power the transmitter in the first solar-powered satellite, Vanguard 1.[7] Vanguard 1 demonstrated that PV solar cells were the only viable power source for satellites; furthermore, it is still in orbit and transmitting around the Earth.

Over the past 60 years, extensive materials research has taken place to improve the conversion efficiencies of solar cells. For commercially available single junction solar cells, conversion efficiencies are in the region of 20%. Over the past 20 years, efforts have been made to reduce the cost of production of solar cells in order for PV to be a cost-effective source of electricity generation for terrestrial usage.

Solar cell technologies can be classified into three generations.[8] The first generation is based on mono or polycrystalline silicon wafers where dopants are diffused into the silicon to create the p-n junction. To reduce the cost and materials utilisation, the second generation was developed based on amorphous or polycrystalline thin film technologies, with full structures being well below 10 μm thick, deposited onto rigid (glass) or flexible (metallic foil or polymer) substrates. The inorganic materials more commonly used are amorphous silicon and chalcogenides, such as copper indium gallium diselenide (CIGS) and cadmium telluride (CdTe). Finally, the third generation encompasses all other technologies that have the potential for higher conversion efficiency, compared with the single junction solar cells used in the first and second generations, and with the potential for lower cost. The challenges in materials used in the second generation and third generation are provided throughout this book. However, an overview of inorganic semiconductors and their electrical and optical properties is given in Section 2.2, followed by an introduction to the single junction solar cells, described in more details in Section 2.3. Finally, the limitations of the single junction and ways of overcoming these limitations are given (Sections 2.3.3 and 2.3.4).

2.2 Fundamentals of PV Materials

2.2.1 Electrical Properties of Inorganic Materials

In terms of electrical conductivity (σ), materials can be classified as metals (or more generally conductors if electrolytes are considered), semiconductors or insulators depending on the band structure of the solid (Figure 2.2). Due to the large number of nuclei in the molecules of a solid (*i.e.* >10^{20}), the discrete energy levels resulting from each molecular orbital combine to form continuous energy bands which are referred to the valence band with the highest occupied energy E_V and the conduction band with the lowest unoccupied energy E_C. By analogy, for organic materials, the valence band corresponds to the highest occupied molecular orbital (HOMO) while the conduction band relates to the lowest unoccupied molecular orbital (LUMO).[9] The bandgap energy (E_g) defines the forbidden zone in the band structure where electrons cannot occupy an energy state (*i.e.* $E_g = E_C - E_V$). In the case of metals, both bands are adjacent or merged and therefore no bandgap exists; the electron can move freely resulting in very good electrical conductivity. In the case of both semiconductors and insulators, an E_g exists and the differentiation between the two material types can be ambiguous and generally referred to a transition where semiconductors that have $E_g \leq 4$ eV, above which, the materials are considered insulators. Materials with a relatively small bandgap are semiconductors with E_g in the range of 0.5–3 eV and can be readily made conducting by introduction of donor or acceptor

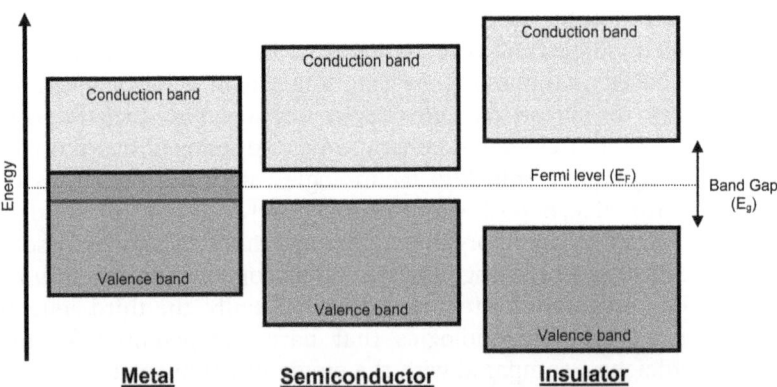

Figure 2.2 Band energy differences between metals, semiconductors and insulators.

impurities. This is the basis of a range of semiconductor devices where photovoltaic cells are based on the p-n junction or 'diode'.

2.2.2 Doping of Semiconductors

In doped semiconductors, the number of electrons and holes is no longer equal, although the total density of carriers remains constant:

$$n_i = (n \cdot p)^{1/2} \tag{2.2}$$

where n_i is the intrinsic carrier concentration, n is the concentration of electrons in the conduction band and p is the concentration of holes in the valence band.

By introducing a donor impurity, an extra valence electron is added to the crystal and $n = n_i + N_D$, where N_D is the donor concentration. Due to the higher number of negative carriers, the semiconductor is now referred to as a n-type semiconductor. In this case, electrons are referred as majority carriers while holes are the minority carriers. Furthermore, as seen in Figure 2.3, as the Fermi level is based on the highest occupancy probability for electrons, and the density n has increased, the position of the new Fermi level is shifted upward. Conversely, for an acceptor impurity, a valence electron is lost and $p = n_i + N_A$, where N_A is the acceptor concentration, resulting in a p-type semiconductor. The reverse of the statement for n-type semiconductors is true as displayed in Figure 2.3. The doping is normally achieved by introducing an impurity (dopant) that will substitute onto a crystal lattice site but with a different valence to the substituted ion. For example, boron (valence of 3) in silicon is an acceptor and phosphorus (valence of 4) is a donor. Thin film PV materials follow the same principles, but being either amorphous or polycrystalline, the

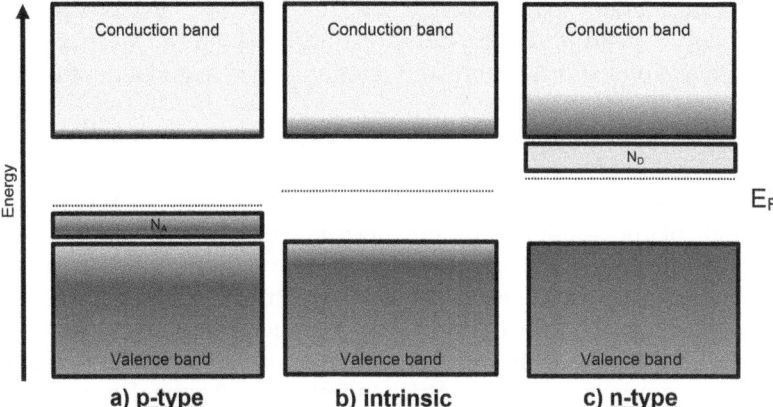

Figure 2.3   Energy band diagram showing the effect of (a) p-type doping, (b) intrinsic & (c) n-type doping of a semiconductor.

electrical properties tend to be dominated by native defects rather than dopant elements.

2.2.3   Band Structure of Solar Absorbers

The properties of the band structure of the p-n junction are important for both the optical and the electrical properties of the PV cell. In its simplest form the bandgap is as an abrupt filter of solar radiation where, for photon energy greater than the bandgap, the radiation will absorb and, for photon energy less than the bandgap, it will transmit through the layer. In reality the absorption is more complex and relates to the band structure of the absorber material and will, in practice, determine the thickness of the absorber required to absorb all the radiation falling on it when the photon energy is greater than the bandgap.

2.2.3.1   Indirect Bandgap Semiconductors

The most widely used absorber material, silicon, is actually a poor absorber because the transition for electrons from the valence band to the conduction band minima requires a momentum exchange at the time of absorption and this reduces the probability for absorption. To understand the absorption process we must look at the energy–momentum space band diagram for silicon. Engineering band diagrams are normally represented in the form of an energy scale on the y-axis and one dimensional space on the x-axis. To understand the absorption of light we have to account for the momentum as well as the energy of an electron in a semiconductor. The momentum is characterised in quantum mechanics by the wave vector (k) which is inversely proportional to the electron wavelength. In free space, this has a simple parabolic relationship with the energy of the electron but in a solid this becomes considerably distorted and hence the curve of energy *versus* k can be complex and depend on the crystal direction in which the electron travels. Further details on band structure and relationship to semiconductor devices can be found in ref. 10.

For simplicity, semiconductor band structures are often considered at zero Kelvin, where a thermal component to the electron energy is neglected. For silicon PV cells, the thermal transfer of energy in the electrons is essential, as the bottom of the conduction band does not occur at the same point in momentum space as the highest energy in the valence band. This is represented in Figure 2.4 which shows the band structure for silicon where the top of the valence band and the bottom of the conduction band are displaced on the k axis. Photons can only create transitions with zero change in the value of k, so some assistance from lattice vibrations (phonons) is required to transfer an electron from the valence band to the conduction band, creating an electron–hole pair. For silicon solar cells, operating at ambient temperature, there is sufficient lattice energy to provide the momentum transfer at

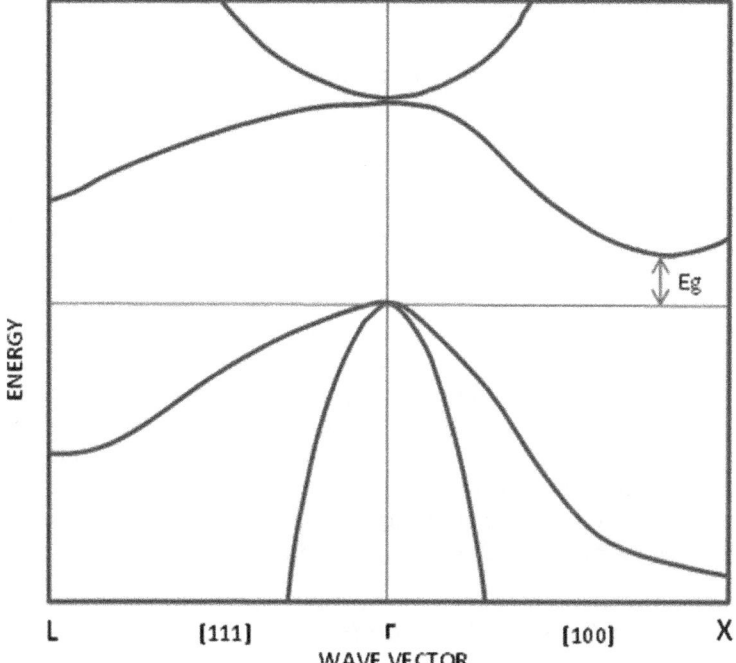

Figure 2.4 Schematic of the *E-k* band diagram for silicon illustrating the indirect transition from the valence band at the zone centre to the conduction band minimum at the zone edge in the [100] crystal direction.

the same time as photon absorption. However, the consequence is that the absorption coefficient of crystalline silicon is less than for a direct bandgap semiconductor where no momentum transfer is required. The transmission of light through an absorbing layer is given by Beer's Law, where reflection losses are ignored:

$$\frac{I(\lambda)}{I_0(\lambda)} = e^{-\alpha(\lambda)d} \tag{2.3}$$

where $I(\lambda)$ is the transmitted intensity of the light at wavelength λ, $I_0(\lambda)$ is the incident light intensity at wavelength λ, $\alpha(\lambda)$ is the absorption coefficient at wavelength λ and d is the thickness of the absorber film.

For crystalline silicon absorbers, the indirect bandgap gives a range of values for $\alpha(\lambda)$ from 1×10^2 cm^{-1} at 1.3 eV ($\lambda = 953$ nm) to 1×10^4 cm^{-1} at 2.5 eV ($\lambda = 496$ nm),[11] so the biggest challenge is to achieve effective absorption at the near infrared part of the solar spectrum. This contrasts with direct bandgap semiconductors that rapidly rise to over 1×10^4 cm^{-1} above the bandgap energy and can increase to over 1×10^5 cm^{-1} for higher photon energy. The significance of $\alpha(\lambda)$ or the materials selection and design of solar cells is in the typical absorption depth for the photons of different wavelengths. As the transmitted intensity is shown from eqn (2.3) to be an exponential decay the typical absorption depth is taken to be when the transmitted intensity reaches $1/e$ of the incident intensity. At this depth $(d_{1/e})$, the ratio $I(\lambda)/I_0(\lambda)$ is 0.37 and therefore 63% of the incident solar radiation at this wavelength has been absorbed. To achieve 86% absorption would require a thickness of $2d_{1/e}$ and so on. The consequence for an effective absorption of light with photon energy of 1.3 eV in silicon is that the $2d_{1/e}$ thickness would be 200 μm, which is the typical thickness of a crystalline silicon solar cell.

For thin film silicon, it is necessary to enhance the absorption of the solar radiation by using amorphous silicon, which behaves as a direct bandgap semiconductor with an absorption thickness of 2 μm at 2 eV. For crystalline silicon thin film layers, the light absorption has to be increased by either increasing d through light scattering in the plane of the thin film or using light absorption enhancement with methods such as plasmonics. The bandgap of amorphous silicon is around 1.7 eV, at the upper end of what would be considered an efficient absorber material from the viewpoint of the Shockley–Queisser limit (discussed in Section 2.3.3).

2.2.3.2 Direct Bandgap Semiconductors

The concept of direct bandgap semiconductors was introduced in the previous section along with the practical consequence of not requiring lattice energy to absorb photons with energy greater than the bandgap. The E-k band diagram for a direct bandgap semiconductor is shown in Figure 2.5. The conduction band has a relatively light effective mass for carriers whereas the hole bands consist of a heavy and light hole band with relatively high density of states compared with the conduction band.[10,12] This band structure is typical for most of the II–VI and III–V classes of direct bandgap semiconductor materials and for the chalcogenides such as copper indium diselenide (CIS). For thin film PV devices, the picture is complicated by the materials being polycrystalline with typical grain size of the order of 1–2 μm. Although this is large enough for the band structure to be the same as for a bulk crystal, the defects associated with the incoherent grain boundaries will significantly affect the electrical properties of the film and are further described in Section 2.4.

The significance of CdTe being a direct bandgap semiconductor is that a thickness of just 1 μm will absorb 90% of the available solar spectrum for photon energy greater than the bandgap. In practice, solar cells are made with CdTe absorber thickness in the region of 5–10 μm, but this has more to do with the electrical properties and non-uniformity in the polycrystalline CdTe film. Theoretical studies by Amin *et al.*[13] have reported the effect of absorber thickness on the optical and electrical performance of a CdTe solar cell, using an absorption coefficient of 2×10^4 cm^{-1} at 800 nm. For a theoretical cell efficiency of 16% at 3 μm absorber thickness, this should only decrease to 15.5% for a 1 μm

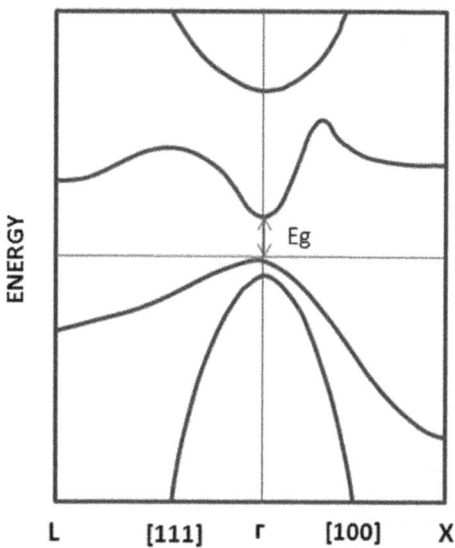

Figure 2.5 Band structure of a direct bandgap semiconductor.

Table 2.1 Absorption coefficients of CdTe polycrystalline thin films for PV cells

	Wavelength/nm		
	405	658	810
Absorption coefficient (cm^{-1})	4.2×10^4	8.8×10^3	4.5×10^3
1/e Absorption depth (nm)	240	1143	2224

absorber thickness. The absorption coefficient measured in thin film CdTe for PV cells over a number of wavelengths from 410 to 810 nm is shown in Table 2.1.[14] It can be seen that absorption measurement in polycrystalline thin films differs significantly from measurements made in bulk material and is strongly wavelength dependent, even with a direct bandgap semiconductor. The consequence is that for absorber thicknesses less than 3 µm, some loss in absorption at the longer wavelengths will occur. Similar characteristics can be expected for the copper based chalcogenide family of absorber materials.

2.2.3.3 Quantum Confined Absorbers

The bandgap of bulk absorbers can be modified by reducing at least one dimension to less than 100 nm. This gives the following three classes of quantum confined absorbers:

1. One-dimensional (1D) – quantum layers as in GaAsP/InGaAs strain-balanced quantum wells[15]

2. Two-dimensional (2D) – nano-rods as in Si nano-rods[16]
3. Three-dimensional (3D) – quantum dots.[17]

   The most developed of these classes of quantum confined absorbers is the 1D confinement where very thin epitaxial layers are grown to modify the band structure perpendicular to the surface of the film. This is well known in opto-electronic emitters such as lasers where very high intensity of light is confined in the region of a quantum well. The laser emission wavelength is controlled by the thickness of the well. In the same way, it is possible to change not only the cut-off wavelength of the absorber but also to create additional levels within the conduction and valence bands. The multi-quantum well structure GaAsP/InGaAs utilises this feature in creating a cascade of electrons from one well to the next *via* a built-in electric field.[15] This is illustrated in Figure 2.6 where the confinement energy E_a determines the longest wavelength for absorption, extending beyond the p and n layer bulk absorptions. The advantage of such structures, over non-quantum confined absorbers, is in the ability to optimise the absorption characteristics by engineering the thickness of the quantum wells and, in similar ways, with the width of the nano-rods or the diameter of quantum dots.

   Nano-rods, made from any of the typical absorber materials, can be formed on any suitable substrate and do not require epitaxial growth as with the 1D confinement. This approach has particular attractions for Si, where the nano-rods structure can also contribute to light capture,[16] and for CdTe where short minority carrier diffusion lengths in bulk polycrystalline films leads to loss of photocurrent. The most challenging but potentially most exciting class of quantum confined absorbers are the quantum dots. In its simplest form, they could replace a dye as the sensitiser in nanoporous titania[18] or as

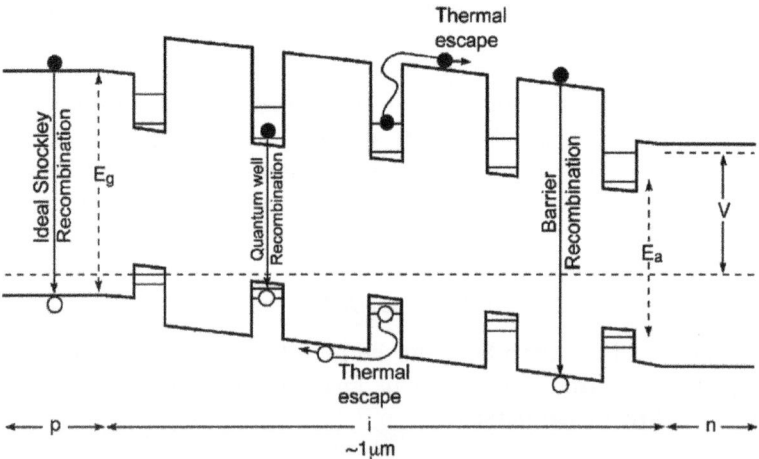

Figure 2.6   Band diagram of strain-balanced III–V MQW (after Mazzer *et al.*[15]).

a hybrid with organic solar cell, such as embedding CdSe quantum dots into conducting polymers. Different sizes of quantum dots would have a different absorption spectrum, so in a suitable matrix could provide efficient multi-band absorption which could exceed the efficiency of a single junction cell (described in Section 2.3.3).

Quantum dots have also been applied to the concept of extremely thin absorber (eta) solar cells where the short minority carrier diffusion length is overcome by creating the absorption close to the junction. High efficiency solar absorption is achieved by creating a large surface area, compared with a smooth surface, either through a structured surface or in the extreme a network of nano-rods. Nano-rod templates, made from a transparent conducting oxide (TCO), can be used to act as the electron-receiving electrode. Zinc oxide (ZnO) is a particularly attractive TCO for this application where nano-rod arrays can be readily formed.[19]

2.3 The pn Junction

If inhomogeneity exists in the carrier concentration of a doped semiconductor, redistribution of the carriers will take place in an attempt to reach equilibrium. In the absence of an externally applied electric field, diffusion of carriers occurs flowing from the high density region to the low density region, generating a diffusion current density (J_{diff}). The diffusion process results in electrostatic charge creating an internal electrical field, which in turn produces a drift current density (J_{drift}) in the opposite direction to J_{diff} resulting in a net zero current flow (*i.e.* $J_{diff} + J_{drift} = 0$).[20] This phenomenon is the basis of a pn junction where a depletion region is created when p-type and n-type semiconductors are brought together, as shown in Figure 2.7(a). An important term here is the Einstein relation, which stipulates that the work done to move a charge carrier must be equal to the energy difference over that region. This relation can be written as follows:

$$\frac{D_n}{\mu_n} = \frac{D_p}{\mu_p} = \frac{k_B T}{q} \tag{2.4}$$

where D_n and D_p are the diffusion coefficient for electrons and holes, q is the elemental charge of a carrier and k_B is the Boltzmann constant. If the mobility of the carrier (μ_n or μ_p) is known at a given temperature T, then D_n and D_p can be calculated.

Note that for hetero-junctions, where two different semiconductor materials are used, such as those used in the second generation of solar cells, Figure 2.7(a) would vary as each semiconductor would have a different bandgap; however, the principle remains the same as long as a good interface between the two semiconductors is made.

The ideal pn junction, also referred to as an ideal diode, is the principle behind most modern electronic components, as it creates a rectifying

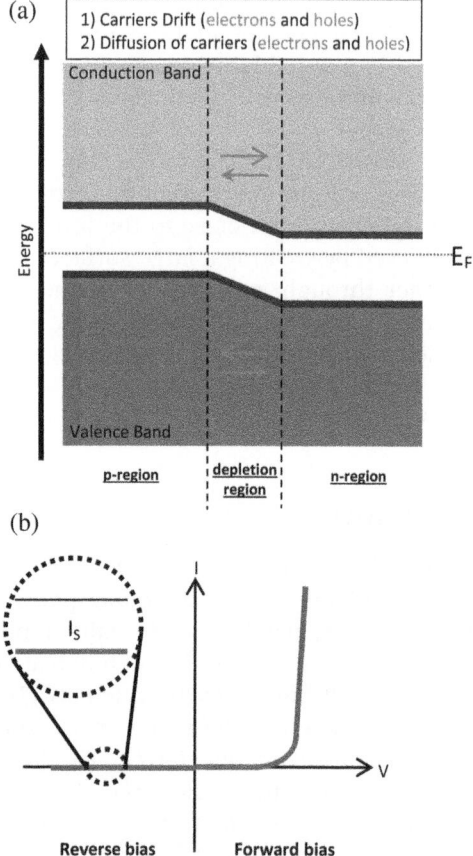

Figure 2.7 (a) Band alignment in a pn homo-junction. (b) Typical I–V characteristic of an ideal diode (not to scale; curve in reverse bias has been magnified to show I_S).

current–voltage (I–V) characteristic, as seen in Figure 2.7(b), which can act as a switch and relates to the following equation:

$$I = I_S\left(e^{qV/k_B T_C} - 1\right) \tag{2.5}$$

where I_S is the reverse saturated current in the pn junction, at a solar cell temperature T_C, which is related to the diffusion of carriers within the depletion region, introduced earlier. I_S can be observed in reverse bias when a reverse applied voltage (V) is equal or lower than the Einstein relation, *i.e.* eqn (2.4). It is noteworthy to remember that current (I) and current density (J) may be used in the literature (*i.e.* $I = J \cdot A$, where A is the surface area covered by the junction).

2.3.1 Fundamentals of Absorption of Solar Radiation in a pn Device

Conduction in an intrinsic semiconductor normally occurs through the thermal generation of electron–hole pairs creating a concentration of free electrons and free holes. Another way of exciting an electron–hole pair is by absorbing photons in the semiconductor as described in Section 2.2. Depending on the context, a photon of light may be described either by its wavelength (λ) or its equivalent energy (E_{ph}). If the E_{ph} of the incident light is equal or higher than E_g, then an electron–hole pair is generated. Due to the presence of an internal electric field in the depletion region of the pn junction, the minority charge carriers generated in either p- or n-type side can flow across the junction, producing the photovoltaic (PV) effect and generating a photocurrent (I_{ph}). In theory, the photocurrent can be generated on either side of the junction. However, this is usually dominant in just one side, with the dominant minority carrier lifetime, and PV devices are designed to take this into consideration.

The maximum amount of photocurrent that may be generated is governed by the optical properties of the semiconductor, namely the refractive index (n) and the extinction coefficient (k). From the solar irradiance at AM1.5G in Figure 2.1, it is possible to calculate the solar spectral flux density $F_{solar}(\lambda)$ at each wavelength over the entire solar spectrum which may penetrate through the atmosphere to the solar cell and contribute to electron–hole pair generation at each wavelength. However, the spectral flux density absorbed $F_{abs}(d, \lambda)$ will depend on the semiconductor used and the spectral flux density losses $F_{loss}(d, \lambda)$ which, based on eqn (2.3), is governed by the absorption characteristic of the semiconductor as follows:

$$F_{loss}(d, \lambda) = F_{solar}(\lambda)e^{-\alpha(\lambda)d} \tag{2.6}$$

where the absorption coefficient can be expressed as a function of $k(\lambda)$:

$$\alpha(\lambda) = \frac{4k(\lambda)\pi}{\lambda} \tag{2.7}$$

and the total absorbed flux density, within the device, can be defined as:

$$F_{abs}(d, \lambda) = F_{solar}(\lambda) - F_{loss}(d, \lambda) = F_{solar}(\lambda)\left(1 - e^{-\alpha(\lambda)d}\right) \tag{2.8}$$

Once $F_{abs}(d, \lambda)$ is known as a function of wavelength, the related number of electron–hole pairs and therefore the maximum photocurrent generated by the solar cell can be calculated by integrating eqn (2.8) with respect to λ; this is further discussed in Section 2.3.3. However, this is where it becomes important to carry out testing of the solar cell and to evaluate its performance against theoretical expectations.

In these measurements, a standard temperature of 25 °C is used. The external and internal quantum efficiencies (EQE and IQE) of the solar cell can

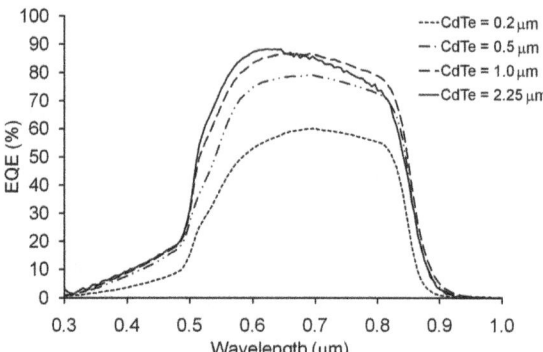

Figure 2.8 EQE curves for CdTe/CdS devices showing the effect of reduced absorption for ultra-thin CdTe absorber layers (after Clayton *et al.*[21]).

be measured using spectral response measurements where the maximum potential photocurrent from the available photon flux is compared with that absorbed in the device (*i.e.* measuring the photocurrent generated in the device), as illustrated in Figure 2.8. This uses a calibrated chopped light source delivered, through a spectrometer arrangement, to the solar cell. A bias light may be used, in these measurements, to fill trap levels, where recombination generally occurs in the junction and provides an 'uninterrupted' path for the charge carriers. This is similar to operating conditions under solar irradiation. The recombination rates in the depletion region are dominated by trap assisted recombination at deep trap levels (*i.e.* $\sim E_g/2$), where the energy transition for electrons and holes are equal. Therefore, the effect of the bias light source will depend on the defect type and trap depths in the depletion region. In contrast to EQE, IQE measurements remove the contributions from the external photon losses such as from reflection and will show more accurately the internal losses either from lack of photon absorption or from carrier recombination.

2.3.2   Electrical Behaviour of a PV Solar Cell

The standard measurement for assessing the performance of photovoltaic solar cells is the current–voltage (I–V) characteristic under both dark and illuminated conditions. The voltage is swept from reverse bias to forward bias (and/or *vice versa*) to reveal a curve, as shown in Figures 2.7(b) and 2.9. The dark response [Figure 2.7(b)] provides information on the solar cell as a diode where I_S can be measured. The measurement under illumination is made with light from a solar simulator, which is filtered to achieve an AM1.5G spectral irradiance. From this, it is possible to determine the photovoltaic device characteristics and the maximum power output. This measurement is used to determine the following parameters [indicated in Figure 2.9(a)]: the maximum power ($P_{max} = I_{max} \times V_{max}$); short-circuit current (I_{sc}); open-circuit voltage (V_{oc}); conversion efficiency (η); and fill factor (FF).

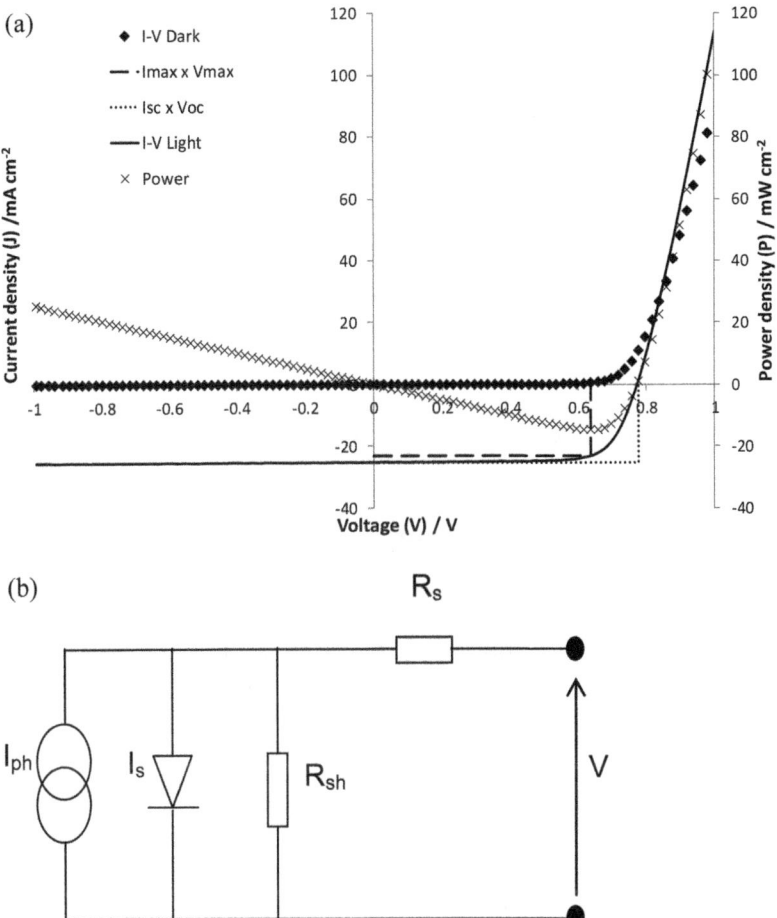

Figure 2.9 (a) Dark and light I–V characteristics of a photovoltaic solar cell, including the power curve, required to determine the cell parameters described in eqn (2.10). (b) Equivalent electrical circuit of the single junction solar cell under illumination.

If the PV device can be represented by an ideal diode, then eqn (2.5) can be modified to include the photocurrent as follows:

$$I = I_{\mathrm{S}}\left(e^{qV/k_{\mathrm{B}}T_{\mathrm{C}}} - 1\right) - I_{\mathrm{ph}} \tag{2.9}$$

The ideal cell parameters can be determined from eqn (2.9) where $I_{\mathrm{ph}} = I_{\mathrm{sc}}$ when $V = 0$, $V = V_{\mathrm{oc}}$ when $I = 0$, while P_{max} is determined from the point on the curve where $I \times V$ is a maximum, as illustrated in Figure 2.9(a). The FF is simply the ratio $I_{\mathrm{max}}V_{\mathrm{max}}/I_{\mathrm{sc}}V_{\mathrm{oc}}$ and is a measure of the 'square-ness' of the I–V curve. However, real devices diverge from the ideal representation of eqn (2.6) due to parasitic resistance, as shown in the circuit of

Figure 2.9(b). The series resistance (R_s) can have a number of contributors, including the resistance of the bulk semiconductor, the contact resistance and the resistance of the conductors, such as the transparent conducting oxide. In addition, there can be a shunt resistance (R_{sh}) that can occur from micro-shorts across the device and would affect the rectifying property in reverse bias. R_s and R_{sh} can be determined from the forward bias and reverse bias characteristics respectively and will modify the I–V characteristics as follows:

$$I = I_s\left(e^{(V-IR_s)q/k_BT_C} - 1\right) + \frac{V - IR_s}{R_{sh}} - I_{ph} \qquad (2.10)$$

The diode equation can be further modified to include other factors according to the physical model used, but the diode eqn (2.10) and corresponding equivalent circuit in Figure 2.9(b) offer a suitable description of PV devices.

2.3.3   Shockley–Queisser Limit

Shockley and Queisser[22] have proposed a maximum conversion efficiency that can be achieved using a single pn junction solar cell. This is referred to as the Shockley–Queisser limit. The fundamental limitation in the conversion efficiency of a single junction solar cell is a balance between the proportion of the solar spectrum that can be absorbed by a semiconductor and the energy that can be converted to electrical energy per photon absorbed. The narrower the bandgap the more photons that can be absorbed at longer wavelengths, but the lower the electrical energy generated per photon absorbed.

The basic assumptions of the model are as follows:

a) Generation of electron–hole pairs for $E_{ph} \geq E_g$, *i.e.* incoming solar photons where absorption is determined by eqn (2.8).
b) The lifetime of minority carriers in the absorber layer is set by the radiative recombination rate.
c) The Sn is treated as a black body at 6000 K and the cell as a black body at 300 K.
d) V_{oc} is a fraction of the bandgap and can only be equal to E_g at $T_C = 0$ K.

The first limit to consider is that every photon with energy equal to or larger than the bandgap will only produce one electric charge (*i.e.* one electron–hole pair). If the absorbed photon has an energy greater than E_g, then the excess energy will be lost through rapid thermalisation (*i.e.* heat generation). On the other hand, any photon with lower energy than the bandgap will not be absorbed by the semiconductor and therefore will not generate an electron–hole pair. Therefore, if we assume that the absorbing semiconductor is thick enough to absorb all of the received photons with energy higher or equal to E_g, eqn (2.8) can be simplified to $F_{abs}(d, \lambda) = F_{solar}(\lambda)$.

Based on these assumptions, Shockley and Queisser predicted that the maximum efficiency for a single junction solar cell was around 30% at 1.5 eV. Treating the Sun as a black body approximates to the AM0 solar spectrum but, for terrestrial applications, the spectrum is significantly modified so a more accurate estimate of the optimum efficiency can be gained by calculating the I–V parameters using the AM1.5G spectrum from Figure 2.1. The theoretical efficiency, V_{oc}, J_{sc} and FF as a function of bandgap energy are shown in Figure 2.10 based on the Shockley–Queisser limit, as follows:

$$J(V, E_g) = q \left(R_r \left(e^{qV/k_B T_C} - 1 \right) - \left(\int_{E_g}^{\infty} N(E)\, dE \right) \right) \qquad (2.11)$$

where $J(V, E_g)$ is the current density for an ideal absorber of bandgap E_g at applied voltage V [note that in Figure 2.10(b), J_{sc} is the maximum photocurrent taken at $V = 0$] and $N(E)$ is the number of photons per photon energy E, derived from $F_{abs}(d, \lambda)$ of eqn (2.8) for AM1.5G. Finally, R_r

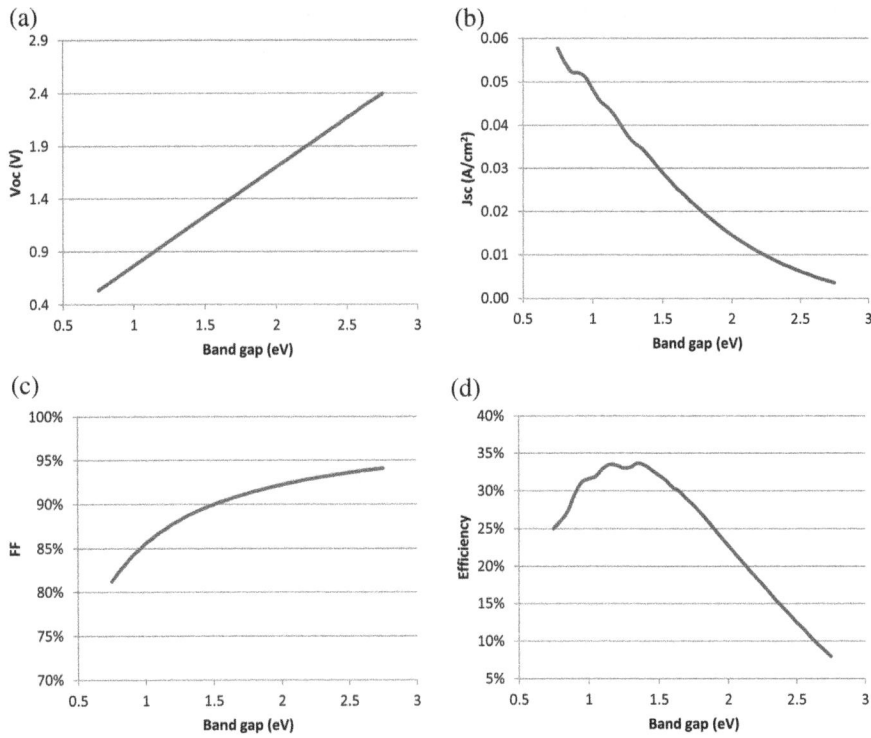

Figure 2.10 Shockley–Queisser limit for AM1.5G spectrum as a function of bandgap energy for (a) V_{oc}, (b) J_{sc}, (c) FF and (d) maximum conversion efficiency.

is the radiative recombination rate, in the absorber, for photon energy equal to or above E_g. The open-circuit voltage, V_{oc}, is calculated from the following:

$$V_{oc}(E_g) = \frac{k_B T_C}{q} \ln\left\{\frac{\int_{E_g}^{\infty} N(E)\, dE}{R_r}\right\} \qquad (2.12)$$

V_{oc} can be seen as the difference between the quasi Fermi levels for the n-type and p-type side of the junction and will be a fraction of the bandgap, increasing as the bandgap increases, as shown in Figure 2.10(a). In contrast, J_{sc} decreases as E_g increases due to the reduction in the above bandgap energy photon density. The theoretical FF increases with E_g, with values up to 90% at 1.5 eV. The decreasing slope, seen in Figure 2.10(c) is caused by its dependence on the decreasing J_{sc} at higher bandgap. The combination of these parameters can be seen in Figure 2.10(d), which shows the expected compromise between high J_{sc} for small bandgap energy and high V_{oc} for high E_g in a single junction solar cell. It can be seen from Figure 2.10(d) that the highest efficiency with AM1.5 illumination is higher than for AM0 with a maximum around 34% in the range 1.2 to 1.4 eV. This provides a target bandgap energy for new solar PV absorbers and for each of the solar cell parameters a means of gauging experimental performance with this maximum Shockley–Queisser limit.

2.3.4 3-G Solar Cells to Beat the Single Junction Limit

The Shockley–Queisser limit shows that there are physical limitations to the maximum conversion efficiency which can be achieved by a single junction solar cell, with a maximum value of 34% for AM1.5. In reality, the actual conversion efficiencies of single pn junction are lowered due to optical losses (*i.e.* reflection at a single air–glass interface is 4%, while at the surface of semiconductor such as Si and GaAs a 30–40% reflection can occur), which can be partly solved by simple anti-reflection coatings or more complex plasmonic light trapping concepts. If the thickness of the absorbing material is not optimum, further optical loss can occur, as shown in eqn (2.8), and may be overcome by applying suitable back reflection arrangements. The manufacturing process of the solar cell also determines the quality of the materials and, therefore, the further losses caused by charge recombination in bulk and at interfaces. These defects might be improved by optimising the manufacturing process. If it can be assumed that the defect structure has been optimised for the remaining discussion, one can try to optimise the conversion efficiencies by one of the following approaches addressing some of the limits discussed in Section 2.3.3.

2.3.4.1 Multi-junction Solar Cells

To avoid the excessive thermalisation losses for photon energies above the bandgap of a single junction, one may design a set of multi-junction solar cells. In this concept, the bandgaps of the individual cells have to be optimised to either match their photocurrents, when connected in series, or their voltage if connected in parallel. In other words, the multi-junction solar cell will perform as well as the sum of the individual cells. The maximum theoretical limit for a triple junction solar cell is 56% (and can be as high as 72% for 36 junctions under 1000 suns).[23] In practice, triple junctions are being produced using III–V semiconductor materials to achieve conversion efficiencies of up to 44.4% under concentrated light of (AM1.5 × 302),[24] which is the current world record produced by Sharp and verified by the Fraunhofer Institute for solar energy systems. A lattice-matched GaInP/GaInAs/Ge structure is generally used with typical bandgaps of 1.86, 1.35 and 0.67 eV, respectively; however, a metamorphic tunnel junction may be used to relax the structure and remove the stringent need for lattice matched layers.[25] These triple junction cells require high quality epitaxial layers monolithically stacked in series and generally manufactured by metal organic vapour phase epitaxy (MOVPE). Due to the high light concentration level used in the concentrator PV (CPV), temperature management must be a carefully designed part of the system to avoid degradation of performance. A more detailed description of multi-junction III–V solar cells is given in Chapter 7.

2.3.4.2 Intermediate Bands

Similarly to multi-junction solar cells, the intermediate band concept[26] addresses the same limitations of thermalisation effect of a single junction for higher energy photons than its bandgap. However, it differs in that the single junction concept is kept, removing the complexity of lattice matched junction and current matching. A wide bandgap semiconductor may be used such as ZnTe[27] to maintain a high V_{oc} and the idea is then to incorporate impurities such as oxygen within the structure in the form of quantum dots or quantum wells to provide additional bands (*i.e.* minimum of three bands inclusive of the conduction and valence band of the main single junction). The bands are creating finite quasi Fermi levels enabling localised electron–hole pairs to be created, using two photons, without excessive thermalisation loss, and therefore increasing the overall photocurrent being generated in a single junction. Although no successful device has been achieved so far, the theoretical conversion efficiency limits are 63% and 72% for three-band and four-band, respectively,[26] which has greater potential compared with multi-junction solar cells.

2.3.4.3 Up/Down Converters

Rather than tuning the active structure (*i.e.* solar cell) to accommodate the broad range of radiation from the solar spectrum, one can think of a 'passive' device or material that would allow the conversion of the solar spectrum to suit single junction devices. Existing single junction solar cells are made from semiconductors having a bandgap between 1 and 2 eV, meaning that a large portion of the ultraviolet (UV) blue and the near infrared (NIR)–far infrared (FIR) are not being utilised. Luminescent materials such as dyes or quantum dots can therefore be used to convert the solar spectrum to be tuned for the specific absorber bandgap. More is described about these processes in Chapter 9.

2.3.4.4 *Hot Carriers*

A large proportion of the photon energy can be lost from the photogenerated electron, defined as the 'hot' carrier, in a non-equilibrium state as it relaxes to the bottom of the conduction band. During that process, energy is being transferred to lattice phonons. For example, a 'blue' photon with energy of 2.6 eV would lose 1.1 eV in a single junction CdTe solar cell simply through thermalisation. This decay process only takes picoseconds to reach equilibrium, which is where the challenges lies: (1) the thermalisation should be reduced within the selected structure; and (2) the selective energy bands, in the structure, should have a narrow energy range to avoid 'cooling' or relaxation of these hot carriers. In both cases, quantum confinement is the solution. Hot carrier solar cells have the potential to reach 68% conversion efficiencies under non-concentrated AM1.5 solar irradiance.[28] This concept has the potential to be simpler than either the multi-junction or intermediate band approaches.

2.4   Defects in Thin Film PV Materials

Thin film PV materials are either polycrystalline or amorphous because of the need for low cost deposition onto cheap substrates. The epitaxial thin film III–V class of semiconductors are in a different category where high quality single crystal Ge or GaAs substrates are used. This has a consequence for the complexity of the defect structure and the effect this has on the device's performance. Consideration of the influence of defects on the performance covers the conversion efficiency of the device and long-term stability. The latter relates to movement of impurities and crystalline defects in the device layers. Once the absorber layer has absorbed a photon, creating an electron–hole pair, there are two possible outcomes—either recombination or diffusion across the junction. Only in the case of movement of the photogenerated minority carrier across the junction will electrical energy be extracted (as illustrated in Figure 2.7). The defects that cause recombination can occur either in the bulk of the film, at the back contact or at the junction. These defects have the potential to reduce both the J_{sc} and V_{oc} of the cell. In the case of thin film crystalline silicon the minority carrier diffusion lengths can be as long as 60 μm[29] and as little as a few hundred nanometres in polycrystalline thin films. For polycrystalline thin films, there is the added complication from recombination at the grain boundaries, as well as at point defects within the grains or at the junction. Unlike single crystal and epitaxial materials the polycrystalline materials can be dominated by a high density of deep levels within the bandgap and will limit the minority carrier diffusion length. All these defect types will contribute to reducing the minority carrier diffusion length and loss of J_{sc}. Specific types of defects in thin film PV devices are now considered.

2.4.1 Staebler–Wronski Effect

The most common type of thin film silicon for PV cells is amorphous silicon (a-Si) due to the high absorption coefficient. An amorphous material will tend to have a high concentration of dangling bonds and these will act as charge centres that can appear within the bandgap and act as traps for minority carriers. This is overcome in a-Si by hydrogenation where the Si dangling bonds are passivated by atomic hydrogen. The Staebler–Wronski[30] effect is the observation of changes in the device performance under illumination from solar radiation. The increase in defect traps on illumination relates to the photo-induced change in bonding; this change is initially very rapid and eventually stabilises. It is important that efficiency measurements quoted for a-Si also quote the light soaking conditions.

Various mechanisms have been proposed for the Staebler–Wronski effect and a common theme is a photo-induced breaking of a weak Si–Si bond and reaction with mobile hydrogen. The process is reversible by heating to 150 °C and is quite different to crystalline silicon at these temperatures where point defects are stable. An engineering solution to the inherent instability in a-Si–H solar cells is to form a tandem cell with microcrystalline silicon. This can be formed by annealing an a-Si film. There is still some degradation but typically less than 10% over the first 100 hours of exposure and relatively stable for longer periods. This is consistent with the Staebler–Wronski effect being associated with weaker Si–Si bonds than appear in a-Si compared with crystalline silicon. The subject of thin film silicon is covered in more detail in Chapter 3.

2.4.2 Minority Carrier Lifetime and Junction Defects

Deep level defects in the bulk of the absorber will act as traps and reduce the minority carrier lifetime of the photogenerated carriers. In the case of crystalline silicon, the trap density is low and recombination within a wafer 200 μm thick is minimal. This compensates for the need to have a thick layer of silicon to absorb the available solar radiation. However, surface recombination becomes more critical and effective surface passivation of the silicon wafer is essential. The direct bandgap in compound semiconductors reduces the need for long minority carrier lifetime and diffusion length. In the case of epitaxial GaAs/GaAsP devices, the bulk lifetime is high but defects forming at the junction due to lattice mismatch will cause recombination within the depletion region. So, choosing layer combinations that not only give the required bandgap for absorption of the radiation but also good lattice matching is essential, hence the importance of the strain-balanced quantum wells described in Section 2.2.3.3.

The minority carrier lifetime can affect device performance through both the J_{sc}, where it will affect the proportion of minority carriers that reach the junction, and the V_{oc}. The effect on V_{oc} can best be described by looking at the relationship between V_{oc} and I_S given by the following equation [a variant to eqn (2.12)]:

$$V_{oc} = \frac{nk_B T_C}{q} \ln\left(\frac{I_{ph}}{I_S}\right) \tag{2.13}$$

where n is the ideality factor and the other terms are as defined previously. I_S is the reverse saturation current given by:

$$I_S = qA\left(\sqrt{\frac{D_{n,p}}{\tau}} \frac{n_i^2}{N_A}\right) \tag{2.14}$$

where $D_{n,p}$ is the minority carrier diffusion coefficient and τ is the minority carrier lifetime. It can be seen from eqn (2.13) and (2.14) that a decrease in τ arising from an increase in deep level traps will cause an increase in I_S which in turn will cause a decrease in V_{oc}. This assumes that the device is essentially homogeneous; the additional effects of inhomogeneities that occur in PV devices are considered in the next section. It is also worth noting here that according to eqn (2.13) and (2.14), an increase in N_A will decrease I_S and hence increase V_{oc}. This is a target in controlling the electrical properties of the thin film PV materials, but current approaches for intrinsic doping make this very challenging.

Polycrystalline thin film PV devices have a more complex defect structure, although the same principles of defects such as dislocations at the junction will apply as for the epitaxial junctions. One advantage of polycrystalline layers is the potential for strain relief within each grain to create a so-called pseudo-morphic junction where the lattice will try to match within the plane of the device. This will constrain the choice of window and absorber layer in hetero-junction devices beyond the constraints of optical absorption in the absorber layer and good transmission in the window layer. The opportunities for minority carrier recombination are shown schematically in Figure 2.11. The grain boundaries create the largest challenge and, without effective passivation, will prevent minority carriers reaching the junction, hence reducing J_{sc}. Referring back to Section 2.2.3.1, we can see that the absorption coefficient is lower for longer wavelength photons, so they will be typically absorbing at depths greater than 1 µm away from the junction and more vulnerable to recombination at unpassivated grain boundaries before they reach the junction. Hence, the EQE curve will show poorer quantum efficiency at longer wavelengths. Grain boundary passivation can be viewed as causing a bending of the bands towards the grain boundary in a way that will

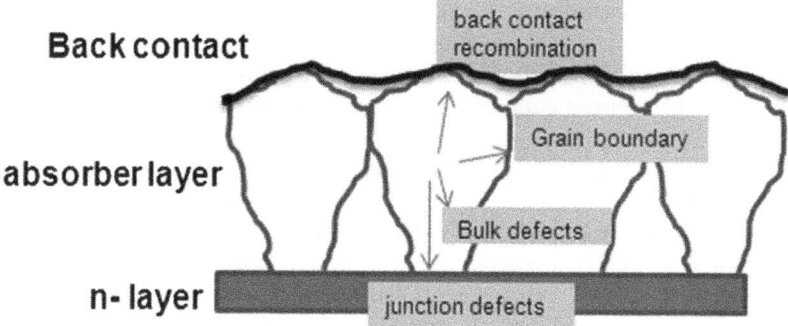

Figure 2.11 Schematic of recombination within a polycrystalline thin film absorber showing recombination within the junction region, bulk of the grain, back contact and grain boundary.

repel the minority carriers. For minority electrons, this will require the conduction band to increase in energy towards the grain boundary. The simplest way of doing this is by doping the grain boundary p-type, but can also be achieved by alloying to increase the bandgap.

The back contact is ideally an ohmic metal contact but in reality it can be a poor Schottky junction. This will be a very defective region of the device and carriers that diffuse to the back contact are likely to recombine. As with the grain boundaries, the minority carrier must be repelled from the back contact either by doping in the case of CdTe or by alloying to change the bandgap as is the case with $Cu(In,Ga)Se_2$.

The recombination of minority carriers at defects can affect the device characteristics in a number of ways, all resulting in a loss of conversion efficiency from the device. The effect on J_{sc}, where recombination occurs before the minority carrier can reach the junction, is described above. However, there is an optical type defect that can also reduce J_{sc}, where the cut-off wavelength of the window will absorb solar radiation in the blue and ultraviolet before it arrives at the junction. The Shockley–Queisser limit assumes that most photons with energy greater than the bandgap will create useful charge generation within the device. The short wavelength cut-off in the EQE spectra such as in Figure 2.8, for CdS/CdTe PV cells, corresponds to the absorption edge of the window layer CdS and shows that the electron-hole pairs, generated by photon absorption in the window layer, do not result in a photocurrent. Hence we are left with a potential efficiency maximum that is less than the ideal Shockley–Queisser limit, depending on the optical properties of the window layer. This is often referred to as the 'blue defect'.

2.4.3 Lateral Non-uniformity of Thin Film PV Devices

A thin film PV module is designed to cover a large area in order to capture the solar radiation. Thin film PV modules are typically monolithically integrated and sub-divided into long thin cells, separated by a three-level scribing process. The deposition process needs to provide uniform films of material over the whole area, which can extend to over one metre. Non-uniformity in the film thickness can result in variations in photon absorption and device performance (*i.e.* across each cell and between cells), resulting in a decrease in the power output of the module.

However, it is not just the macroscopic uniformity that can affect the cell or module performance; microscopic variations in morphology, grain size, composition and pinholes can all have an influence. The types of non-uniformity are influenced by the type of thin film cell and their deposition techniques. For example, the characteristics of grain size and variation in composition are not applicable to amorphous silicon which is generally easier than polycrystalline CdTe or CIGS to deposit with sufficient uniformity over large areas. The polycrystalline thin film PV devices tend to have better long-term stability, not having the dangling bonds; however, passivation of grain boundaries is important. Compositional uniformity variations in ternary and quaternary thin film absorbers can alter the cut-off wavelength of the devices and charge collection efficiency; hence, this must be carefully controlled in materials such as CIS and CIGS. The slight grading of Ga composition in CIGS towards the junction is an advantage in creating a small built-in electric field to sweep carriers towards the junction.

The effect the density distribution of pinholes across the module has been studied by Koishiyev and Sites,[31] who showed that the distribution of microshunts across the cells can have a large effect on the reduction in PV module efficiency. A range of characterisation techniques from optical mapping through to laser beam induced current (LBIC) can be used to survey modules and cells for spatially distributed defects. The appropriate characterisation techniques for each of the material systems covered in the book are described in the following chapters. The nature of polycrystalline thin film materials presents a challenge to achieving uniform properties necessary to make high efficiency PV modules.

2.5   Conclusions

This chapter has introduced a number of topics that will be expanded in later chapters of this book. The emphasis has been on the properties of the different layers in a thin film solar cell and the relationship with device performance. For a single junction cell there is a fundamental limit to solar cell efficiency which is defined by the Shockley–Queisser limit. In practice, thin film solar cells do not get close to this limit but understanding the fundamental limitations to performance makes a start in understanding the performance of the materials in thin film solar cells. These practical limitations will become much clearer in later chapters and where the challenge of new materials developments could get us closer to the fundamental limits. This chapter has also introduced the reader to new concepts in extending the range of solar capture through bandgap engineering or through photonic materials achieving improved conversion of photon energy into electrical energy. Again, these themes will be expanded in chapters 8 and 9.

Acknowledgements

This research was supported by the RCUK Energy Programme through the PV SUPERGEN (PV Materials for the 21st Century) collaborative research project. We also acknowledge financial support from the Low Carbon Research Institute (LCRI) and the European Regional Development Fund (ERDF) for the SPARC Cymru collaborative industrial research project. Members of both collaborative research consortia are thanked for their contribution, along with members of the Centre for Solar Energy Research (CSER) team.

References

1. EIA, Table 13: 'OECD and non-OECD net electricity generation by energy source, 2010–2040 (trillion kilowatthours)', in *International Energy Outlook 2013*, US Energy Information Administration, Washington, DC, 2013.
2. V. Smil, 'Energy at the crossroads', paper presented at OECD Global Science Forum, Paris, 17–18 May 2006.
3. American Society for Testing and Materials (ASTM), reference solar spectral irradiance AM0 (ASTM E490) and AM1.5G (ASTM G173).
4. A. E. Becquerel, *C. R. Acad. Sci.*, 1839, **9**, 561.
5. C. E. Fritts, *Proc. Am. Assoc. Adv. Sci.*, 1883, **33**, 97.
6. D. M. Chapin, C. S. Fuller and G. L. Pearson, *J. Appl. Phys.*, 1954, **25**, 676.
7. R. L. Easton and M. J. Votaw, *Rev. Sci. Instrum.*, 1959, **30**, 70–75.
8. M. A. Green, in *Third Generation Photovoltaics: Advanced Solar Energy Conversion*, Springer-Verlag, Berlin, 2003, pp. 1–6.
9. P. W. Atkins, *Physical Chemistry*, Oxford University Press, Oxford, 5th edn, 1994, pp. 461–505.
10. S. M. Sze and K. K. Ng, *Physics of Semiconductor Devices*, John Wiley and Sons, Hoboken, NJ, 3rd edn, 2007, ch. 1.
11. G. Beaucarne, *Adv. OptoElectron.*, 2007, DOI: 10.1155/2007/36970.
12. D. J. Chadi, J. P. Walter, M. L. Cohen, Y. Petroff and M. Balkanski, *Phys. Rev.*, 1972, **B5**, 3058.
13. N. Amin, K. Sopian and M. Konagai, *Sol. Energy Mater. Sol. Cells*, 2007, **91**, 1202.
14. W. S. M. Brooks, S. J. C. Irvine and V. Barrioz, *Energy Procedia*, 2011, **10**, 232.
15. M. Mazzer, K. W. J. Barnham, I. M. Ballard, A. Bessiere, A. Ionnides, D. C. Johnson, M. C. Lynch, T. N. D. Tibbits, J. S. Roberts, G. Hill and C. Calder, *Thin Solid Films*, 2006, **511**, 76.
16. B. M. Kayes, H. A. Atwater and N. S. Lewis, *J. Appl. Phys.*, 2005, **97**, 114302.

17. S. Mirabella, R. Agosta, G. Franzo, I. Crupi, M. Miritello, R. Lo Savio, M. A. Di Stefano, S. Di Marco, F. Simone and A. Terrasi, *J. Appl. Phys.*, 2009, **106**, 103505.

18. M. Samadpour, A. I. Zad, N. Taghavinia and M. Molaei, *J. Phys. D: Appl. Phys.*, 2011, **44**, 045103.

19. M. Willander, P. Klason, L. L. Yang, S. M. Al-Hilli, Q. X. Zhao and O. Nur, *Phys. Status Solidi C,* 2008, **5**, 3076.

20. F. H. Mitchell Jr. and F. H. Mitchell Sr., in *Introduction to Electronics Design*, Prentice-Hall International, Upper Saddle River, NJ, 2nd edn, 1992, ch. 3, pp. 93–135.

21. A. J. Clayton, S. J. C. Irvine, E. W. Jones, G. Kartopu, V. Barrioz and W. S. M. Brooks, *Sol. Energy Mater. Sol. Cells*, 2012, **101**, 68.

22. W. Shockley and H. J. Queisser, *J. Appl. Phys.*, 1961, **32**, 510.

23. J. M. Olson, D. J. Friedman and S. Kurtz, in *Handbook of Photovoltaic Science and Engineering*, ed. A. Luque and S. Hegedus, Wiley, New York, 2003, ch. 9, p. 359.

24. M. A. Green, K. Emery, Y. Hishikawa, W. Warta and E. D. Dunlop, *Prog. Photovoltaics: Res. Appl.*, 2013, **21**, 827.

25. J. F. Geisz, D. J. Friedman, J. S. Ward, A. Duda, W. J. Olavarria, T. E. Moriarty, J. T. Kiehl, M. J. Romero, A. G. Norman and K. M. Jones, *Appl. Phys. Lett.*, 2008, **93**, 123505.

26. (a) A. Luque and A. Martí, *Phys. Rev. Lett.*, 1997, **78**(26), 5014; (b) A. Martí and A. Luque, *Adv. Sci. Tech.*, 2010, **74**, 143.

27. B. Lee and L.-W. Wang, *Appl. Phys. Lett.*, 2010, **96**, 071903.

28. R. T. Ross and A. J. Nozik, *J. Appl. Phys.*, 1982, **53**, 3813.

29. K. Ohdaira, H. Takemoto, K. Shiba and H. Matsumura, *Appl. Phys. Express,* 2009, **2**, 061201.

30. A. Kolodziej, *Opto-Electron. Rev.*, 2004, **12**, 21.

31. G. T. Koishiyev and J. R. Sites, *Mater. Res. Soc. Symp. Proc.*, 2009, **1165**, 203.

CHAPTER 3

Crystalline Silicon Thin Film and Nanowire Solar Cells

HARI S. REEHAL* AND JEREMY BALL

Department of Engineering and Design, London South Bank University, 103 Borough Road, London SE1 0AA, UK
*E-mail: reehalhs@lsbu.ac.uk

3.1 Introduction

Thin film silicon has a long history in photovoltaics. Effort has mostly focused on plasma-deposited amorphous hydrogenated silicon (a-Si:H) and microcrystalline silicon (μc-Si:H) which is a mixture of a-Si:H and Si nano-crystallites (nc-Si).[1,2] These materials lie at the heart of current thin film silicon-based photovoltaic (PV) manufacturing. The devices use the p–i–n structure with the intrinsic (i) layer being the absorber. Hydrogenated amorphous silicon–germanium alloys have also been used as the absorber layer.[3] The light-induced degradation problems in a-Si:H are well-known and microcrystalline Si was introduced to improve stability and cell performance. This has led to the so-called 'micromorph' technology which is a tandem a-Si:H and microcrystalline Si structure. Stabilised efficiencies of small area micromorph cells exceeding 12% have been reported.[4] Other multi-junction approaches include the triple junction a-Si:H/μc-Si:H/μc-Si:H and a-Si:H/a-SiGe:H/μc-Si:H designs with reported small area stabilised efficiencies of 13.6% and 13.3%, respectively.[4]

Fully crystalline forms of thin Si (without any amorphous content) offer the promise of stable device operation as well as the other attributes of crystalline Si. These include its non-toxicity, high abundance in the Earth's crust

RSC Energy and Environment Series No. 12
Materials Challenges: Inorganic Photovoltaic Solar Energy
Edited by Stuart J C Irvine
© The Royal Society of Chemistry 2015
Published by the Royal Society of Chemistry, www.rsc.org

and, importantly, a high efficiency potential despite its indirect bandgap and poor optical absorption properties. Single junction efficiencies exceeding 15% have been suggested for many years as being possible in μm scale thicknesses by using effective light trapping.[5] Conventional, fully crystalline, thin forms consist of thin film polycrystalline Si (with grain sizes ≥1 μm) and monocrystalline silicon produced either epitaxially[6] or by a variety of etching or lift-off processes. Lift-off approaches generally yield fairly thick layers (>10–20 μm) and have been studied for many years. Thinner (~1 μm) monocrystalline layers have also recently been demonstrated using lift-off.[7]

This review covers thin film polycrystalline Si (TF poly-Si) solar cell technology on glass. Some recent work on higher temperature substrates is also discussed, together with thin (<5 μm), monocrystalline Si planar forms produced by epitaxy or by lift-off from crystalline Si wafers. Developments in plasmonic concepts relevant to thin crystalline Si cells are described. Finally, progress in Si nanowire solar cells is discussed.

3.2 Planar Thin Film Crystalline Silicon Technology

The main approaches to the preparation of TF poly-Si since 2000 fall into two broad categories. The first is crystallisation of as-deposited amorphous layers using solid phase crystallisation and liquid phase crystallisation. The second uses a two-step process of forming a thin crystalline Si seed layer with large grains which is then epitaxially thickened. Temperatures have to be generally kept below ~600 °C for glass substrates, although excursions up to ~900 °C are needed to remove defects, improve crystallinity and optimise dopant activation. This means the use of borosilicate or aluminosilicate substrates rather than the cheaper soda lime glass. Research in these areas is summarised below together with some recent work on epitaxy and lift-off approaches for thin layers.

3.2.1   Crystallisation of Amorphous Silicon

3.2.1.1   *Solid Phase Crystallisation*

In terms of device performance, solid phase crystallisation (SPC) of a-Si:H has until recently been the most successful TF poly-Si formation technique. The process consists of thermally annealing the deposited films at ~600 °C for a time period of up to several tens of hours. Film thicknesses are typically in the range 1–3 μm. The average grain size of device grade SPC films is in the region of 1 μm with a high density of intra-grain defects as well as grain boundaries. Sanyo reported a 9.2% efficient cell in 1996.[8] Subsequently the process was developed by several groups, most notably CSG Solar using its novel point contact crystalline silicon on glass device technology[9,10] (see Figure 3.1).

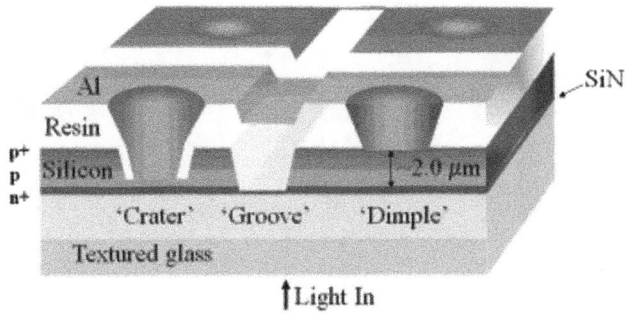

Figure 3.1 CSG Solar point contact crystalline Si on glass technology. After ref. 10.

In the CSG Solar technology, amorphous Si layers (~2 μm thick) with an n^+pp^+ structure were deposited by plasma-enhanced chemical vapour deposition (PECVD) onto textured and silicon nitride coated glass sheets. The coated sheets were then heated to 600 °C for about 24 hours to achieve SPC. This was followed by a short defect anneal at ~900 °C using rapid thermal annealing (RTA). Hydrogen plasma defect passivation was then carried out followed by contact patterning. An efficiency of 10.4% was realised in 2007 in a 94 cm^2 minimodule using a Si thickness of 2.2 μm with an open-circuit voltage (V_{oc}) of 492 mV per cell.[10] CSG Solar successfully demonstrated batch processing to mitigate the slow nature of SPC, but the limited V_{oc} of this technology prevented large-scale commercialisation and the company became insolvent in 2011.

In recent years electron beam (e-beam) evaporation has been investigated as a higher deposition rate (up to ~1 μm min^{-1}) alternative to PECVD for amorphous Si growth prior to SPC.[11,12] Using this technique, Sontheimer *et al.*[13] achieved an efficiency of 7.8% in minimodules fabricated on planar SiN-coated glass, confirming the electronic quality of e-beam deposited layers. These authors also reported faster SPC on aluminium-doped zinc oxide (ZnO:Al) coated substrates, which offer the potential of simpler device fabrication, due to a significant reduction in the activation energy of steady state nucleation from 5.0 eV on SiN to 2.9 eV on ZnO:Al. The SPC of a-Si on large grained poly-Si seed layers prepared by aluminium-induced crystallisation (see Section 3.2.2.1) has also been reported.[14] The presence of the crystalline seed layer leads to the transfer of its structure to the SPC layer. However, the performance of these cells has been limited with V_{oc} values below ~450 mV. Various other aspects of SPC technology have been studied. Examples include the influence on crystallisation of the annealing temperature[15] and the RTA process.[16]

3.2.1.2 Liquid Phase Crystallisation

Alternative techniques for a-Si crystallisation have been studied for many years. They include zone-melting crystallization processes using light sources and electron beams. An earlier review was provided by Bergmann.[17] Earlier

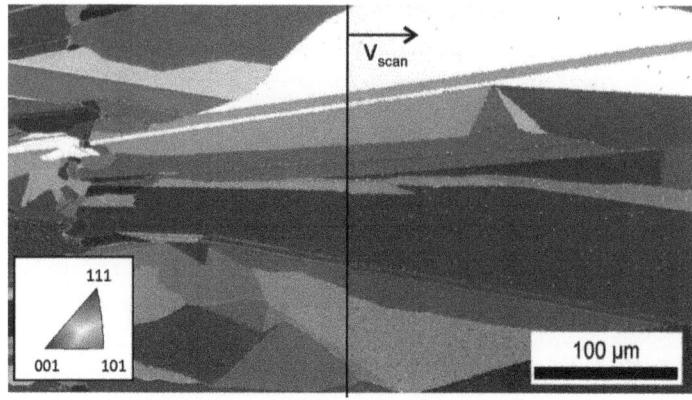

Figure 3.2 Electron backscatter diffraction images of poly-Si thin films on glass crystallised by a line-shaped electron beam (scan direction as indicated). The inset shows the grain orientation map. After ref. 20.

work generally focused on high temperature substrates such as ceramics. An example of recent work using e-beam zone-melting crystallization of 8–13 μm thick Si films is the report by Amkreutz *et al.*[18] The films were deposited by low-pressure chemical vapour deposition (LPCVD) at 670 °C on SiC_x coated high temperature glass substrates (Corning Eagle XG). Solar cells formed using a-Si hetero-emitters yielded efficiencies up to 4.7%. A V_{oc} of up to 545 mV was achieved showing the high electronic quality of the absorber layers. The SiC_x interlayer can crack during crystallisation, so in more recent work a combined interlayer consisting of 200 nm SiO_x followed by 20 nm SiC_x was used to resolve this issue whilst suppressing impurity diffusion from the substrate. The interlayers were sputtered whilst high rate electron beam evaporation was used to grow the 10 μm thick absorber layer. The V_{oc} was improved to 582 mV, although the overall efficiency was 4.3% due to a lower short-circuit current and fill factor.[19] The poly-Si absorber layers exhibit large grains up to ~100 μm wide and ~1 cm in length with low defect densities as shown by the electron backscatter diffraction (EBSD) image of Figure 3.2.[20]

Significant progress has been made in recent years in crystallizing a-Si layers on glass substrates using laser crystallisation (LC) employing line-focused diode lasers operating at 808 nm. Long crystal grains similar to the e-beam crystallised films are obtained upon solidification. Using this technique, Dore *et al.*[21] reported an initial efficiency of 11.7% with a V_{oc} of 585 mV for a 1 cm^2 area cell on borosilicate glass. The Si film was 10 μm thick and deposited by e-beam evaporation using a sputtered $SiO_x/SiN_x/SiO_x$ intermediate buffer stack. The structure of the basic cell is shown in Figure 3.3. The rear surface of the silicon was chemically etched to achieve a random texture to promote light trapping giving a short-circuit current density (J_{sc}) of 27.6 mA cm^{-2}. The efficiency and V_{oc} values are the highest reported to date for a TF poly-Si on glass solar cell. However, a degradation in efficiency was

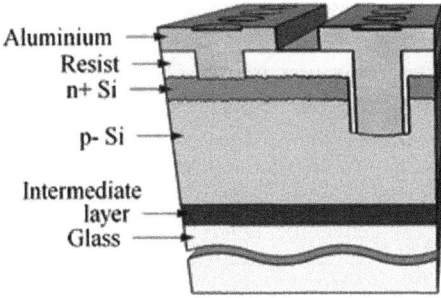

Aluminium
Resist
n+ Si

p- Si

Intermediate
layer
Glass

Figure 3.3 Device schematic for the basic cell structure of Dore *et al.* using a laser crystallised absorber layer. The cell is operated in a superstrate orientation. After ref. 21.

observed which has been attributed to the poor contact between the sputtered Al and the lightly doped Si absorber layer. A selective p^+ metallization scheme has been devised which was reported to eliminate the degradation but with a reduced efficiency of 10.4%.[21]

The use of high irradiance (several tens of J cm^{-2}) Xe flash lamp annealing has also been reported for the crystallisation of a-Si films several μm thick on quartz glass substrates. The work is at an early stage with a reported solar cell efficiency of 1.37% in a 2×2 mm^2 crystallised device.[22]

In summary, SPC films have small grain sizes and high levels of intra-grain defects which have limited device performance to around 10%. Liquid phase crystallisation yields better quality material with large (cm scale) grain sizes leading to higher efficiencies. It offers the potential for a successful crystalline Si thin film technology on glass if it can be implemented cost-effectively.

3.2.2 Seed Layer Approaches

The concept is based on first forming a thin layer of Si, the seed layer, with good crystalline quality and large grains. This is then epitaxially thickened to form the solar cell absorber layer with the crystalline structure of the underlying seed. The main techniques investigated for seed layer formation are aluminium induced crystallisation and laser crystallisation.

3.2.2.1 Seed Layer Formation by Aluminium Induced Crystallisation and Layer Exchange

Aluminium induced crystallisation (AIC) is based on the layer exchange of thin (<500 nm), adjacent a-Si or μc-Si and Al films when they are heated to below the eutectic temperature of Al (577 °C).[23] The subject has been studied extensively[24–26] and only the salient points are described below.

The starting layer sequence of substrate/Al/a-Si is the one most commonly employed. Both evaporated and sputtered Al layers have been successfully used. E-beam evaporation, sputtering and PECVD have been the techniques

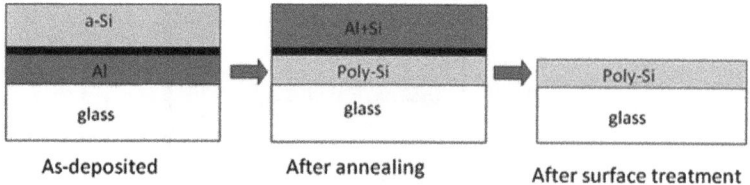

As-deposited   After annealing
                                                       After surface treatment

Figure 3.4   AIC process schematic where the thick black line represents the permeable interface between the Al and Si layers. Adapted from Fuhs *et al.*[27]

most widely used for Si deposition. The use of hot wire CVD (HWCVD) has also been reported. Upon annealing, Si atoms diffuse into the Al film through a thin permeable membrane at the Al–Si interface (usually aluminium oxide formed by atmospheric exposure). This is followed by nucleation of Si grains at the interface and grain growth within the Al layer until adjacent Si grains impinge. The Al is displaced towards the substrate leading to layer exchange and the transition of the a-Si into poly-Si which becomes p$^+$ doped by the Al to its solid solubility limit in Si of ~18 cm^{-3}. The poly-Si film thickness is defined by the thickness of the starting Al layer. The final layer sequence is substrate/poly-Si/Al+Si where the Al and Si phases segregate in the top Al rich layer. A schematic of the process is shown in Figure 3.4.

The poly-Si films generally exhibit a preferential (100) orientation of about 60–70% with average grain sizes in the region of ~10 µm, although grains exceeding 50 µm have been observed. The permeable membrane at the interface plays a critical role in controlling these parameters. Schneider *et al.*[28] have contributed to the theoretical understanding of the process. A property of the AIC process is that Si islands form on top of the poly-Si layer.[26] For subsequent epitaxial thickening of the seed layers, the Al and the Si islands have to be removed to leave a smooth surface. This is a critical factor for epitaxial thickening, especially at low temperatures. Various approaches to this have been implemented including wet chemical etching followed by mechanical abrasion,[26] chemical–mechanical polishing[27] and selective wet etching of Al combined with reactive ion etching of the remaining Si islands.[29]

Solar cells based on seed layers directly on glass require relatively complex contacting schemes due to the absence of a highly conducting back contact. To overcome this drawback, Gall *et al.*[26] successfully demonstrated seed layer formation by AIC on transparent conducting oxide (TCO) coated glass substrates where the TCO is ZnO:Al. The grain size was somewhat reduced (*e.g.* average size decreased from 7 to 5 µm under the conditions used) but, as on bare glass, the preferential (100) orientation was ~60%. Overall the properties were quite similar to layers formed on bare glass.

The reverse of the structure described above (R-AIC), with the starting layer sequence of substrate/a-Si/Al, has also been studied.[27,30] This yields the final layer sequence after layer exchange of substrate/Al+Si/poly-Si with the Al rich,

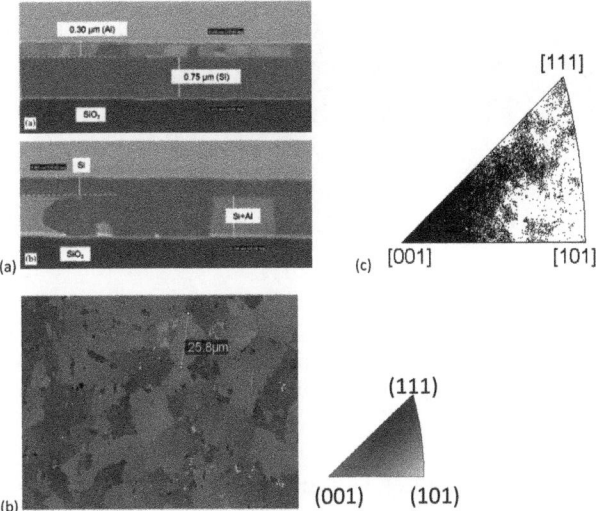

Figure 3.5 (a) Focused ion beam images showing the layer exchange process for the R-AIC geometry: top – before annealing; bottom – after annealing. (b) EBSD map of poly-Si grains after AIC with grain orientation colour key. (c) Inverse pole figure showing preferred (100) orientation. The poly-Si layer is ~0.3 μm thick. After ref. 30.

Al+Si mixed phase left underneath the poly-Si surface. Therefore, this geometry requires no Al etching step to expose the poly-Si layer and the Si islands form underneath the poly Si layer to leave a relatively smooth surface. In addition, the bottom Al+Si layer offers the possibility of forming a low resistance back contact. A cross section of the R-AIC process is shown in Figure 3.5(a). Figure 3.5(b) and (c) show an electron backscatter diffraction (EBSD) map and the corresponding inverse pole figure, respectively, confirming large grains with a preferred (100) orientation. Although Si island formation is avoided in this scheme, the as-formed poly-Si surface can still exhibit significant roughness and requires further treatment for successful absorber layer deposition. Excimer laser processing has been shown to improve surface morphology.[31]

3.2.2.2 Seed Layer Formation by Laser Crystallisation

Laser crystallisation of thin amorphous Si layers on glass has been an active field of study for many years driven by the thin film transistor (TFT) market, with pulsed excimer lasers being the dominant technology. The utility of this technique to crystalline Si solar cells has been studied. The fact that excimer lasers operate in the ultraviolet (UV) and vacuum ultraviolet (VUV) (wavelengths between 351 and 157 nm) restricts the Si layer thickness that can be crystallised to below ~100 nm. This is acceptable for seed layer formation but the grains are generally up to ~1 μm in size at best and possess a mixed

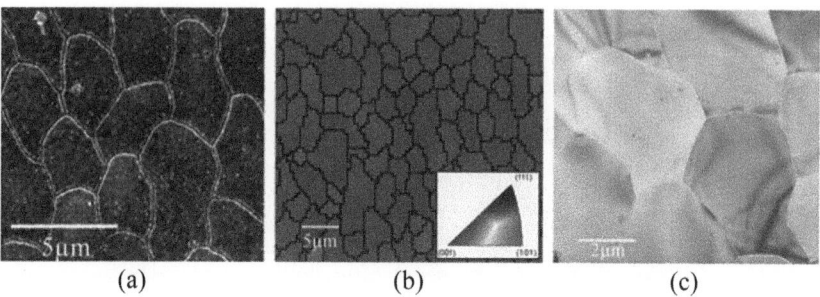

Figure 3.6   (a) SEM of Schimmel defect etched MPS sample. (b) EBSD map showing (100) texture. (c) Transmission electron micrograph (TEM) showing the grains are mostly devoid of intra-grain defects. After ref. 36.

orientation.[32,33] Pulsed copper vapour lasers operating at 511 or 578 nm enable the crystallisation of thicker layers. For example, *via* sequential lateral solidification using a copper vapour laser, Bergmann *et al.* obtained Si crystallites of several tens of μm in length in 400 nm thick films on glass.[34] More recently, Falk and collaborators[35] have shown that large grained poly-Si seed layers on glass substrates can be produced by scanning the line focused beam of a continuous wave (CW) diode laser operating at 806 nm, or of a green pulsed laser emitting at 515 nm. Grains exceeding 100 μm in size were formed in 400 nm thick amorphous Si starting layers by the CW laser, whereas the green laser yielded grains with sizes in the 10 μm range in 60 nm thick layers. Both approaches led to mixed orientation grains as evidenced by EBSD analysis.

New approaches studied include multiple-exposure mixed-phase solidification (MPS) using a scanning laser beam. Promising results have been reported with large grain size (~5 μm), nearly 100% (100) oriented, low intra-grain defect density poly-Si films being formed on quartz substrates as shown in Figure 3.6.[36] The data is for a 130 nm thick a-Si film on quartz processed using a CW laser operating at 532 nm.

3.2.2.3   Absorber Growth on Seed Layers and Device Results

Direct epitaxial growth has been widely reported for growing absorber layers on seed layers using a number of low to medium temperature techniques. Solid phase epitaxy of a-Si films deposited on the crystallised seed layers has also been investigated. To date, most of the work on solar cells on seed layers has used simple structures without any complex light trapping schemes thereby leading to modest short circuit currents.

The first successful epitaxial growth of Si on AIC seed layers used ion-assisted deposition (IAD) under non-UHV conditions.[37] The technique is based on e-beam evaporation and subsequent creation of Si ions which are accelerated towards the substrate by a low voltage (typically 20 V) to increase surface adatom mobilities. To obtain several µm thick smooth poly-Si films requires temperatures in the vicinity of 600 °C or above. Seed layers with a preferred (100) orientation were found to be advantageous for the epitaxial thickening. However, the development of solar cell structures using these layers has been challenging with typical as-grown devices having V_{oc} values in the region of 100 mV. A high density of intra-grain defects and impurities have been identified as the key factors limiting performance. Post deposition treatments such as rapid thermal defect annealing at ~1000 °C and remote plasma hydrogenation lead to a significant improvement in performance. However, performance has remained poor with best V_{oc} values of ~420 mV and efficiencies of ~2% in small area cells with ~2 µm thick absorber layers.[13]

Plasma deposition techniques such as electron cyclotron resonance CVD (ECRCVD) can also be used to provide low energy ion bombardment of the growth surface to increase adatom surface mobilities and reduce epitaxial growth temperatures to below 600 °C. Epitaxial thickening of AIC seed layers by ECRCVD has been reported by several authors including Gall *et al.*[25] and Ekanayake *et al.*[38] The best quality epitaxial layers below 600 °C are obtained on (100) oriented Si wafers. Typically, epitaxy breaks down above a thickness of ~2.5 µm on (100) Si and at much lower thicknesses for other orientations. This has consequences for the epitaxial thickening of AIC seed layers with their mixed grain orientation. At best, epitaxial thickening of ~70–80% of the surface has been reported.[27]

Like the IAD deposited absorber case, the performance of solar cells fabricated from ECR thickened seed layers has been limited. Best reported efficiencies in small area (4 × 4 mm^2) cells on glass, with inter-digitated contacts and a 2 µm thick absorber layer, were in the region of 1% with a V_{oc} of 397 mV. This was achieved after post-deposition defect annealing using RTA at 900 °C and hydrogen plasma passivation.[39] Reference cells grown on p$^+$ (100) Si wafers exhibited an efficiency exceeding 4.2% without light trapping, with a V_{oc} of 458 mV before defect annealing or passivation. Analysis of the layers using Seeco etching has shown that they exhibit a very high density ($\geq 10^9$ cm^{-2}) of extended defects and growth regions of different structural quality which limits device performance.[40] Interestingly, in a study of cells prepared on AIC seed layers with the reverse structure (R-AIC) formed on a silver–indium tin oxide (ITO) back contact, Jaeger *et al.* reported cell efficiencies exceeding 5% using 2 µm thick non-epitaxial absorber layers grown by PECVD at 180 °C.[41]

Epitaxial thickening of AIC seed layers on glass using high rate e-beam evaporation under non-UHV conditions has been systematically studied by the group at Helmholtz–Zentrum, Berlin.[42] As for the ECRCVD films, the layer structural quality depends strongly on the orientation of the underlying substrate, with (100) orientation producing the best results and pointing to the importance of a high preferential (100) orientation of the seed layers. Layers grown at 600 °C on Si (100) substrates exhibited no extended defects whereas defects were present on other orientations. Films grown on seed layers do exhibit defects and it has been suggested that these originate from imperfections at the seed layer surface. The defect densities are lower than in ECRCVD films. This translated to superior performance in solar cells. The best poly-Si cell on glass was reported to have an efficiency of 3.2% using an absorber thickness of 2.2 μm. The V_{oc} was 407 mV after RTA defect annealing and hydrogen plasma passivation. A 1.8 μm thick reference cell grown on a p^+ (100) Si wafer at 650 °C without any light trapping exhibited an efficiency of 5.86%, with a V_{oc} of 570 mV. The same group also reported TF poly-Si cells grown on AIC seed layers formed on ZnO:Al coated glass. This allows for a simpler contacting scheme and light trapping. Solar cells using 2 μm thick, e-beam evaporated epitaxial absorbers achieved a V_{oc} of 389 mV and an efficiency of 2%.[43]

Thickening of AIC seed layers has also been reported using HWCVD to produce solar cell structures on glass. For example, Wang *et al.*[44] reported an initial efficiency of 5.6% with a V_{oc} of 470 mV for a cell where the AIC seed layer was formed on a Ti contact layer. The absorber layer had a p–i–n configuration and a microcrystalline structure with a crystalline fraction of 93%.

Laser crystallised seed layers have been investigated to see if they offer better structural quality and hence better performance poly-Si on glass solar cells. Andra *et al.* studied layered laser crystallisation (LLC) whereby the laser crystallised seed layer was epitaxially thickened by simultaneous deposition of a-Si using e-beam evaporation and repeated pulses of an excimer laser. A V_{oc} of 517 mV was achieved with an efficiency of 4.2% using a 2 μm thick absorber layer, without an anti-reflective coating (ARC) or light trapping.[45] However the approach is difficult to scale up. Schneider *et al.* reported solid phase epitaxial (SPE) thickening of e-beam deposited a-Si on CW laser crystallised seed layers.[46] The CSG Solar contacting technology was used and an efficiency of 4.9% reported in a minimodule (12 series connected cells). Cell results based on excimer laser crystallised seed layers with grain size up to 1100 nm have been reported.[35] SPE of amorphous films on the seed layers gave higher V_{oc} values (up to 443 mV) compared with e-beam deposited epitaxial absorber layers.

The best performance in epitaxially thickened AIC seed layer cells has been reported using higher temperature substrates and thermal CVD at ~1100 °C by IMEC. IMEC used alumina and transparent glass ceramic substrates and achieved a record efficiency of 8.5% on alumina using a heterojunction emitter and an inter-digitated contacting scheme.[47] A 2–3 μm thick absorber layer was grown on a 250 nm thick seed layer followed by remote plasma hydrogen defect passivation and plasma texturing. The textured surface improved light trapping, and together with an ITO ARC, led to a J_{sc} of 21.6 mA m^{-2}, a V_{oc} of 523 mV and a fill factor (FF) of 75.8%, giving an efficiency of 8.54%. An efficiency of 6.4% was reached on the glass ceramic. The structure of the 8.54% efficient device and light *I–V* curves are shown in Figure 3.7. Detailed analysis showed that the V_{oc} of the cells was almost independent of the grain size and that a high density (~10^9 cm^{-2}) of electronically active

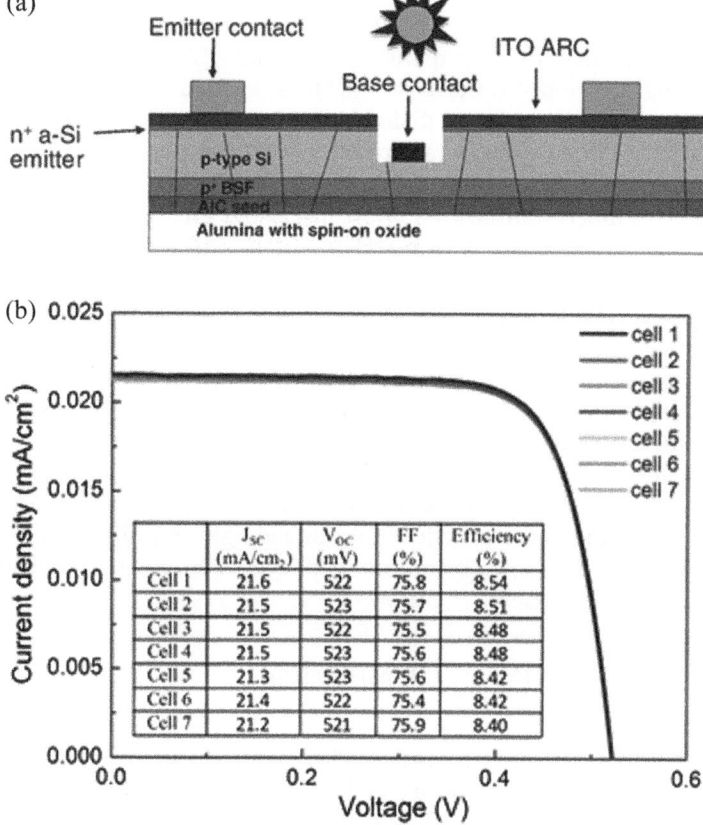

Figure 3.7 (a) Device structure and (b) light *I–V* characteristics of an 8.5% efficient cell grown on an AIC seed layer using thermal CVD. After ref. 47.

intra-grain defects in the epitaxially thickened AIC layers was limiting cell performance.[48]

The details of the AIC process, and the status and potential of AIC solar cell research, were reviewed recently by Van Gestel *et al.*[49] The high density of intra-grain defects and mixed grain orientation are key factors limiting performance. They concluded that further material improvement and the use of advanced light trapping schemes could lead to J_{sc} values above 30 mA cm^{-2}. V_{oc} values will need to improve significantly to >580 mV if efficiencies in the region of 14% are to be achieved. This is a challenging requirement and it is not clear if these improvements can be realised in a cost-effective process to compete with laser and e-beam crystallisation approaches.

The use of monocrystalline Si seed layers has been investigated by Branz and co-workers using a heterojunction cell structure with an a-Si emitter.[50] They used HWCVD at 730 °C to epitaxially thicken 450 nm thick monocrystalline Si (100) layers, oxide bonded to Corning EAGLE XG display glass. The best device had a V_{oc} of 460 mV and an efficiency of 4.8% using an absorber thickness of 2.5 μm and without any RTA, defect passivation or light trapping. Reference wafer based devices reached an efficiency of 6.7% with a V_{oc} of 570 mV. High dislocation densities were reported to be a limiting factor due to the lower growth temperature used compared with thermal CVD.

The use of monocrystalline seed layers has also been reported by IMEC.[7] Its seed layers were created by transferring 300 nm thick, (100) oriented monocrystalline Si layers onto glass ceramic substrates using Corning's anodic bonding and implant-induced separation technology.[51] Seed layer size is limited by the size of the starting Si wafer so that wafer scale cells are, in principle, possible. Epitaxial thickening of the wafers by high temperature thermal CVD followed by plasma texturing led to an efficiency of 10.8% in a 1 cm^2 inter-digitated device with an absorber thickness of ∼8 μm. V_{oc} values exceeding 600 mV were recorded illustrating the good layer quality (defect density reduced to ∼10^5 cm^{-2}). The J_{sc} values were still relatively low (24.3 mA cm^{-2}) due to the absence of advanced light trapping schemes. The good V_{oc} values using monocrystalline seed layers demonstrate the potential of the seed layer approach.

3.2.3   Lift-Off and Epitaxy Approaches

The preparation of thin monocrystalline Si wafers by lift-off from a parent wafer has been extensively reported. There are two main approaches: (a) induced cleaving; and (b) porous Si based methods.[52] Cleaving methods generally lead to layer thicknesses in the range of 20–50 μm and are not discussed further. Porous Si based methods involve the creation of two porous Si layers in the top surface of the wafer. A high porosity layer sits beneath a lower porosity upper layer which can support the growth of epitaxial silicon by high temperature CVD. The mechanically weaker, high porosity layer enables the epitaxial layer to be detached from the substrate which can be re-used. Using this method an efficiency of 19% in a 43 μm thick cell with an aperture area of 3.98 cm^2 has been reported.[53]

Progress on thinner layers has been limited. IMEC has reported the production of ultra-thin (∼1 μm) monocrystalline silicon layers using lift-off by the so-called 'epifree' approach.[7] This relies on the formation of a uniform array of cylindrical micropores in the surface of a mono-crystalline wafer, followed by annealing at high temperature. The annealing leads to merging of the pores into a wide, plate-like void under a thin film which can subsequently be detached. An efficiency of up to 4.1% was reported in simple, 'proof of concept', solar cells fabricated on films with a thickness of ∼1 μm. The cell fill factor was 75%, but V_{oc} and J_{sc} were low at 426 mV and 12.8 mA cm^{-2}, respectively, suggesting the need for significant improvements in passivation and thicker films. The epifree process is illustrated in Figure 3.8.

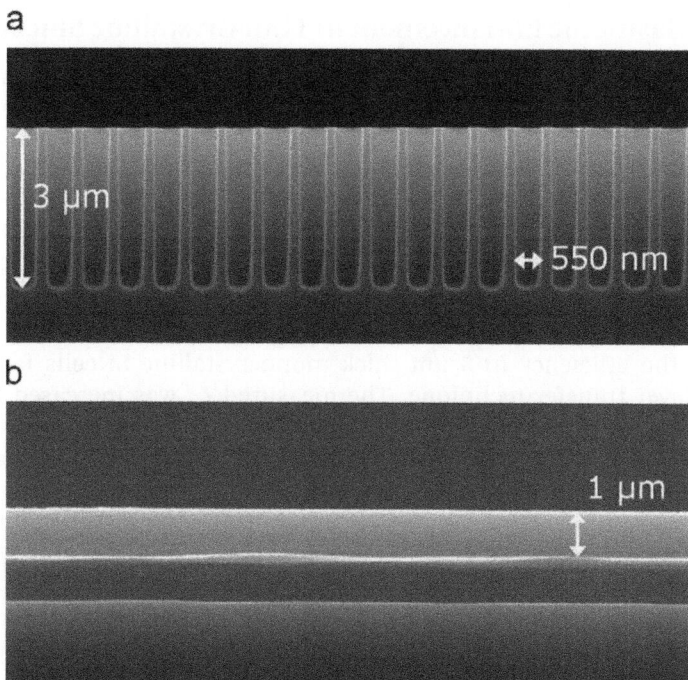

Figure 3.8 (a) Regular pores with a diameter of 550 nm and a pitch of 800 nm obtained by deep UV lithography and reactive ion etching. (b) Transformation of pores into a single void with an overlaying film of 1 μm thickness after annealing for 60 min at 1150 °C in hydrogen. After ref. 7.

In a development of the IMEC work, Hernandez *et al.*[52] reported a multiple 'epifree' process (which they term 'silicon millefeuille') whereby several crystalline silicon layers with thickness in the ∼1–7 μm range can be produced from a single wafer in a single technological step. The work is at an early stage and the performance of the layers in devices has not yet been reported.

Turning to the direct production of crystalline layers, Cariou *et al.*[54] reported an approach to produce epitaxial silicon films by RF-PECVD at a very low growth temperature of 165 °C. These epitaxial layers have been used as absorbers in heterojunction solar cells. For a 2.4 μm thick absorber, a V_{oc} of 546 mV, FF of 77%, J_{sc} of 16.6 mA cm^{-2} and an efficiency of 7% was achieved in a cell with an area of 4 cm^2 without any light trapping features. The same group has reported a process based on low temperature (200 °C) RF-PECVD to produce ultra-thin crystalline silicon films (0.1–1 μm) on flexible substrates.[55] It was suggested that, by optimising the flow of H_2 gas during processing, very highly crystalline films can be grown on an interface mainly composed of micro-cavities to facilitate lift-off. Device results on these layers are awaited.

3.2.4   Plasmonic Enhancement in Thin Crystalline Silicon Cells

Due to the weak optical absorption of crystalline silicon near its band edge, effective light trapping is a requirement to achieve high performance in thin film crystalline silicon solar cells. Traditional light trapping schemes used in wafer based silicon technology involve pyramid texturing on the micron scale and are not feasible in very thin structures.

Several new approaches to light trapping in thin film silicon solar cells have been proposed in recent years. These include the use of photonic crystals as back reflectors. A combined grating and one-dimensional photonic crystal as a distributed Bragg reflector has been shown to enhance the efficiency in 5 µm thick monocrystalline Si cells fabricated using a layer transfer technique. The measured J_{sc} was increased by 19% compared with a theoretical prediction of 28%.[56] Becker *et al.* have summarised approaches based on replacing the planar absorber layer by a periodically structured nanophotonic thin film formed using nano-imprint lithography (NIL), which is a promising technology for fabricating submicron light trapping textures on large areas.[20] They reported a significant increase of absorption over the range 400 to 1100 nm in a solid phase crystallised 2 µm periodic, poly-Si microhole array with a nominal thickness of 2.1 µm fabricated using NIL. Results on how these structures perform in devices are awaited.

Another method for achieving light trapping in thin film solar cells that has attracted considerable attention in recent years, and is the focus for the remainder of this section, is the use of metallic nanostructures that support surface plasmons.[57] These are excitations of the conduction electrons at the interface between a metal and a dielectric. Both localised surface plasmons (LSPs) excited in metal nanoparticles and surface plasmon polaritons (SPPs) propagating at the metal–semiconductor interface can contribute to light trapping in thin cells. Plasmonic nanoparticles have a very strong interaction with light near the resonance frequency. Incident light can be either absorbed or scattered, and the contribution from each mechanism depends on the size, shape and composition of the particle as well as the surrounding medium.[58] The nanoparticles can be applied to a planar semiconductor layer and remove the need for rough textured surfaces. Theoretically, plasmonic light trapping schemes can outperform conventional light trapping schemes based on surface texturing.[59]

Excited LSPs can decay radiatively resulting in scattering, or nonradiatively which gives rise to absorption. Absorption dominates for small particles <50 nm in size, whereas larger particles up to ~100 nm are more efficient scatterers.[58] For thin film poly-Si solar cells, low absorption losses across the visible and near infrared (NIR) region and large scattering cross-sections are required, particularly in the NIR where transmission losses are more significant. Silver has been studied widely as a nanoparticle as its resonance wavelength is in the visible and can be tuned towards NIR wavelengths. It provides the highest scattering and lowest absorption for Si solar cells compared with other metals such as Au, Cu and Al.[58] Al nanoparticles have been shown to be unsuitable for polycrystalline Si solar cells due to the presence of an interband region in the NIR. The opposite is found for Au nanoparticles, which feature an interband threshold region in the visible that makes their optical properties suitable for crystalline and polycrystalline silicon solar cells.[60] However, Au is a well-known killer centre in crystalline Si. Review of the field have been given by Catchpole and Polman[61] and Pillai and Green.[62]

Arrays of metal nanoparticles can be applied to the front or rear of thin film solar cells. Many designs have used Ag nanoparticles deposited by self-assembly (annealing of nm scale thick films) on the front, illuminated surface of fully fabricated solar cells. This approach led to an increase in photocurrent of up to a factor of 16, at a wavelength of 1050 nm, in 1.25 µm thick silicon-on-insulator solar cells.[63] However, these designs frequently suffer from absorption losses in the nanoparticles in the blue end of the spectrum.[64] Locating the nanoparticles at the rear of the cell offers the advantage of short wavelength light being absorbed in the cell before reaching the particles, while the long wavelength light reaching the rear of the cell can be scattered back and trapped. This is supported by numerical simulations based mostly on finite difference time domain (FDTD) methods.[65,66] The role of very thin dielectric spacer layers between the Si and the rear located nanoparticles has been investigated. The scattering cross-section was found to increase dramatically with a decrease in thickness of the dielectric spacer layer where this layer was <10 nm thick.[67]

Experimental efforts on applying particles to the rear of functioning TF poly-Si cells are at an early stage and largely come from research at the University of New South Wales in Australia. They include the work of Ouyang *et al.*[68] who prepared 2 µm thick polycrystalline Si cells on glass by solid phase crystallisation of e-beam evaporated a-Si precursor films. The cells had a superstrate structure with inter-digitated contacts as shown in Figure 3.9. The planar borosilicate glass substrates were coated with a silicon nitride (SiN_x) layer which served as a diffusion barrier and an anti-reflection layer. Random arrays of Ag nanoparticles were formed using self-assembly by annealing a thin Ag film at ~200 °C on the rear surface of the cell, with and without a thin dielectric spacer layer. The particle size was of the order of 150–250 nm, with a coverage of ~50%. LSPs excited in the nanoparticles were shown to increase light absorption in the Si films and enhance the spectral response and efficiency. Nanoparticles formed directly on Si enhanced the J_{sc} by 29% and the efficiency by 23%, more than double the enhancement provided by the particles on a 30 nm thick SiO_2 spacer layer. A combination of the Ag nanoparticles directly on the Si surface with a detached white paint back surface reflector further improved the J_{sc} and efficiency enhancement to 38% and 31%, respectively.

In subsequent work on similar devices a J_{sc} enhancement of 44% was achieved when a nanoparticle/magnesium fluoride/diffuse paint back surface reflector structure was employed.[69] Again, the optimum arrangement

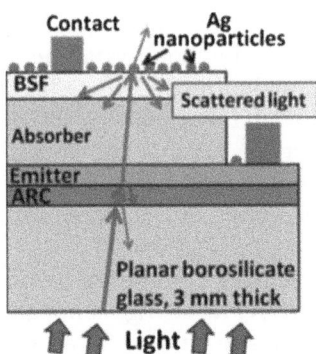

Figure 3.9 Plasmonic thin film poly-Si solar cell structure with back surface field (BSF). After ref. 68.

was when the Ag nanoparticles were formed directly on the rear Si surface of the cells without using the thin spacer layer.[70] Rao *et al.*[71] investigated the change of the optical properties of the silver nanoparticles when overcoated with different dielectric layers (MgF_2, Ta_2O_5 and TiO_2). They found that TiO_2 (highest refractive index) provided the highest absorption enhancement of 75.6%. However, the highest J_{sc} enhancement of 45.8% was achieved with the lower refractive index MgF_2 coating. Optimizing the thickness of the latter, together with the use of a diffuse white paint back reflector, gave a final J_{sc} enhancement of 50.2%.

One of the challenges with plasmonic solar cells is the large area fabrication of nanoparticle arrays having the desired size, shape and distribution. Self-assembly by thermal annealing provides poor control over these parameters whilst particle distribution is a limitation of techniques using colloidal suspensions. Electron beam lithography is limited to small area research devices. Alternatives with scale-up potential such as nanoimprint lithography are being explored but the focus so far has been on applying these to amorphous Si cells. An example is the work of Ferry *et al.*[72] who fabricated cm^2 scale a-Si:H solar cells using soft nanoimprint lithography to incorporate plasmonic nanostructures in ordered arrays. A schematic diagram, photograph of a finished solar cell substrate and scanning electron microscope (SEM) images of the patterned substrate and substrate cross section are shown in Figure 3.10. The structures were printed into a sol–gel silica layer which was coated with 200 nm of Ag followed with the growth of a 130 nm ZnO:Al layer, n–i–p a-Si:H solar cell and 80 nm ITO top contact. When compared with planar control cells the plasmonic structures exhibited higher photocurrents. The maximum performance for a cell thickness of 160 nm was seen for 250 nm diameter particles at a pitch of 500 nm which showed a 50% higher J_{sc} compared with the flat reference cells, and 10% higher than randomly textured Asahi cells. The best efficiency was 6.6%.

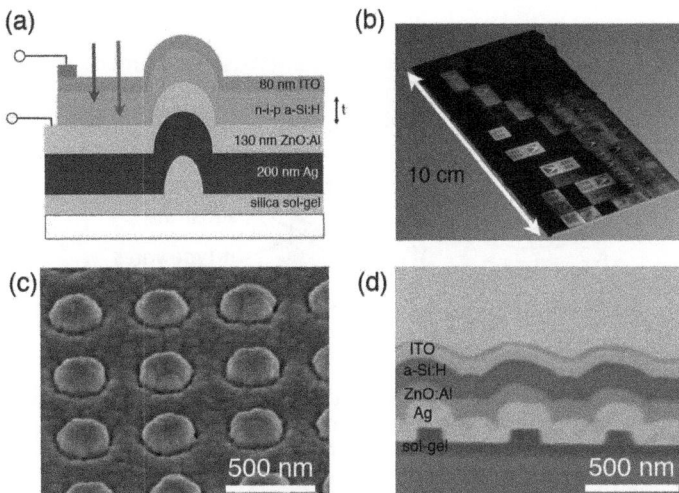

Figure 3.10 Plasmonic light trapping a-Si:H solar cell design: (a) schematic cross section; (b) photograph of finished imprinted patterned solar cell substrate where each square is a separate device with different particle diameter and pitch; (c) SEM image of Ag over coated patterns showing 290 nm diameter particles with 500 nm pitch; and (d) SEM image of a cross-section of a fabricated cell, cut using focused ion beam milling. From ref. 72.

In summary, promising results showing J_{sc} enhancements have been seen in plasmonic thin film Si solar cells. However, reported improvements in absolute efficiency are limited and generally in the region of 1% or below at best.

3.3 Silicon Nanowire Solar Cells

Single junction solar cell geometries based on crystalline Si nanowires (SiNW) with radial junctions are attracting considerable interest as they provide new opportunities for enhanced light trapping and increased performance, as well as using significantly less silicon compared with wafer cells.[73,74] The radial geometry is illustrated in Figure 3.11. The direction of solar radiation is parallel to the wire axis such that light is absorbed parallel to the p–n junction. Therefore absorption and carrier separation are orthogonalised, with minority carrier generation close to the junction. This means carrier diffusion lengths required are much shorter than in wafer Si cells which relaxes the requirements on material purity. Measurements on single wire test structures and optoelectronic simulations suggest that large area SiNW solar cells have the potential to exceed 17% energy conversion efficiency.[75]

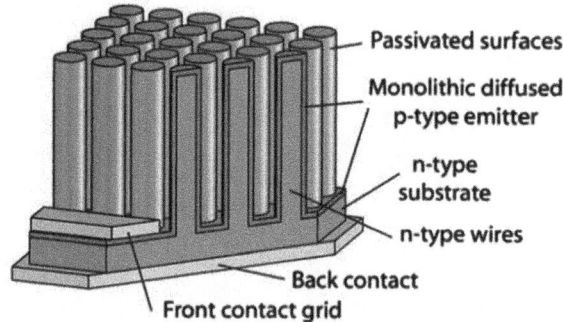

Passivated surfaces

Monolithic diffused
p-type emitter

n-type
substrate

n-type wires

Back contact

Front contact grid

Figure 3.11 Radial cell geometry showing the short diffusion distance inherent in a simple SiNW cell design. From ref. 75.

Methods of wire formation fall into two main categories, bottom–up and top–down. The vapour–liquid–solid (VLS) method is the most widely studied bottom–up approach wherein wire growth takes place from the vapour phase using a source of Si. Many techniques for supplying the Si and synthesising wires have been studied,[76] with chemical vapour deposition (CVD) and related approaches being the most common. Consequently, our discussion is restricted to CVD based approaches with wire diameters above the region where quantum confinement effects come into play (<10 nm). In the latter regime the Si bandgap increases and SiNWs have been considered for the top cell of an all Si tandem structure.[77]

3.3.1   SiNW Growth using the Vapour–Liquid–Solid Method

The VLS method has attracted considerable attention. It was first reported by Wagner and Ellis in 1964[78] and involved placing a small particle of Au on a Si (111) growth substrate followed by CVD using $SiCl_4$ and H_2 at 950 °C. SiNW growth was observed at the site of the Au particles with the particles being present at the tip of the wires during growth. It was suggested that the Au particle formed a liquid alloy droplet with the substrate material whilst undergoing heating and acted as a catalyst for arriving Si atoms, resulting in saturation of the droplet under continual gas flow. The excess Si, containing a small concentration of Au impurity, was expelled causing the droplet to rise from the substrate on the top of the wire as shown in Figure 3.12(a). SiNWs with diameter and length up to ~0.25 μm and ~2 μm, respectively, were reported. The VLS effect is suggested to take place in the area of the Au–Si phase diagram to the right and above of the eutectic point and below the liquidus line, identified as 'core growth' in Figure 3.12(b).[79] Thickening the core with a shell is possible if deposition conditions are varied as shown in Figure 3.12(b).

The favourable phase diagram of the Au–Si system and the inertness of Au has resulted in it being the most widely used catalyst for SiNW growth. Unfortunately, as mentioned previously, Au forms deep levels in the Si energy

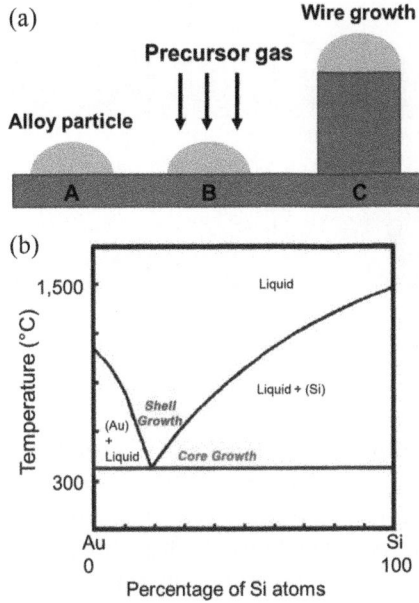

Figure 3.12 (a) Model of the VLS effect: (A) metal alloy droplet; (B) droplet acting as sink or catalyst for precursor gas carrying Si atoms; and (C) Si is expelled from the droplet in a continuous process causing the droplet to rise on top of the wire. (b) Binary phase diagram of Au–Si material system with areas of core and shell growth. From ref. 80.

bandgap and is detrimental to PV performance. As a result many other catalyst metals have been investigated. They include Ag and Al which have similar phase diagrams. However, Ag also forms deep levels in the bandgap and the sensitivity of Al to oxidation is an issue. Low Si solubility metals such as In, Ga and Sn have also been studied and are attractive due to their low eutectic temperatures, but the low surface tensions of these metals and the low solubility are impediments. However, successful wire growth using these metals has been reported by PECVD.[23] Successful SiNW growth has also been achieved using silicide forming metals such as Cu and Ni but the growth temperatures required are high, in excess of ~800–900 °C. The role of different catalyst metals has been discussed by Schmidt *et al.*[76] in a wide ranging and comprehensive review of SiNW growth. Their review also covers the mechanisms of wire growth including thermodynamics aspects.

As the SiNW solar cell concept requires wide area arrays of SiNWs, the position and morphology of the wires is of interest. The growth site and diameter are influenced by the initial position and diameter of catalyst particles.[81] SiNWs have been grown from colloidal Au spheres deposited on the growth substrate[82,83] giving a degree of diameter control. Alternative methods include self-organisation whereby particles are formed from deposited thin metal films. In this method, the heated metal film de-wets

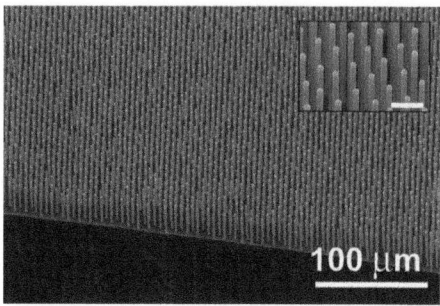

Figure 3.13 Tilted SEM views of a Cu-catalysed Si wire array over a 1 cm^2 area. The scale bar in the inset is 10 µm. From ref. 89.

from the surface forming metal–substrate material alloy particles[84,85] with a pseudo random position and diameter. This is reflected in the grown wire arrays. Particle formation is influenced by factors including annealing temperature,[86] metal layer thickness[87] and substrate orientation.[88] Well-ordered arrays of SiNWs have been demonstrated by confining the catalyst metal within patterned oxide masks produced using photolithography and metal lift off techniques.[89] An example of high fidelity wire growth using the latter technique is given in Figure 3.13. Arrays of vertically oriented Si wires with diameters of 1.5 µm and lengths of up to 75 µm were grown on Si (111) over areas >1 cm^2 using CVD and SiCl$_4$ as the precursor gas. Cu was used as the catalyst metal with optimal growth occurring between 1000 and 1050 °C.[89]

3.3.1.1 Wire Structure, Growth Direction and Crystallography

Typically, silicon wires grown *via* the VLS effect on single crystal Si substrates by CVD grow epitaxially and are single crystal in structure.[76,78,90] The growth direction and crystal orientation are affected by the underlying growth substrate. Three growth directions, $\langle 111 \rangle$, $\langle 110 \rangle$ and $\langle 112 \rangle$, have been observed on single crystal wafer substrates. Wagner[91] has suggested that wire growth direction is influenced by a 'single lowest free energy solid liquid interface', this being parallel to the (111) plane. It has been observed by Wu *et al.*[81] and by Schmidt *et al.*[92] that the wire growth direction is affected by diameter. In general, wires with diameters in excess of ~50 nm on (111) oriented silicon substrates grow in a $\langle 111 \rangle$ direction and hence normal to the substrate surface. Due to this preferential growth direction, wires grown on substrate orientations other than (111) exhibit different growth angles. On Si (100) growth is angled at 35.3° in four different directions at 90° to each other. The substrate dependant growth orientation has been suggested by Fortuna and Li[93] to only be present on surfaces free from native oxide.

While the initial substrate orientation has a large influence on wire growth direction, changing growth conditions can modify the final growth direction giving rise to kinking. The importance of temperature and precursor partial pressure on kinked growth has been shown by Westwater *et al.*[94] Using SiH_4 as a source gas it was observed that the higher the temperature and lower the partial pressure, the greater the kink free growth regime and larger the SiNW diameter. Schmidt *et al.*[76] have discussed the morphology of SiNWs including defects. The presence of defects has been found to become more pronounced in wires grown at lower growth temperatures using plasma assistance with low solubility catalysts such as Sn, Ga and In.[87,95–98] Interestingly, these wires also exhibit significant levels of tapering.

3.3.1.2 Optical Properties

SiNW arrays exhibit enhanced absorption compared to a Si layers of the same thickness. A considerable degree of simulation and modelling work has been carried out in this area.[99–101] The analysis by Hu and Chen[102] of a modelled square array of SiNWs showed a low reflection from the wire arrays of <5% over the studied wavelength range. At wavelengths below ~500 nm, levels of absorption were in the region of 90–95% compared with 40–60% for a planar thin silicon film. Disorder in SiNW arrays has been shown by Lagos *et al.*[103] to increase absorption. Bao and Ruan[104] have used finite difference time domain (FDTD) simulations to study this using a 4 × 4 ordered square array as a control. This had SiNWs of 100 nm diameter, length of 2 μm and a centre to centre spacing of 200 nm. The effects of disorder in wire length, position and diameter were studied. Compared with the control, arrays with wires in random positions had a slightly improved absorption but a similar reflection. When length was randomised absorption was improved and reflection reduced, while randomised diameter increased and broadened the absorption spectrum. The enhanced absorption was attributed to inter-wire scattering and resonance effects.

Optical characterisation of as grown wire arrays has been reported by several groups including Stelzner *et al.*,[105] Tsakalakos *et al.*,[106] Convertino *et al.*,[107] Kuo *et al.*[108] and Muskens *et al.*[109] All observed low reflectivity ranging between ~1% and 20% dependent on wavelength. When the wire arrays were grown on glass,[105,106] the measurement of transmission across a range of wavelengths typically resulted in results in the 1–2% range up to ~700 nm. The absorption results presented by Tsakalakos *et al.*[106] calculated as (1-T-R) indicated very high levels of absorption, typically over 90% in the visible. The properties of tapered nanowires and nanocones have also been investigated. These structures provide both efficient anti-reflection properties and absorption enhancement over a broadband spectrum and a wide range of angles of incidence.[110,111]

3.3.1.3 SiNW Solar Cells Grown Using Au Catalyst

Radial junction SiNW solar cells grown *via* the VLS effect generally employ the p–n or p–i–n geometry with intrinsic and doped shells formed over the nanowire core. Early SiNW solar cells were pioneered by the Lieber group who grew p-type wires and then, with reduced pressure and increased temperature, deposited phosphine doped n-type shells.[112,113] Tian et al. grew p–i–n SiNWs.[80] Test results on a single SiNW cell yielded a V_{oc} of 260 mV and a FF of 55% yielding efficiency values between 2.3 and 3.4%. In 2007 Tsakalakos and co-workers grew a SiNW cell with a radial junction on a stainless steel substrate by VLS growth using CVD.[114] A 40 nm thick n-type a-Si:H shell was deposited by PECVD to create a radial p–n junction on p-type SiNWs 109 ± 30 nm in diameter and ~16 μm long. Optical reflectance was reduced significantly compared with planar cells but device performance was poor, with devices of area 1.8 cm^2 having a V_{oc} of ~130 mV, a short-circuit current (I_{sc}) of 3 mA and an FF of ~only 28%. Reasons for the poor performance were stated as nanowire geometry, regions of localised shunting, high contact resistance and the use of the Au catalyst. Using a p–i–n radial junction on a p-type wafer, the same group improved the efficiency to 1.18% with a V_{oc} of 348 mV, a J_{sc} of 7.12 mA cm^{-2} and an FF of 47.6%.[115]

Gunawan and Guha[116] have reported a SiNW cell grown by the VLS method with n-type wires on an n-type monocrystalline substrate. The wire diameter was 80–100 nm with lengths between 1.5 − 2 μm. A p-type emitter was deposited on the wires to form the junction together with an Al front contact and an In–Ga back contact. The cell was on a 10 mm × 5 mm, 3 μm deep mesa structure. The highest efficiency reached was ~0.9% with a V_{oc} of ~300 mV and a J_{sc} of ~11 mA cm^{-2}. The presence of the Au impurity in the wires and surface recombination were identified as key limiting factors. The use of a conformal Al_2O_3 film as a surface passivation layer increased efficiency to 1.8%.

Several authors, including Perraud et al.,[117] have reported cells consisting of n-type SiNWs grown on p-type wafers The 40–50 nm diameter, 1 μm long SiNWs were imbedded in a spin on glass (SOG) matrix. The surface was planarised for contacting and a front contact finger grid of Ni/Al on ITO applied together with and an Al back contact. An efficiency of 1.9% was reported with a V_{oc} of 250 mV, J_{sc} of 17 mA cm^{-2} and an FF of 44%. Kuo et al.[118] reported an efficiency of 3.47% in a SiNW on a p-type substrate cell with a V_{oc} of 500 mV, a J_{sc} of 11.62 mA cm^{-2} and an FF of 59.7%. The nanowires were doped with phosphorus after growth to form the p–n junction.

SiNW solar cells grown on glass have been reported by Andra et al.[119] A 200–400 nm thick boron-doped a-Si film was deposited on the glass and subsequently crystallised by a diode laser to produce a polycrystalline layer with grains in the region of 100 μm. N-type wires were grown on this layer by thermal CVD using Au colloids with diameters in the 30–150 nm range. The device was completed by embedding poly(methyl methacrylate) (PMMA) into the SiNW carpet and providing Al contacts. Rectifying and photovoltaic behaviour was observed due to a junction formed between the nanowires and the base layer, but the performance was very poor with a V_{oc} in the region of ~100 mV.

In summary, the performance of SiNW cells constructed from wires grown using Au catalyst particles has been limited. Aspects such as high levels of surface recombination, doping, etc. are issues common to all SiNW designs, but many authors have linked the poor performance in Au catalysed wires to the well-known fact that Au forms a deep trap level in Si with a corresponding detrimental effect on minority carrier lifetime. Interestingly, Kempa *et al.*[120] have recently demonstrated V_{oc} values up to 480 mV in single nanowire Au-catalysed solar cells with a core/multi-shell geometry. They suggest that Au impurities do not significantly lower V_{oc} and that the overall quality of the core/multi-shell structure is the most important driver of good electrical performance.

3.3.1.4 Nanowire Solar Cells Grown from Non-Au Metals

As an alternative to Au, other catalyst metals have been examined for SiNW solar cell growth. These include Sn, Bi, Cu and Al[121–124] but work in this area is limited. The temperature of the eutectic point in the Si–metal binary phase system has an influence on wire growth temperature. For example, growth from the Si–Sn system takes place at <400 °C whereas temperatures >900 °C are required for Si–Cu.

In 2009 Jeon and Kamisako[125] grew SiNWs on a single crystal p-type wafer using Sn as the catalyst and a microwave plasma assist to fabricate the seed particles and the wires. Sn is isoelectronic with Si and does not form deep trap levels in Si. Tapered SiNWs were grown having base and top diameters of $\sim$60 nm and $\sim$10 nm, respectively, with a length of $\sim$1.5 µm. Doping was carried out after wire synthesis using a spin-on phosphorous source to form the solar cell structure which was essentially a SiNW emitter on a p-type base. An encouraging V_{oc} of 520 mV was measured in a 1 cm^2 device but the J_{sc} and FF were poor due, amongst other reasons, to shunting and high series resistance.

A hybrid amorphous Si/crystalline SiNW core radial junction solar cell on top of a SnO$_2$ or ITO coated glass substrate has been reported by Yu *et al.*[126] The undoped SiNWs were grown by RF-PECVD using Sn catalyst particles prepared by a H$_2$ plasma to reduce the TCO layer. A thin, intrinsic a-Si:H layer followed by a top n$^+$ a-Si layer were conformally deposited on the SiNW core. Photovoltaic action was demonstrated but the V_{oc} was 240 mV, with a J_{sc} and FF of 2.6 mA cm^{-2} and 41%, respectively. Device performance has subsequently been significantly improved in these structures using the n–i–p configuration where the SiNW core is doped p-type during growth (Figure 3.14). Using a ZnO back contact to grow the SiNWs, Cho *et al.*[127] reported an efficiency of 4.9%, with a V_{oc} of 800 mV and J_{sc} of 12.4 mA cm^{-2}. Further work by the same group to optimise the SiNW density has led to a V_{oc} of 800 mV, a J_{sc} of 16.1 mA cm^{-2}, an FF of 62.8% and an efficiency of 8.14%.[128] The high V_{oc} reflects the a-Si nature of the active layer of the device.

An all-crystalline microwire array radial junction cell using Cu as the catalyst metal was reported by the Atwater group in 2010.[129] The wires had a diameter of $\sim$2.5 µm with a length of $\sim$60 µm and were grown by CVD on heavily doped p$^+$ Si (111) wafers. BCl$_3$ was used as a dopant to achieve p-type wires. The fabrication process is shown in Figure 3.15. Polydimethylsiloxane

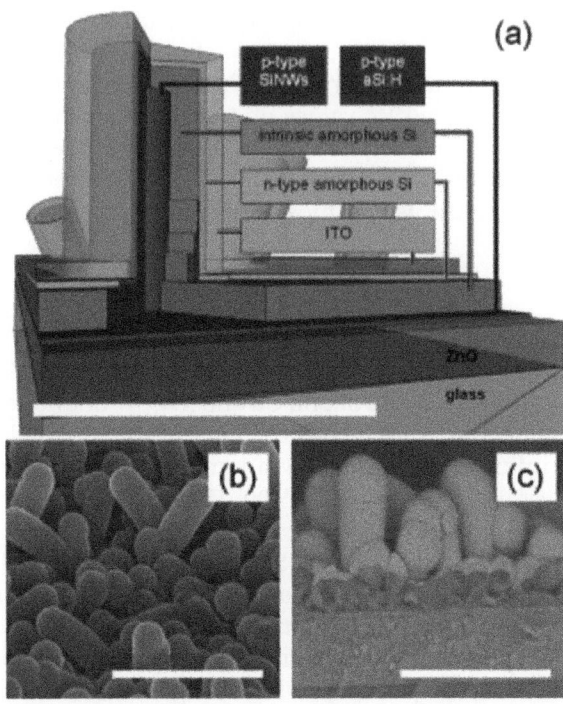

Figure 3.14   (a) Schematic of the solar cell structure of Cho *et al.*[127] A sputtered ITO
                   contact completes the device. (b) SEM of p-type silicon nanowires
                   covered with intrinsic a-Si:H, n-type a-Si:H. (c) Cross-sectional SEM
                   image. Scale bars: 1 μm.

(PDMS) was used as an etch barrier for the thermal oxide located at the
bases of the wires (Figure 3.15(C)). The best cell gave a V_{oc} of 498 mV, a J_{sc} of
24.3 mA cm^{-2}, an FF of 65.4% and an efficiency of 7.92%. It utilised an
a-SiN$_x$:H antireflection/passivation layer, a Ag back reflector and Al$_2$O$_3$
particles embedded between the wires to scatter light. This appears to be the
highest efficiency reported for a bottom–up grown Si radial junction array
cell to date. Up to 9.0% apparent PV, a V_{oc} of 600 mV and over 80% FF has
been reported by the same group in a single-wire radial p–n junction solar
cell fabricated using the same approach.[2] This has led the authors to suggest
that large-area Si wire-array solar cells have the potential to exceed 17%
energy conversion efficiency.

3.3.2   Etched SiNWs and Solar Cells

Top–down approaches to wire fabrication involve the wet or dry etching of Si.
This has the major advantage that the wire internal structure and material
composition can be well regulated but scaleability is an issue. The patterns
for etching are typically formed by processes such as conventional

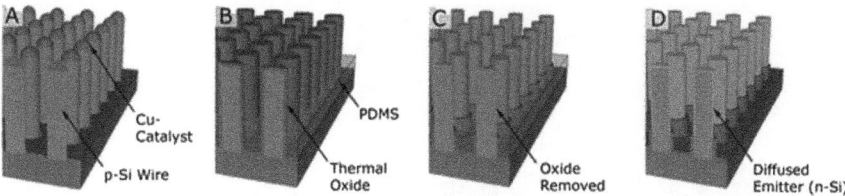

Figure 3.15 Schematic of the radial p–n junction fabrication process: (A) VLS-grown, p-Si microwire array; (B) microwire array after catalyst removal, growth of a thermal oxide and deposition of a PDMS layer; (C) removal of the unprotected thermal oxide; and (D) removal of the PDMS and subsequent phosphorus diffusion to complete the fabrication of a radial p–n junction. After ref. 129.

lithography, e-beam lithography and nanosphere lithography. Self-organised arrays of wires can be formed with a metal assisted chemical etch (MAC etch). Also known as electroless etching, one common method involves etching in silver nitrate/hydrofluoric acid/water solution as described by Peng *et al.*[130] During the etching, Ag particles form and deposit on the surface of the Si wafer. These particles catalyse oxidation of the Si locally with the resulting SiO_2 removed by the HF resulting in nanohole formation and nanowires after prolonged etching (Figure 3.16).

The majority of the work reported on top–down SiNW cells is on bulk Si wafer substrates where photogeneration from the substrate most likely contributes to device performance but has rarely been quantified. These are not 'thin film' structures but are included here for completeness. In some cases the nanowires are heavily doped and utilised purely for their anti-reflective properties in a Si wafer cell. An example of this is the report by Oh *et al.*[131] of a wafer [300 μm thick, p-type (100)] based cell with silver nitrate etched nanowires. A junction was formed *via* conventional phosphorus diffusion. Auger recombination caused by excessive doping, and not simply surface recombination associated with the high surface area of the wires, was shown to limit photogenerated charge collection and efficiency. By suppressing Auger recombination an efficiency of 18.2% was obtained without any additional anti-reflective coating. The J_{sc} was 36.45 mA cm^{-2}, with a V_{oc} of 628 mV and an FF of 79.6%. Based on this work, design rules for core-shell radial junction cells were proposed suggesting that in addition to excellent surface passivation, Auger recombination needs to be suppressed for efficient devices.

Work on SiNW devices on standard Si wafer substrates where diffusion doping of the wires is employed includes Kumar *et al.*[132] who used MAC etching to fabricate wire arrays of length ~4 μm on p-type (1–5 Ω cm) wafers followed by phosphorous diffusion for junction formation. This structure realised a J_{sc} of 37 mA cm^{-2}, a V_{oc} of 544 mV and an efficiency of 13.7%. However, it was not clear whether the p–n junction was formed coaxially in the SiNWs or the SiNWs were converted fully to n-type after diffusion. The use

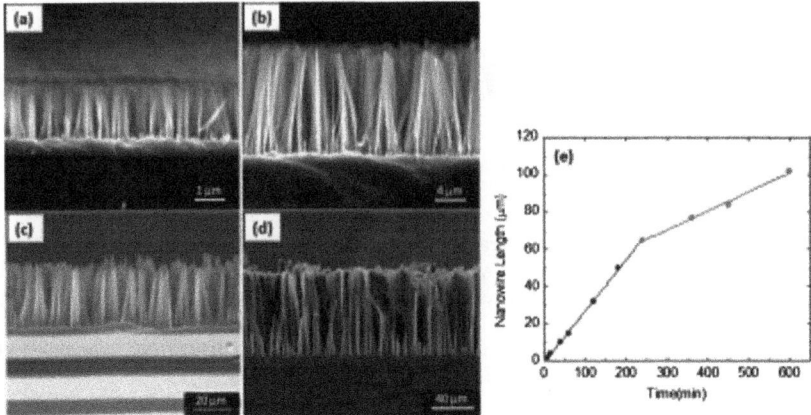

Figure 3.16 Cross-sectional SEM images of the vertically standing SiNW arrays
obtained by electroless etching at 40 °C in a solution containing
4.6 M HF/0.02 M AgNO$_3$ for: (a) 8 min; (b) 40 min; (c) 120 min; and
(d) 360 min. (e) Variation of nanowire length with etching time. Line
is a visual aid. From ref. 130.

of conventional lithography in conjunction with MAC etching has been
demonstrated by Lee *et al.*[133] Periodic photoresist dots 4 μm in diameter with
a period of 12 μm were patterned on a p-type Si wafer. A 15 nm thick Au film
was then deposited followed by MAC etching. The resulting 12 μm long wires
were diffusion doped giving a cell with a J_{sc} of 18.8 mA cm^{-2}, V_{oc} 432 mV, FF
of 38% and an efficiency of 3.2%.

The impact of doping concentration on microwire cells fabricated by
photolithography and deep reactive ion etching (DRIE) on a 380 μm thick p-
type wafer has been investigated by Mallorqui *et al.*[134] The hexagonally
packed wires were 45 μm long with diameters of 1.86, 2.4 and 3.1 μm spaced
7, 8 and 10 μm apart, respectively. The n$^+$-shell was formed by diffusing
POCl$_3$. The most efficient cell was obtained with the 3.1 μm diameter wires
and reached 9.7% with a J_{sc} of 24.9 mA cm^{-2}, a V_{oc} of 530 mV and an FF of
68%. It was concluded that the doping of the wire core should be kept low to
avoid bulk recombination. It was also suggested that a thin n-layer would
reduce emitter losses and scaling the wire diameter with depletion width was
important.

Work has also been published on SiNW/organic hybrid core-shell heterojunction solar cells which offer the advantage that all processing can be
carried out at low temperatures. An example is the report by He *et al.*[135] who
fabricated vertically aligned, single-crystal SiNW arrays on 2–4 Ω cm, n-type
Si (100) wafers by chemical etching. The wire diameter was in the range
30–150 nm with length up ∼5 μm. A thin layer of the p-type small molecule
2,2′,7,7′-tetrakis(*N*,*N*′-di-*p*-methoxyphenylamine)-9,9′-spirobifluorene (spiro-
OMeTAD) was spin coated on the SiNWs followed by a highly conducting

layer of the polymer poly(3,4-ethylenedioxythiophene) polystyrene sulfonate (PEDOT:PSS) and a silver grid to complete the device. The highest efficiency of 10.3% was observed in a cell with a wire length of 0.35 μm with a J_{sc} of 30.9 mA cm^{-2}, a V_{oc} of 0.57 V and an FF of 58.8%. Although the SiNWs significantly enhanced the light absorption and the photocurrent, absorption in the Si substrate contributed to the photocurrent.

We next consider core-shell devices on standard Si wafer substrates using amorphous Si shells deposited by PECVD. MAC etching was used in 2008 by Garnett and Yang[136] to etch wires ~18 μm long in an n-type wafer. The nanowires were coated with p-type amorphous silicon yielding radial junctions with efficiencies up to 0.5%. The limited performance was ascribed to interfacial recombination and high series resistance. Jia *et al.*[137] have reported TCO/a-Si/SiNW heterojunction cells on arrays also prepared by MAC etching of an n-type silicon wafer. Wires with diameters ranging up to 300 nm and lengths of ~900 nm were produced. The intrinsic and p-type a-Si layers shells were deposited by PECVD followed by a TCO layer to complete the device. The mesa-structured 7 mm^2 area cells yielded a conversion efficiency of 7.29% with a V_{oc} of 476 mV, J_{sc} of 27 mA cm^{-2} and FF of 56.2%. Electron beam induced current (EBIC) measurements demonstrated that the nanowires played an active part in the cell's PV response.

Kim *et al.*[138] have fabricated p-type Si microwires with a diameter, spacing and length of 3 μm, 7 μm and 9 μm, respectively, in Si (100) wafers by photolithography and DRIE. The devices featured CVD grown poly-crystalline i- and n- layers for p–n and p–i–n structures. When compared with a planar control, all the microwire devices exhibited superior efficiency. The p–i–n structure was given an 80 nm thick a-SiN:H passivating and anti-reflective layer which improved efficiency from 8.9% to 10.6%. The best efficiency achieved was 11.0% ($V_{oc} = 0.58$ V, $J_{sc} = 29.2$ mA cm^{-2}, FF $= 0.694$).

Gharghi *et al.*[139] also fabricated microwires by lithographically defined patterns and DRIE. Single crystal, n-type upgraded metallurgical grade Si wafers (~200 μm thick) were used as substrates. The wires had radii ranging from 1.5 to 50 μm with lengths of 22 μm and centre-to-centre distances of ~2.7 times the radius. Amorphous silicon i- and p-layers were deposited by PECVD at <150 °C to form the radial heterojunction structures. The highest efficiency obtained was 12.2% with a J_{sc} of 31.1 mA cm^{-2}, V_{oc} of 591 mV and FF of 66%, despite the very short carrier lifetime of <1 μs in the starting wafer. Laser mapping of the photocurrent showed that a significant part of the response was obtained from the planar regions of the wafer between the wires.

Next we describe work on thin film or thin film equivalent structures which uses glass substrates or thin Si epilayers on heavily doped Si wafers where the substrate does not contribute to photogeneration. Gunawan *et al.*[140] used Si substrates consisting of a thin (2.3 μm) p-epitaxial layer grown on a highly doped p$^+$ wafer and nanosphere lithography (NSL) for SiNW array formation.

The NSL technique involves deposition of an ordered monolayer of submicron polystyrene, latex or silica beads onto the substrate to be patterned. Typical methods of bead deposition include spin coating and the Langmuir–Blodgett technique.[141] The spheres can be plasma etched to reduce bead size and used as a mask for dry etching to form nanowires with high aspect ratios.[142,143] The wires of Gunawan *et al.* were etched in the p-epitaxial layer and had a length of 1.1 μm with diameters of 0.44, 0.85 and 1.65 μm. $POCl_3$ diffusion was used to form the n-type emitters in a radial configuration. Capacitance-voltage measurements were used to show that only the 1.65 μm diameter wires formed fully radial junction devices with an active p-type region which served as an absorber. The best efficiency obtained was 4.9% with a V_{oc} of 563 mV and a J_{sc} of 12.7 mA cm^{-2}.

Garnett and Yang[144] also used nanosphere lithography and DRIE to form nanowires ~390 nm in diameter and length ~5 μm. The substrates were heavily doped n$^+$ wafers with lightly doped epitaxial absorber layers on top with thicknesses of 8 and 20 μm. A radial p-n junction was formed by boron diffusion. Average efficiencies of 4.83% and 5.30% were obtained for the 8 μm and 20 μm thick absorber layers, respectively, without any surface passivation. By using optical transmission and photocurrent measurements on arrays fabricated with different lengths on different thickness absorber layers, the authors showed the SiNWs can increase the path length for incident solar radiation by up to a factor of 73, above the Lambertian limit. Also, the enhanced absorption can dominate over surface recombination, even without any surface passivation for the 8 μm thick absorber layers.

Work on etched wires on glass substrates is limited. Sivakov *et al.*[145] used MAC etching to fabricate non-radial junction SiNW cells. A planar, 2.7 μm thick multicrystalline p$^+$–n–n$^+$ stack was formed by e-beam evaporation and diode laser crystallisation. This was then etched with the best cell yielding an open circuit voltage of 450 mV, J_{sc} 40 mA cm^{-2} and efficiency of 4.4%. A strong broadband optical absorption (>90% at 500 nm) was observed. The same group has recently reported an efficiency of 8.8% with a V_{oc} of 530 mV in core-shell radial junction cells on glass.[146] The SiNWs were formed by MAC etching of an n-type, 8 μm thick laser crystallised multi-crystalline Si film. PECVD was used to deposit an a-Si heteroemitter around the nanowires followed by an Al_2O_3 passivation layer. The space between the wires was filled with TCO, ZnO:Al. The device concept is shown in Figure 3.17.

In summary, the full potential of SiNW and microwire cells has yet to be realised. The best performance in VLS grown wire cells is ~8% in a microwire array as well as a nanowire array using a-Si active layers. Top–down cells on Si wafer substrates have better performance reflecting the superior wire material quality. Efficiencies in the region of 10–12% have been demonstrated in microwire cells but photogeneration in the Si substrate is likely to be contributing in many cases. The performance of top–down etched nanowire cells on glass has reached a promising level of ~9%.

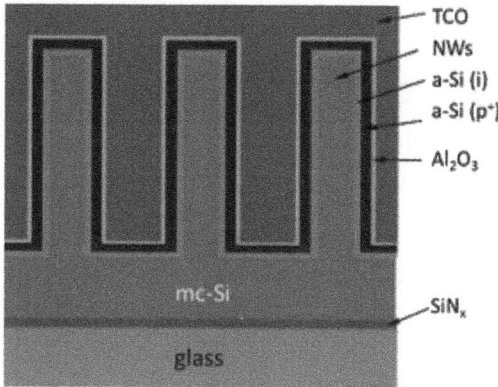

Figure 3.17 Concept of the SiNW thin film solar cell formed on glass using a multicrystalline Si layer (mc-Si) produced by diode laser crystallisation. After ref. 146.

3.4   Conclusions

Significant progress has been made in recent years in thin film crystalline silicon solar cell research. Crystallisation of amorphous silicon on glass to form polycrystalline thin films has been intensively studied using solid and liquid phase approaches. Solid phase crystallisation on glass was developed to the commercial stage but the limited V_{oc} (~500 mV) of this technology and manufacturing throughput issues proved a barrier to large-scale commercialisation. Liquid phase crystallisation is showing more promise with efficiencies and V_{oc} values approaching 12% and 600 mV, respectively, due to superior material quality. Using high rate e-beam evaporation combined with laser processing, this technology offers excellent prospects for a competitive thin film Si technology, exceeding 15%, particularly if viable light trapping schemes based on nanophotonics can be implemented.

Seed layer approaches on glass can yield large grain sizes (>10 μm) but efficiencies have struggled to exceed a few percentage due to material quality issues. Efficiencies of up to 8.5% have been achieved using aluminium induced crystallisation on high temperature compatible substrates and thermal CVD for absorber layer deposition. However, a high density of intragrain defects and mixed grain orientation are key factors that have limited V_{oc} and device performance. It is not clear if adequate improvements in material quality can be achieved using cost-effective processes to compete with other approaches.

Plasmonic enhancement of thin film cells has attracted considerable interest in recent years and promising results showing improvements in short-circuit current have been seen in thin film Si solar cells. However, reports of improvements in efficiency are limited with absolute gains in the region of 1% or below at best. Many challenges remain and the field remains open for further research.

Work on Si nano/micro wire solar cells is at an early stage. So far efficiencies of small area laboratory cells are in the region of 10% or below. Improvements need to be made in several areas including wire quality for bottom–up approaches. Common factors needing attention for all approaches include minimising surface and bulk recombination, control of dopant concentration and distribution and cell design including efficient contacts for carrier extraction. These challenges need to be met to realise the full potential and benefits of this approach.

In summary, the crystalline Si thin film and SiNW fields continue to provide significant research challenges and opportunities. Considerable progress has been made. Further developments in materials quality and device engineering will enable the full potential of these approaches to be realised. Integrating both the crystallisation and nanowire concepts could be beneficial.

References

1. A. V. Shah, J. Meier, E. Vallat-Sauvain, N. Wyrsch, U. Kroll, C. Droz and U. Graf, *Sol. Energy Mater. Sol. Cells,* 2003, **78**, 469–491.
2. S. Hänni, G. Bugnon, G. Parascandolo, M. Boccard, J. Escarré, M. Despeisse, F. Meillaud and C. Ballif, *Prog. Photovoltaics,* 2013, **21**, 821–826.
3. S. Guha, J. Yang and A. Banerjee, *Prog. Photovoltaics,* 2000, **8**, 141–150.
4. A. V. Shah, E. Moulin and C. Ballif, *Sol. Energy Mater. Sol. Cells,* 2013, **119**, 311–316.
5. B. Bergmann and T. J. Rinke, *Prog. Photovoltaics,* 2000, **8**, 451–464.
6. P. Rosenits, F. Kopp and S. Reber, *Thin Solid Films,* 2011, **519**, 3288–3290.
7. I. Gordon, F. Dross, V. Depauw, A. Masolin, Y. Qui, J. Vaes, D. Van Gestel and J. Poortmans, *Sol. Energy Mater. Sol. Cells,* 2011, **95**, S2–S7.
8. T. Matsuyama, N. Terada, T. Baba, T. Sawada, S. Tsuge, K. Wakisaka and S. Tsuda, *J. Non-Cryst. Solids,* 1996, **198–200**, 940–944.
9. M. A. Green, P. A. Basore, N. Chang, D. Clugston, R. Egan, R. Evans, D. Hogg, S. Jarnason, M. Keevers, P. Lasswell, J. O. Sullivan, U. Schubert, A. Turner, S. R. Wenham and T. Young, *Sol. Energy,* 2004, **77**, 857–863.
10. M. J. Keevers, T. L. Young, U. Schubert and M. A. Green, in *Proceedings 22nd European Photovoltaic Solar Energy Conference,* 2007, pp. 1783–1790.
11. A. G. Aberle, *J. Cryst. Growth,* 2006, **287**, 386–390.
12. O. Kunz, Z. Ouyang, S. Varlamov and A. G. Aberle, *Prog. Photovoltaics,* 2009, **17**, 567–573.
13. T. Sontheimer, C. Becker, F. Ruske, C. Klimm, U. Bloeck, S. Gall, O. Kunz, T. Young, R. Egan, J. Hupkes and B. Rech, in *Proceedings 35th IEEE Photovoltaic Specialists Conference,* 2010, pp. 000614–000619.

14. S. He, J. Wong, D. Inn, B. Hoex, A. G. Aberle and A. B. Sproul, *Thin Solid Films*, 2010, **518**, 4351–4355.

15. Y. Tao, S. Varlamov, J. Wong, O. Kunz and R. Egan, in *Proceedings 35th IEEE Photovoltaic Specialists Conference*, 2010, pp. 000620–000625.

16. B. Rau, T. Weber, B. Gorka, P. Dogan, F. Fenske, K. Y. Lee, S. Gall and B. Rech, *Mater. Sci. Eng. B,* 2009, **159–160**, 329–332.

17. R. B. Bergmann, *Appl. Phys. A,* 1999, **69**, 187–194.

18. D. Amkreutz, J. Muller, M. Schmidt, T. Hanel and T. F. Schulze, *Prog. Photovoltaics,* 2011, **19**, 937–945.

19. J. Haschke, L. Jogschies, D. Amkreutz, L. Korte and B. Rech, *Sol. Energy Mater. Sol. Cells,* 2013, **115**, 7–10.

20. C. Becker, D. Amkreutz, T. Sontheimer, V. Preidel, D. Lockau, J. Haschke, L. Jogschies, C. Klimm, J. J. Merkel, P. Plocica, S. Steffens and B. Rech, *Sol. Energy Mater. Sol. Cells,* 2013, **119**, 112–123.

21. J. Dore, D. Ong, S. Varlamov, R. Egan and M. A. Green, *IEEE J. Photovoltaics,* 2014, **4**, 33–39.

22. K. Ohdaira, T. Nishikawa, S. Ishii, *et al.* in *Proceedings 25th European Photovoltaic Solar Energy Conference*, 2010, pp. 3546–3548.

23. O. Nast and S. R. Wenham, *J. Appl. Phys.,* 2000, **88**, 124–132.

24. P. I. Widenborg and A. G. Aberle, *J. Cryst. Growth,* 2002, **242**, 270–282.

25. S. Gall, J. Schneider, J. Klein, K. Hubner, M. Muske, B. Rau, E. Conrad, I. Seiber, K. Petter, K. Lips, M. Stoger-Pollach, P. Schattschneider and W. Fuhs, *Thin Solid Films,* 2006, **511–512**, 7–14.

26. S. Gall, C. Becker, K. Y. Lee, T. Sontheimer and B. Rech, *J. Cryst. Growth,* 2010, **312**, 1277–1281.

27. W. Fuhs, S. Gall, B. Rau, M. Schmidt and J. Schneider, *Sol. Energy,* 2004, **77**, 961–968.

28. J. Schneider, A. Sarikov, J. Klein, M. Muske, I. Sieber, T. Quinn, H. S. Reehal, S. Gall and W. Fuhs, *J. Cryst. Growth,* 2006, **287**, 423–427.

29. D. Van Gestel, I. Gordon, A. Verbist, L. Carnel, G. Beaucarne and J. Poortmans, *Thin Solid Films,* 2008, **516**, 6907–6911.

30. G. Ekanayake, T. Quinn and H. S. Reehal, *J. Cryst. Growth,* 2006, **293**, 351–358.

31. T. Quinn and H. S. Reehal, in *Proceedings 24th European Photovoltaic Solar Energy Conference*, 2009, pp. 2517–2520.

32. S. Summers, H. S. Reehal and G. J. Hirst, *J. Mater. Sci.: Mater. Electron.,* 2000, **11**, 557–563.

33. M. Weizman, C. Klimm, M. Nittel, M. Kastner, C. Hernandez, N. H. Nickel, T. Sontheimer and B. Rech, in *Proceedings 25th European Photovoltaic Solar Energy Conference*, 2010, pp. 2828–2831.

34. R. B. Bergmann, J. Kohler, R. Dassow, C. Zaczek and J. H. Werner, *Phys. Status Solidi A,* 1998, **166**, 587–602.

35. G. Andra, J. Bergmann, A. Gawlik, I. Hoger, T. Scmidt, F. Falk, B. Burghardt and G. Eberhardt, in *Proceedings 25th European Photovoltaic Solar Energy Conference*, 2010, pp. 3538–3542.

36. J. S. Im, M. Chahal, P. C. Van der Wilt, U. J. Chung, G. S. Ganot, A. M. Chitu, N. Kobayashi, K. Ohmori and A. B. Limanov, *J. Cryst. Growth*, 2010, **312**, 2775–2778.
37. N. P. Harder, T. Puzzer, P. I. Widenborg, S. Oelting and A. G. Aberle, *Cryst. Growth Des.*, 2003, **3**, 767–771.
38. G. Ekanayake, T. Quinn, H. S. Reehal, B. Rau and S. Gall, *J. Cryst. Growth*, 2007, **299**, 309–315.
39. B. Rau, E. Conrad and S. Gall, in *Proceedings 21st European Photovoltaic Solar Energy Conference*, 2006, pp. 1418–1421.
40. B. Rau, K. Petter, I. Sieber, M. S. Pollach, D. Eyidi, P. Schattschneider, S. Gall, K. Lips and W. Fuhs, *J. Cryst. Growth*, 2006, **287**, 433–437.
41. C. Jaeger, T. Matsui, M. Takeuchi, M. Karasawa, M. Kondo and M. Stutzmann, *Jpn. J. Appl. Phys.*, 2010, **49**, 112301.
42. S. Gall, C. Becker, E. Conrad, P. Dogan, F. Fenske, B. Gorka, K. Y. Lee, B. Rau, F. Ruske and B. Rech, *Sol. Energy Mater. Sol. Cells*, 2009, **93**, 1004–1008.
43. K. Y. Lee, P. Dogan, F. Ruske, B. Gorka, S. Gall, B. Rech and J. Hupkes, in *Proceedings 3rd European Photovoltaic Solar Energy Conference*, 2008, pp. 2261–2264.
44. J.-H. Wang, S.-Y. Lien, C.-F. Chen and W.-T. Whang, *IEEE Electron Device Lett.*, 2010, **31**, 38–40.
45. G. Andra, J. Plentz, A. Gawlik, E. Ose, F. Falk and K. Lauer, in *Proceedings 22nd European Photovoltaic Solar Energy Conference*, 2007, pp. 1967–1970.
46. J. Schneider, J. Dore, S. Christianson, F. Falk, N. Lichtenstein, B. Valk, R. Lewandowska, A. Slaoui, X. Maeder, J. Labar, G. Safran, M. Werner, V. Naumann and C. Hagendorf, in *Proceedings 25th European Photovoltaic Solar Energy Conference*, 2010, pp. 3573–3576.
47. Y. Qiu, O. Kunz, S. Venkatachalam, D. Van Gestel, R. Egan, I. Gordon and J. Poortmans, in *Proceedings 25th European Photovoltaic Solar Energy Conference*, 2010, pp. 3633–3637.
48. I. Gordan, L. Carnel, D. Van Gestel, G. Beaucarne and J. Poortmans, *Thin Solid Films*, 2008, **516**, 6984–6988.
49. D. Van Gestel, I. Gordan and J. Poortmans, *Sol. Energy Mater. Sol. Cells*, 2013, **119**, 261–270.
50. D. L. Young, K. Alberi, C. Teplin, I. Martin, P. Stradins, M. Shub, C. Beall, E. Iwaniczko, H. Guthrey, M. J. Romero, T. Chuang, E. Mozdy and H. M. Branz, in *Proceedings 35th IEEE Photovoltaic Specialists Conference*, 2010, pp. 000626–000630.
51. D. Dawson-Elli, C. Kosik Williams, J. Couillard, J. Cites, R. Manley, G. Fenger and K. Hirschman, *ECS Trans.*, 2007, **8**(1), 223.
52. D. Hernandez, T. Trifonov, M. Garin and R. Alcubilla, *Appl. Phys. Lett.*, 2013, **102**, 172102.
53. J. H. Petermann, D. Zielke, J. Schmidt, F. Haase, E. Garralaga Rojas and R. Brendel, *Prog. Photovoltaics*, 2012, **20**, 1–5.

54. R. Cariou, M. Labrune and P. Roca i Cabarrocas, *Sol. Energy Mater. Sol. Cells*, 2011, **95**, 2260–2263.
55. M. Moreno and P. Roca i Cabarrocas, *PV Direct*, 2010, **1**, 10301.
56. L. Zeng, P. Bermel, Y. Yi, B. AAlamariu, K. A. Broderick, J. Liu, C. Hong, X. Duan, J. Joannopoulos and L. C. Kimerling, *Appl. Phys. Lett.*, 2008, **93**, 221105.
57. H. A. Atwater and A. Polman, *Nat. Mater.*, 2010, **9**, 205–213.
58. T. L. Temple and D. M. Bagnall, *Prog. Photovoltaics*, 2013, **21**, 600–611.
59. P. N. Seata, V. E. Ferry, D. Pacifici, J. N. Munday and H. A. Atwater, *Opt. Express*, 2009, **17**, 20975–20990.
60. T. Temple and D. M. Bagnall, *J. Appl. Phys.*, 2011, **109**, 084343.
61. K. Catchpole and A. Polman, *Opt. Express*, 2008, **16**, 21793.
62. S. Pillai and M. A. Green, *Sol. Energy Mater. Sol. Cells*, 2010, **94**, 1481–1486.
63. S. Pillai, K. R. Catchpole, T. Trupke and M. A. Green, *J. Appl. Phys.*, 2007, **101**, 093105.
64. T. Temple, G. D. K. Mahanama, H. S. Reehal and D. Bagnall, *Sol. Energy Mater. Sol. Cells*, 2009, **93**, 1978–1985.
65. F. J. Beck, A. Polman and K. R. Catchpole, *J. Appl. Phys.*, 2009, **105**, 114310.
66. A. Centeno, B. Ahmed, H. Reehal and F. Xie, *Nanotechnology*, 2013, **24**, 415402.
67. F. J. Beck, S. Mokkapati, A. Polman and K. R. Catchpole, *Appl. Phys. Lett.*, 2010, **96**, 033113.
68. Z. Ouyang, S. Pillai, F. Beck, O. Kunz, S. Varlamov, K. R. Catchpole, P. Campbell and M. A. Green, *Appl. Phys. Lett.*, 2010, **96**, 261109.
69. Z. Ouyang, X. Zhao, S. Varlamov, Y. Tao, J. Wong and S. Pillai, *Prog. Photovoltaics*, 2011, **19**, 917–926.
70. S. Pillai, F. J. Beck, K. R. Catchpole, Z. Ouyang and M. A. Green, *J. Appl. Phys.*, 2011, **109**, 073105.
71. J. Rao, S. Varlamov, J. Park, S. Dligatch and A. Chtanov, *Plasmonics*, 2013, **8**, 785–791.
72. V. Ferry, M. A. Verschuuren, H. B. T. Li, E. Verhagen, R. J. Walters, R. E. I. Schropp, H. A. Atwater and A. Polman, *Opt. Express*, 2010, **18**, A237–A245.
73. M. D. Kelzenberg, M. A. Filler, B. M. Kayes, M. C. Putnam, D. B. Turner-Evans, N. S. Lewis and H. A. Atwater, in *Proceedings 33rd IEEE Photovoltaic Specialists Conference*, 2008, 6 pages.
74. B. M. Kayes, N. S. Lewis and H. A. Atwater, *J. Appl. Phys.*, 2005, **97**, 114302.
75. M. D. Kelzenberg, D. B. Turner-Evans, M. C. Putnam, S. W. Boettcher, R. M. Briggs, J. Y. Baek, N. S. Lewis and H. A. Atwater, *Energy Environ. Sci.*, 2011, **4**, 866.
76. V. Schmidt, J. V. Wittemann and U. Gosele, *Chem. Rev.*, 2010, **110**, 361.

77. J. K. Mann, R. Kurstjens, G. Pourtois, M. Gilbert, F. Dross and J. Poortmans, *Prog. Mater. Sci.,* 2013, **58**, 1361–1387.
78. R. S. Wagner and W. C. Ellis, *Appl. Phys. Lett.,* 1964, **5**, 89.
79. B. Tian, T. J. Kempa and C. M. Lieber, *Chem. Soc. Rev.,* 2009, **38**, 16.
80. B. Tian, X. Zheng, T. J. Kempa, Y. Fang, N. Yu, G. Yu, J. Huang and C. M. Liber, *Nature,* 2007, **449**, 885.
81. Y. Wu, Y. Cui, L. Huynh, C. J. Barrelet, D. C. Bell and C. M. Lieber, *Nano Lett.,* 2004, **4**, 433–436.
82. S. Hofmann, C. Ducati, R. J. Neill, S. Piscanec, A. C. Ferrari, J. Geng, R. E. Dunin-Borkowski and J. Robertson, *J. Appl. Phys.,* 2003, **94**, 6005.
83. T. Kawashima, T. Mizutani, H. Masuda, T. Saitoh and M. Fujii, *J. Phys. Chem.,* 2008, **112**, 17121.
84. B. Ressel, K. C. Prince, S. Heun and Y. Homma, *J. Appl. Phys.,* 2003, **93**, 3886.
85. N. Ferralis, R. Maboudian and C. Carraro, *J. Am. Chem. Soc.,* 2007, **130**, 2681.
86. P. J. F. Harris, *Int. Mater. Rev.,* 1995, **40**, 97.
87. J. Ball, L. Bowen, B. G. Mendis and H. S. Reehal, *Cryst. Eng. Comm.,* 2013, **15**, 3808.
88. J. Ball and H. S. Reehal, *Thin Solid Films,* 2012, **520**, 2467.
89. B. M. Kayes, M. A. Filler, M. C. Putnam, M. D. Kelzenberg, N. S. Lewis and H. A. Atwater, *Appl. Phys. Lett.,* 2007, **91**, 103110.
90. R. Q. Zhang, Y. Lifshitz, D. D. D. Ma, Y. L. Zhao, T. Frauenheim, S. T. Lee and Y. S. Tong, *J. Chem. Phys.,* 2005, **123**, 14403.
91. R. S. Wagner, in *Wisker Technology*, ed. A. P. Levitt, Wiley, New York, 1970.
92. V. Schmidt, S. Senz and U. Gosele, *Nano Lett.,* 2005, **5**, 931.
93. S. A. Fortuna and X. Li, *Semicond. Sci. Technol.,* 2010, **25**, 024005.
94. J. Westwater, D. P. Gosain, S. Tomiya and S. Usui, *J. Vac. Sci. Technol.,* 1997, **15**, 3.
95. S. Misra, L. Yu, W. Chen and P. Roca i Cabarrocas, *J. Phys. Chem.,* 2013, **117**, 17786–17790.
96. S. K. Chong, B. T. Goph, C. F. Dee and S. A. Rahman, *Thin Solid Films,* 2013, **529**, 153–158.
97. I. Zardo, S. Conesa-Boj, S. Estrade, L. Yu, F. Peiro, P. Roca i Cabarrocas, J. R. Morante, J. Arbiol, A. Fontcuberta and I. Morral, *Appl. Phys. A,* 2010, **100**, 287–296.
98. M. J. Hernandez, M. Cervera, E. Ruiz, J. L. Pau, J. Piqueras, M. Avella and J. Jimenez, *Nanotechnology,* 2010, **21**, 455602.
99. J. Li, H. Yu and Y. Li, *Nanotechnology,* 2012, **23**, 194010.
100. C. Battaglia, C.-M. Hsu, K. Soderstrom, J. Escarre, F.-J. Haug, M. Charriere, M. Boccard, M. Despeisse, D. Alexander, M. Cantoni, Y. Cui and C. Ballif, *ACS Nano,* 2012, **6**, 2790.
101. J. Li, H. Yu, S. M. Wong, G. Zhang, X. Sun, P. G.-Q. Lo and D.-L. Kwong, *Appl. Phys. Lett.,* 2009, **95**, 033102.
102. L. Hu and G. Chen, *Nano Lett.,* 2007, **7**, 3249.

103. N. Lagos, M. M. Sigalas and D. Niarchos, *Photo. Nano. Fund. Appl.,* 2011, **9**, 163.
104. H. Boa and X. Ruan, *Opt. Lett.,* 2010, **35**, 3378.
105. Th Stelzner, M. Pietsch, G. Andra, F. Falk, E. Ose and S. Christiansen, *Nanotechnology,* 2008, **19**, 295203.
106. L. Tsakalakos, J. Balch, J. Fronheiser, M.-Y. Shih, S. F. LeBoeuf, M. Pietrzykowski, P. J. Codella, B. A. Korevaar, O. Sulima, J. Rand, A. Davuluru and U. Rapol, *J. Nanophotonics,* 2007, **1**, 013552.
107. A. Convertino, M. Cuscuna and F. Martelli, *Nanotechnology,* 2010, **21**, 355701.
108. C. Y. Kuo, C. Gau and B. T. Dai, *Sol. Energy Mater. Sol. Cells,* 2011, **95**, 154.
109. O. L. Muskens, J. G. Rivas, R. E. Algra, E. P. A. M. Bakkers and A. Lagendijk, *Nano Lett.,* 2008, **8**, 2638.
110. J. Zhu, Z. Yu, S. Fan and Y. Cui, *Mater. Sci. Eng. R,* 2010, **70**, 330–340.
111. J. Li, H. Yu, S. M. Wong, G. Zhang, G.-Q. Lo and D.-L. Kwong, *J. Phys. D: Appl. Phys.,* 2010, **43**, 255101.
112. L. J. Lauhon, M. S. Gudiksen, D. Wang and C. M. Lieber, *Nature,* 2002, **420**, 67.
113. G. Zheng, W. Lu, S. Jin and C. M. Lieber, *Adv. Mater,* 2004, **16**, 1890.
114. L. Tsakalakos, J. Balch, J. Fronheiser and B. A. Korevaar, *Appl. Phys. Lett.,* 2007, **91**, 233117.
115. L. Tsakalakos, J. Balch, J. Fronheiser, B. A. Korevaar, O. Sulima and J. Rand, in *Proceedings 23rd European Photovoltaic Solar Energy Conference,* 2008, pp. 11–16.
116. S. Gunawan and O. Guha, *Sol. Energy Mater. Sol. Cells,* 2009, **93**, 1388.
117. S. Perraud, S. Poncet, S. Noel, M. Levis, P. Faucherand, E. Rouviere, P. Thony, C. Jaussaud and R. Delsol, *Sol. Energy Mater. Sol. Cells,* 2009, **93**, 1568.
118. C. Y. Kuo, C. Gau and B. T. Dai, *Sol. Energy Mater. Sol. Cells,* 2011, **95**, 154.
119. G. Andra, M. Pietsch, V. Sivakov and F. Falk, in *Proceedings of the 24th European Photovoltaic Solar Energy Conference,* 2009, pp. 418–420.
120. T. J. Kempa, R. W. Day, S.-K. Kim, H.-G. Parkc and C. M. Lieber, *Energy Environ. Sci.,* 2013, **6**, 681.
121. M. Jeon and K. Kamisako, *Mater. Lett.,* 2008, **62**, 3903.
122. L. Yu, F. Fortuna, B. O'Donnell, T. Jeon, M. Foldyna, G. Picardi and P. Roca i Cabarrocas, *Nano Lett.,* 2012, **12**, 4153.
123. M. C. Putnam, D. B. Turner-Evans, M. D. Kelzenberg, S. W. Boettcher, N. S. Lewis and H. A. Atwater, *Appl. Phys. Lett.,* 2007, **91**, 103110.
124. Y. Wang, V. Schmidt, S. Senz and U. Gosele, *Nat. Nanotechnol.,* 2006, **1**, 186.
125. M. Jeon and K. Kamisako, *Mater. Lett.,* 2009, **63**, 777.
126. L. Yu, B. O'Donnell, P.-J. Alet and P. Roca i Cabarrocas, *Sol. Energy Mater. Sol. Cells,* 2010, **94**, 1855.
127. J. Cho, B. O'Donnell, L. Yu, K.-H. Kim, I. Ngo and P. Roca i Cabarrocas, *Prog. Photovoltaic,* 2013, **21**, 77–81.

128. S. Misra, L. Yu, M. Foldyna and P. Roca i Cabarrocas, *Sol. Energy Mater. Sol. Cells,* 2003, **118**, 90.
129. M. C. Putnam, S. W. Boettcher, M. D. Kelzenburg, D. B. Turner-Evans, J. M. Spurgeon, E. L. Warren, R. M. Briggs, N. S. Lewis and H. A. Atwater, *Energy Environ. Sci.,* 2010, **3**, 1037.
130. K. Peng, Y. Xu, Y. Wu, Y. Yan, S.-T. Lee and J. Zhu, *Small,* 2005, **1**, 1062.
131. J. Oh, H.-C. Yuan and H. M. Branz, *Nat. Nanotechnol.,* 2012, **7**, 743.
132. D. Kumar, S. K. Srivastava, P. K. Singh, M. Husain and V. Kumar, *Sol. Energy Mater. Sol. Cells,* 2011, **95**, 215.
133. E. Lee, Y. Kim, M. Gwon, D.-W. Kim, S.-H. Baek and J.-H. Kim Sol, *Energy Mater. Sol. Cells,* 2012, **103**, 93.
134. A. D. Mallorqui, F. M. Epple, D. Fan, O. Demichel and A. Fontcuberta i Morral, *Phys. Status Solidi A,* 2012, **8**, 1588.
135. L. He, C. Jiang, Rusli, D. Lai and H. Wang, *Appl. Phys. Lett.,* 2011, **99**, 021104.
136. E. C. Garnett and P. Yang, *J. Am. Chem. Soc.,* 2008, **130**, 9224.
137. G. Jia, M. Steglich, I. Sil and F. Falk, *Sol. Energy Mater. Sol. Cells,* 2012, **96**, 226.
138. D. R. Kim, C. H. Lee, P. M. Rao, I. S. Cho and X. Zheng, *Nano Lett.,* 2011, **11**, 2704.
139. M. Gharghi, E. Fathi, B. Kante, S. Sivoththaman and X. Zhang, *Nano Lett.,* 2012, **12**, 6278.
140. O. Gunawan, K. Wang, B. Fallahazad, Y. Zhang, E. Tutuc and S. Guha, *Prog. Photovoltaics,* 2011, **19**, 307–312.
141. C. L. Cheung, R. J. Nikolic, C. E. Reinhardt and T. F. Wang, *Nanotechnology,* 2006, **17**, 1339.
142. S. W. Schmitt, F. Schechtel, D. Amkreutz, M. Bashouti, S. K. Srivastava, B. Hoffmann, C. Dieker, E. Spiecker, B. Rech and S. H. Christiansen, *Nano Lett.,* 2012, **12**, 4050.
143. F. Laermer and A. Schilp, *US Pat.,* 5501893, 1996.
144. E. Garnett and P. Yang, *Nano Lett.,* 2010, **10**, 1082.
145. V. Sivakov, G. Andrä, A. Gawlik, A. Berger, J. Plentz, F. Falk and S. H. Christiansen, *Nano Lett.,* 2009, **9**, 1549.
146. G. Jia, A. Gawlik, J. Bergmann, B. Eisenhawer, S. Schonherr, G. Andra and F. Falk, *IEEE J. Photovoltaic,* 2014, **4**, 28–32.

CHAPTER 4

A Review of NREL Research into Transparent Conducting Oxides

TIMOTHY J. COUTTS[*†], JAMES M. BURST, JOEL N. DUENOW, XIAONAN LI, AND TIMOTHY A. GESSERT

National Renewable Energy Laboratory, Golden, Colorado 80401, USA
*E-mail: tcoutts1229@comcast.net

4.1   Introduction

In a previous publication we discussed remaining materials challenges in the development of transparent conducting oxides (TCOs) for thin film photovoltaics.[1] We pointed out that researchers should aim to achieve film resistivities of about 5×10^{-5} Ω cm and optical transmittances of at least 85% across the entire range of wavelengths of the solar spectrum. For a film thickness of 0.5 μm, the corresponding sheet resistance of 1 Ω $\square^{-1}$ is far less than is typically achieved either in manufacture or even in research, and it is a value that would benefit thin film solar cells by increasing their current and fill factor. It is also desirable to avoid mismatch of the conduction band edges of the TCO and n-CdS, the latter typically being the second layer encountered by incoming photons. Unless this can be achieved, a barrier to the transport of excess minority charge can result. Finally, there are many

[†]NREL Fellow Emeritus.

RSC Energy and Environment Series No. 12
Materials Challenges: Inorganic Photovoltaic Solar Energy
Edited by Stuart J C Irvine
© The Royal Society of Chemistry 2015
Published by the Royal Society of Chemistry, www.rsc.org

practical challenges that must be considered in the further development of TCOs. A recent review discusses many aspects of TCOs and summarizes recent progress.[2]

Transparent conducting oxides are used in all thin film solar cells, currently at an advanced state of development. The role of the TCO is to reduce resistive losses incurred in the collection of photogenerated excess charge. In some thin film devices and modules, a collection grid is also used and the sheet resistance of the TCO determines the spacing of the grid fingers. A reduction in the resistivity of the TCO offers the opportunity to increase the spacing of the grid fingers, thereby reducing optical losses due to shadowing. In circumstances such that individual subcells of a module are formed by scribing and then connected in series and parallel strings, the number of scribe lines is determined by the sheet resistance of the TCO. Reductions in TCO sheet resistance again offer significant advantages because the number of scribe lines and subcells may be reduced.

The critical parameter in optimizing the properties of TCOs is the mobility—or more fundamentally, the relaxation time—of the electrons. It is important to make higher-quality material by adjusting the deposition conditions, or by using alternative materials with reduced charge scattering. In reality, both options have been vigorously investigated in recent years and we shall discuss some of the progress we have made in our own work, as well as that of others. It is debatable whether significant advances toward the goals discussed earlier have been made by adjusting fabrication conditions.

Because the TCO is the first thin film layer of a thin film multilayer device encountered by incident photons, one of the key requirements is that it is highly transmissive. The TCO film must allow all photons in the solar spectrum, which can be absorbed usefully by the electrically active layers, to be transmitted to the region of the electrostatic junction. In general, TCO films have a transmittance window that extends over the entire solar spectrum, but the transmittance is usually significantly less than 100%. Typically, the range of high transmission is from about 0.3–0.4 µm to about 1.5–2.0 µm. The former range is determined mainly by the fundamental bandgap of the semiconducting TCO, whereas the latter range is determined by the density of free electrons, their effective mass and, far less extensively discussed in the literature, the high frequency permittivity of the TCO. The topic of TCOs has been the subject of extensive research at NREL for many years, and in this chapter we discuss our work on some of the materials studied. The review concludes with a discussion of the development of high permittivity TCOs, which could lead to a new family of materials that can be readily synthesized. For each material considered, we indicate remaining challenges, the solution of which will help its adoption in thin film solar cells.

Before discussing the more fundamental aspects of TCOs and their properties, we begin by considering some of the practical challenges that must be overcome if novel materials are to be acceptable to industry.

4.2　Practical Challenges Facing TCOs

In addition to the difficulty of making highly transparent, low-resistivity material, there are several challenges for the TCO in a thin film module that are often not discussed. Some of these may represent significant challenges at the practical level.

4.2.1　Elemental Abundance and Cost

Claims have been made that the element indium is scarce and that novel TCOs must be developed that do not include it. Because of its putative scarcity, it is also costly and its cost per unit mass is rather variable.[3] The claim of scarcity is unproven and, in practice, there seems to be a strong link between demand and price. The major market for indium has historically been for indium tin oxide (ITO) electrodes for flat panel displays.[3] In the event of a supply shortage of indium, it seems probable that this market could afford to pay the increasing costs of indium, whereas the photovoltaic (PV) market, which needs module costs to be reduced, may find difficulty in remaining competitive if there are severely increased material costs because of the need to meet the goals for levelized cost of electricity.[4,5]

In partial response to claims of cost, note that the thickness of a typical TCO is limited to about 0.5 μm, whereas the thickness of the absorber layer in a copper indium gallium selenide (CIGS) solar cell is about 2 μm. The equivalent cost of the indium in a CIGS solar cell is about 5 cents per watt, so the justification for seeking a replacement for indium-containing TCOs in thin film PV seems marginal, because the cost is only about 1 cent per watt. Continuation of supply is a separate issue and will depend on factors such as reserves of indium, competing technologies and rate of production. Ensuring that prices remain stable may require manufacturers of indium-containing TCOs and CIGS solar cells to place long-term contracts with refiners of indium.

4.2.2　Toxicity

Cadmium stannate (Cd_2SnO_4 or CTO) has been shown to be one of the most promising TCOs and was used in a recent world record CdTe solar cell.[6] However, criticisms of cadmium-containing materials are commonplace and often made without regard to their usage. Although cadmium is certainly one of the most toxic elements known, this is not necessarily true of cadmium-containing compounds and alloys. Both CdTe and CTO are likely far less toxic than cadmium alone, and their stability was demonstrated by Wu *et al.*[7] Fthenakis *et al.* demonstrated that the amount of cadmium in a type-C battery (Ni–Cd) is about the same as that in one kilowatt of CdTe modules.[8] In addition, fire tests showed that, even under exposure to intense heat, the CdTe remained stable and the amount of Cd emitted from the modules was minimal.

Hence, it appears that claims regarding the risks of toxicity are overstated. Nevertheless, criticisms will doubtless continue, particularly as PV technology begins to compete with conventional sources of electricity in years to come. The PV industry, or that part of it that uses cadmium-based semiconductors, must therefore be ready to answer such criticisms swiftly and aggressively.

4.2.3   Ease of Deposition

TCOs must have the following qualities: be simple to deposit rapidly, using low-cost deposition equipment; not damage previously deposited layers in the thin film stack; not require high-temperature processing steps during or after deposition; be free of strain and of tendencies to crack or delaminate; and maintain excellent optical and electrical properties. It must also be straightforward to coat large areas with uniform films both laterally and in depth. Several techniques meet most or all of these requirements, including chemical vapor deposition and sputtering. Others, such as sol-gel coating, are not well developed for TCOs, but appear to have the potential to meet requirements.[9]

4.2.4   Stability

One of the most important properties of a TCO in a thin film cell or module is that it is stable against internal mechanisms such as interdiffusion between adjoining layers and against external mechanisms such as electrochemical corrosion. There is ample evidence that the degradation in performance of CIGS modules is due to moisture-related corrosion of the ZnO TCO bilayer.[10] This problem represents a major challenge for the CIGS technology and compounds such as amorphous $Zn_xIn_{1-x}O$ ($0.6 < x < 0.8$) are being investigated because of their remarkable resistance to moisture and their excellent electro-optical properties.[11]

4.2.5   Contacting

The purpose of the TCO in a thin film module is to collect the photo-generated current. However, occasionally an external metallic contact is used and grids or bus bars are used for this purpose. Consequently, it must be simple to make electrical contact to the TCO. TCOs usually have a very high concentration of free electrons and it should not be difficult to make reliable metallic contact, but the specific contact resistance must be minimal and stable. This implies that the interfacial region must be chemically stable and free of potentially high resistance oxides. The TCO contact to a CIGS cell or module consists of a bilayer of ZnO. The first layer of ZnO is in contact with the n-CdS layer of the junction and is not deliberately doped. However, its carrier concentration is still on the order of 10^{18} cm^{-3}. The second layer is usually doped with aluminum and has a carrier concentration of about

5×10^{20} cm^{-3}. When a CIGS module is exposed to standard environmental testing conditions of humidity and temperature (80% relative humidity, 80 °C) for a period of hundreds of hours, the performance deteriorates slightly, primarily due to corrosion of the TCO contact. This is one of the main issues facing the future development of CIGS as a technology and the development of a corrosion-resistant TCO, with suitable electronic and optical properties, remains a challenge for the CIGS technology.[12]

4.3   Background Science

The range of useful power density in the solar spectrum for most thin film PV applications extends from about 0.4–1.2 µm, and the peak in the solar power occurs within this range.[13] Generally, TCO transmittance in this range of wavelengths is high, and reflection and absorption losses are low—generally around 0–5% for research-quality material. Reflection losses can be reduced by use of an anti-reflection coating, although this may not be desirable in large-volume manufacture of low-cost, thin film solar cells. However, given the large flux, absorption losses in this range ($\sim$0.6–1.0 µm) can amount to a significant loss of current, which is much more difficult to reduce. This range of energies ($\sim$1.2–2.0 eV) is far less than the fundamental bandgap of any TCO used in thin film solar cells, so the absorption mechanism responsible must involve near-mid-gap states.

In this section, we review efforts at NREL to develop improved TCOs and present results that suggest they may benefit PV devices. Nevertheless, further improvements in their properties may still be realized. One of the two main functions of a TCO is to permit as much incident solar radiation as possible to pass unimpeded into the electrically active region of the solar cell. To achieve this, the TCO must have low absorptance across the entire solar spectrum. In general, a TCO transmits across a limited range of wavelengths, referred to as the transmittance window. The 'width' of the window is determined by fundamental constraints, although it may be adjusted by changing the properties of the materials. First, we review the basic physical mechanisms that determine the short wavelength cut-off and then those that affect the cut-off in the near-infrared (NIR) region.

4.3.1   The Transmission Window

The TCO transmits short wavelength light above its bandgap relatively freely, but it absorbs sub-bandgap wavelengths strongly. The bandgap of a typical TCO lies between about 3 and 4 eV, which roughly corresponds to a wavelength range of 0.3–0.4 µm. The solar irradiance is relatively small at such wavelengths, but solar cell designers would still prefer to capture photons in this range to generate additional current, even if small.[13] The range of wavelengths transmitted by a TCO depends on the Burstein–Moss effect in the short wavelengths ($\sim$0.35 µm), and on free electron absorption in the NIR in the wavelength range ($\sim$1 µm).[14] We first discuss the shift of the optical gap of the semiconductor due to its degeneracy, as originally demonstrated by Burstein[15] and Moss,[16] and then summarize our efforts to increase the fundamental bandgap by alloying.

4.3.1.1 The Burstein–Moss Effect

A semiconductor is defined as degenerate when the Fermi energy is above the minimum in the conduction band for n-type material; for higher electron concentrations, the Fermi level moves further into the conduction band. The degeneracy is the energy difference between the conduction band minimum and the Fermi level, and a typical TCO becomes degenerate when the free carrier concentration is about 10^{18} cm^{-3}. All electron states beneath the Fermi energy but above the bottom of the conduction band are filled, which means that electrons excited from the valence band will not be absorbed unless they have an energy at least equal to the energy gap plus the degeneracy. The manifestation of this effect, known as the Burstein–Moss shift, is a shift of the optical gap of the TCO to shorter wavelengths, with potential benefit to the device. The Burstein–Moss formulation gives an estimate of the shift of the optical gap (as distinct from the fundamental gap) with the carrier concentration. For a typical TCO, with a rather flat valence band and a large hole effective mass, the reduced effective mass and the electron effective mass are approximately equal and generally about 0.35 m_e, where m_e is the free electron mass.

The magnitude of the effect depends primarily on the curvature of the conduction band and a semiconductor with an effective mass of about 0.35 m_e becomes degenerate for a carrier concentration of about 10^{18} cm^{-3}. Large shifts (up to about 1 eV) in the optical bandgap are found for concentrations in the mid 10^{20} cm^{-3} range.

It has previously been shown that CdO, a material with a fundamental bandgap of about 2.2 eV, which is too small for application in solar cells, has an optical gap that can be increased by about 1 eV with increasing degeneracy.[17] Although increasing the carrier concentration may appear to provide an obvious benefit by enabling lower TCO resistance, there may be disadvantages. For example, electron transport properties may deteriorate, long wavelength transmittance may suffer and transmittance in the visible range of wavelengths may decrease. In this case, the challenge is to ensure that exploitation of the Burstein–Moss effect—bearing in mind that the increase in current is necessarily small—does not adversely impact other aspects of device performance.

4.3.1.2 Free Electron Absorption

The free electron effect often adequately accounts for the optical and electrical properties of TCO materials in the NIR region of the spectrum.[14,18] The essence of the theory is that the sign of the real part of the dielectric permittivity changes from positive to negative at the plasma wavelength. The plasma wavelength is determined by the concentration of free carriers, the effective mass of conduction electrons, and other quantities. In the region of the plasma wavelength, the optical properties of the TCO change from being highly transmissive (at shorter wavelengths) to highly reflective (at longer wavelengths). The reflectance and transmittance do not equal 0.5 (in the range from 0 to 1) at their point of intersection, meaning that there is finite absorption in the vicinity of the plasma wavelength. An absorption band (the free carrier absorption band) is formed and it can also impair device performance. To minimize the width and height of the absorption band, it is crucial to maximize the electron mobility or, more fundamentally, the electron relaxation time.[18] Most research and development into TCOs needs to be focused on optimization of the relaxation time. It is a far better strategy to increase the relaxation time than the free-carrier concentration because this will lead to an improvement of both the optical and electrical properties of the TCO.

4.4 Binary Compounds

4.4.1 ZnO

ZnO-based TCO is technologically important because of its abundance and non-toxicity, its deposition by industrially scalable processes such as sputtering or chemical vapor deposition, and the reasonable performance achieved when depositing films without intentional heating. ZnO films, both extrinsically doped and undoped, have been studied extensively since the 1970s.[19] Most dopants have been selected to provide a cation substitution for Zn (+2), although doping with fluorine, as an anionic dopant, has also been investigated.[20-25] Minami[26] and Gordon[27] have compiled lists of TCOs and extrinsic dopants that have been investigated. Al and Ga, each with the valence state +3, have proven to be among the most effective substitutional dopants for zinc (valence state +2).[19,26,28-36] ZnO carrier concentration values as high as 10^{21} cm^{-3} have been achieved, but maximum mobility values reported for ZnO thin films have rarely exceeded 50 cm^2 V^{-1} s^{-1}.[26,28,37] Values of 20 cm^2 V^{-1} s^{-1} or less are typical in commercial ZnO:Al films. Mobility and resistivity improvements have continued for doped ZnO, whereas other common TCOs such as indium tin oxide (ITO) and SnO$_2$ have shown little recent change.[26] This suggests that future gains in ZnO material properties may still be available and further investigations are worthwhile.

4.4.1.1 Properties of ZnO

ZnO films are typically thermodynamically stable in the hexagonal wurtzite (B4) ABAB lattice structure, with lattice constants of $a \sim 0.3245$ nm and $c \sim 0.5206$ nm.[31] Polycrystalline ZnO films are typically oriented in the ZnO (0002) direction, with the c-axis oriented normal to the substrate. Calculations have been performed by different methods to determine the ZnO band structure.[38-41] Recent calculations show that ZnO has an s-like lowermost conduction band consisting of zinc 4s states, while the valence bands include oxygen 2p and zinc 3d states.[39,40] The bandgap of ZnO was calculated to be 3.4 eV.

ZnO is intrinsically n-type. Oxygen vacancies (V_O) and zinc interstitials (Zn_i) have been suggested as providers of intrinsic free carriers in ZnO.[42-45] Others have calculated that V_O are deep donors and Zn_i have a low activation energy for diffusion from their interstitial sites, suggesting the possibility that neither of these mechanisms provides a substantial number of free carriers in ZnO.[46-49] In recent years, hydrogen, an impurity common to most film deposition methods, has been suggested as an alternate source of free carriers in ZnO, where hydrogen acts exclusively as a donor.[50-56] Which of these mechanisms dominates remains unclear.

4.4.1.2   Recent Studies of ZnO-Based TCOs at NREL

Typically, ZnO films have been grown at room temperature by sputtering in pure argon; doping is usually required to achieve adequate conductivity. A dopant concentration of 2.0 wt% aluminum is commonly used, although it is not obvious that this is optimal for PV applications. In addition, non-traditional metal dopants in ZnO may offer advantages over aluminum. At NREL, we have studied various doping levels (using fully oxidized targets containing 0.05, 0.1, 0.2, 0.5, 1.0 and 2.0 wt% Al_2O_3 in ZnO), less common dopants including zirconium (valence state +4) and vanadium (+2, 3, 4, 5) that have received less attention, and the role of oxygen and hydrogen incorporation in the argon radio-frequency (r.f.) sputtering ambient.[57,58] We found that reduced aluminum doping can enhance film properties by reducing the amount of electron scattering from extrinsic defects. The electrical properties of ZnO:V appear promising, although further work is required to improve its optical properties.

Based on our initial studies of heavily doped films deposited in pure argon, a substrate temperature of 200 °C was chosen for these studies. Films were subsequently deposited at room temperature and were found to be of equal, if not superior, performance. We also investigated the effects of adding oxygen or hydrogen to the argon sputtering gas. Gas flows were controlled with needle valves and measured using an ion gauge. The chamber pressure was throttled to 15 mtorr during film deposition.

When small amounts of oxygen are added to the argon ambient, the carrier concentration and mobility both decrease sharply [Figure 4.1(a) and (b)]. The quantity of oxygen is expressed as a percentage of the flow rate relative to that of argon. The most significant decrease occurs for the lightest doped, or undoped, ZnO because it has a low initial carrier concentration. Films with the lowest aluminum doping (0.1 wt% Al_2O_3 shown here) retain a slightly higher carrier concentration in an oxygen partial pressure ambient than the undoped ZnO films, whereas the more heavily Al-doped films experience a smaller proportional decrease in carrier concentration. Conversely, ZnO:V

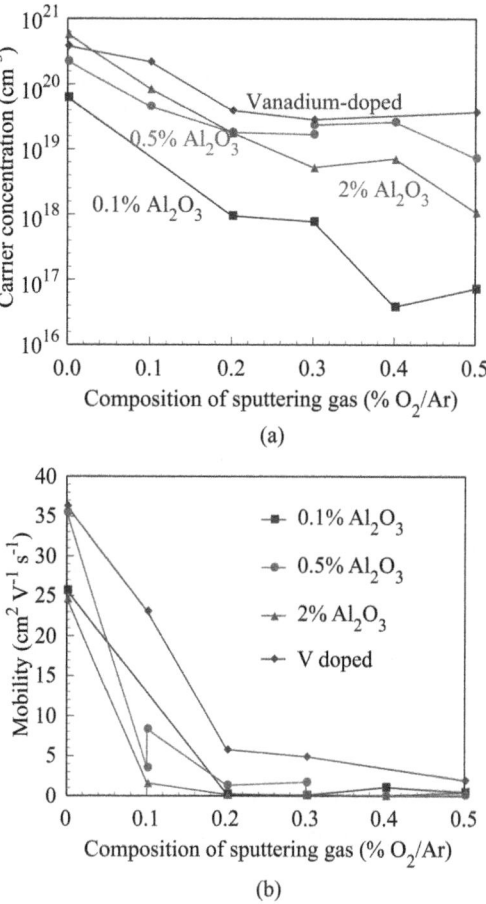

Figure 4.1 (a) Variation of carrier concentration with oxygen percentage in argon sputtering gas. The *x*-axis refers to the percentage of the flow rate of oxygen to argon. (b) Variation of mobility with oxygen percentage in argon sputtering gas.[59]

exhibits the greatest tolerance for oxygen and it retained the highest carrier concentration in oxygen of all the dopants and doping levels examined in this study. Hall mobility values also decrease significantly when small amounts of oxygen are added to the ambient, regardless of aluminum doping level. ZnO:V again appears to be the most tolerant to the adverse effects of oxygen, retaining a mobility value of $\sim$24 cm^2 V^{-1} s^{-1} at 0.1% O$_2$/Ar, whereas the next highest (belonging to the 0.5 wt% Al$_2$O$_3$ film) was only $\sim$8 cm^2 V^{-1} s^{-1}.

These decreasing values of carrier concentration and mobility in oxygen could have a number of possible causes. Because the films are poly-crystalline, with an average grain size of $\sim$30 nm, the grain boundary density is high.[59] Oxygen may adsorb on the grain boundaries and could remove free carriers from the grains, thereby reducing the measured carrier concentration.[60] Trapped charge, due to the adsorbed oxygen, would also establish electrostatic potential energy barriers that could inhibit carrier transport between grains, thereby reducing the measured Hall mobility.[61] Other mechanisms may also describe this behavior. Oxygen vacancies and zinc interstitials have been suggested as donors in ZnO.[43,45] Addition of oxygen to the ambient could reduce the concentration of these defects, thereby decreasing the number of free carriers. Others have suggested that hydrogen may be a significant source of free carriers in ZnO.[47,53] The addition of oxygen to the growth ambient could remove beneficial effects of any residual hydrogen in the growth chamber.

Controlled amounts of hydrogen in the sputtering gas increase both carrier concentration and mobility, as shown in Figure 4.2(a) and (b). The quantity of hydrogen is expressed as a percentage of the flow rate relative to

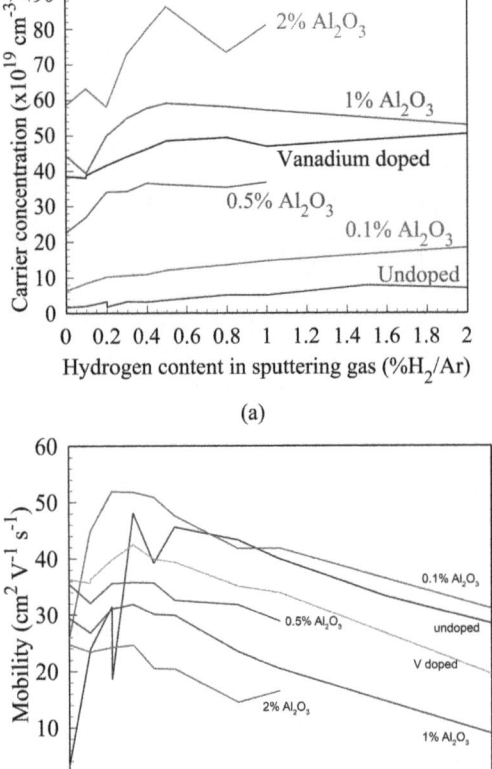

Figure 4.2 (a) Variation of carrier concentration with hydrogen percentage in argon sputtering gas. The *x*-axis refers to the percentage of the flow rate of hydrogen to argon. Variation of mobility with hydrogen percentage in argon sputtering gas.[64]

that of argon. Carrier concentration values of ZnO:Al films grown in a hydrogen-rich ambient increase systematically from 1 to 8×10^{20} cm^{-3} as the Al$_2$O$_3$ content of the target increases from 0.05 to 2.0 wt%, respectively. The highest carrier concentration of the ZnO:V films is 5.0×10^{20} cm^{-3}. A small amount of hydrogen (optimal H$_2$/Ar $\sim$0.3%) added to the argon sputtering ambient was particularly important in obtaining the highest Hall mobilities for undoped and lightly Al-doped films. Remarkably, films containing 0.05–0.2 wt% Al$_2$O$_3$ exceed the mobility of undoped ZnO, reaching values greater than 50 cm^2 V^{-1} s^{-1} near 0.3% H$_2$/Ar. The highest mobility of ZnO:Zr films is 24 cm^2 V^{-1} s^{-1}, offering no improvement over well-established ZnO:Al, while having a lower corresponding carrier concentration. ZnO:V films have a peak mobility of 42 cm^2 V^{-1} s^{-1} for 0.3% H$_2$/Ar. Combined with their reasonable carrier concentration values, ZnO:V films offer the lowest film resistivity (not shown) in this study in nearly all hydrogen and oxygen partial pressure ambients. The ZnO:Al films containing 2.0 wt% Al$_2$O$_3$ offer only slightly lower resistivities than ZnO:V films in the 100% Ar and 0.1% H$_2$/Ar ambients. We observe that mobility values of films doped with vanadium decrease for H$_2$/Ar ratios greater than 0.3%, similarly to those doped with aluminum.

Hydrogen in the sputtering ambient could form complexes with adsorbed oxygen on ZnO grain boundaries, thereby passivating negative oxygen ions and returning carriers to the conduction band. The height of the electrostatic barriers between grains would also be reduced, increasing the measured Hall mobility as observed. Alternatively, the use of hydrogen creates a reducing environment that could be conducive to forming donor defects, V$_O$ and Zn$_i$. The decrease in mobility with larger amounts of Al$_2$O$_3$ and excessive hydrogen content (>0.3% H$_2$/Ar) is likely due to increased ionized or neutral impurity scattering because of Al$^+$ and H$^+$ and neutral scattering centers.

To obtain further information about hydrogen doping of ZnO, we performed a secondary-ion mass spectrometry (SIMS) study to quantify the amount present in the ZnO films as a function of the H$_2$/Ar ratio of the deposition ambient.[59] Hydrogen concentrations were obtained at mid-film depth ($\sim$200 nm) by sputter depth-profiling to avoid surface contamination effects. Hydrogen incorporation in the films was found to increase linearly with the H$_2$/Ar ratio in the growth ambient from 8×10^{19} cm^{-3} at 100% Ar to 4×10^{21} cm^{-3} at 2.0% H$_2$/Ar. These data and the carrier concentration are shown for these films in Figure 4.3. Although the corresponding carrier concentration increases with H$_2$/Ar ratio, the carrier concentration represents only $\sim$1.5% of the hydrogen content in the films. Thus, a maximum of only $\sim$1.5% of the hydrogen atoms present in the ZnO films appear to contribute carriers. Interestingly, the saturation concentration in single crystal ZnO is much lower ($\sim$10^{17}–10^{18} cm^{-3}) than the concentration observed in our films and in previous film studies ($\sim$10^{21} cm^{-3}).[62] This possibly supports our suggestion that surfaces and grain boundaries play an important role in providing locations for hydrogen in ZnO films.

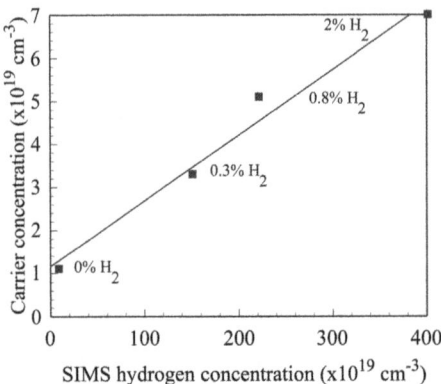

Figure 4.3   Measured carrier concentration of ZnO films as a function of the hydrogen concentration of these films. The hydrogen concentration was measured using secondary-ion mass spectrometry.[64]

Optical absorptance data for the highest-mobility ZnO TCO films of each dopant and doping level in this study are shown in Figure 4.4. In the visible wavelengths, the undoped and Al-doped films all exhibit low absorptance of less than 3%. The ZnO:V film, however, has higher visible absorptance and is not yet suitable for PV devices. In the infrared region, the peak absorptance shifts to shorter wavelengths as carrier concentration increases due to free carrier effects. The absorptance peak is evident for films with the highest carrier concentrations, but the peaks for the undoped ZnO and 0.1 wt% Al_2O_3 film occur at longer wavelengths and are of little relevance to solar cells. Low absorptance up to ~1100 nm is required for optimal current generation in $Cu(In,Ga)Se_2$ PV devices and significant gains may be feasible by using lightly doped ZnO.

ZnO has satisfactory electrical and optical properties, and is suitable for use with some types of solar cell. Production advantages such as inexpensive, non-toxic starting material and the possibility of room temperature deposition are also beneficial. The primary challenges for ZnO-based TCOs lie in their limited heat and moisture tolerance. Lightly doped films deposited in hydrogen remain stable up to temperatures of ~250 °C, but temperature-programmed desorption and annealing experiments indicate that hydrogen begins to desorb from the films at higher temperatures, leading to decreased Hall mobility and carrier concentration.[63] Films with higher aluminum concentration were found to be stable to temperatures of ~400 °C.[64] ZnO generally is not used for CdTe solar cells because it is not stable at the high temperature used for deposition of the CdTe; other TCOs, such as SnO_2 and Cd_2SnO_4, are more tolerant of high temperatures. Moisture tolerance is another significant concern for ZnO.[65,102] Although some TCOs, such as SnO_2, offer excellent damp-heat tolerance, ZnO generally has been found to be poor in this regard. Using good engineering practices to guard against moisture ingress, ZnO films can be implemented in PV modules successfully.

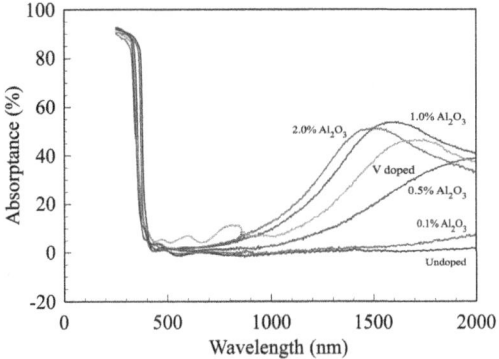

Figure 4.4 Absorptance *vs.* wavelength for the highest mobility film at each Al_2O_3 level or dopant, deposited at 200 °C in 0.3% H_2/Ar (except for the 2.0 wt% Al_2O_3 film, deposited in 100% Ar). The bandgap of CIGS is shown for comparison.[64]

4.4.2 In₂O₃-Based TCOs

Doped In_2O_3 (IO) films represent some of the highest quality TCOs presently available because of high electron mobility at high carrier concentration, low optical absorption and low moisture sensitivity. IO films can be doped using several Group IVa and IVb dopants, including Ti, Zr (IZrO) and Sn (ITO).[66] However, most reports have suggested that the optimal balance between electrical and optical quality is produced with Sn doping.[67] IO-based TCO films can be deposited by a range of techniques including evaporation, sputtering, spraying and CVD using a wide range of deposition temperatures (room temperature to ~500 °C), and yet can maintain high film quality during equally high post-deposition treatments. Present applications of IO-based TCOs are largely in high-value consumer electronic products where the volatility of indium price is not a major issue (*e.g.* liquid crystal, light-emitting diode and plasma displays). However, because of their robust mechanical/chemical/electrical properties, applications also include touch panels, antistatic shielding, hot mirrors, de-icing layers and PV solar cells.

4.4.2.1 Brief History of In₂O₃-Based TCOs

Although IO-based TCOs are among the longest-studied TCOs, an all-encompassing description of how to produce high-quality films using different deposition techniques and processes has not been established. For many high-throughput applications, films are produced at relatively low temperature by reactive evaporation or sputtering from metal sources. Although this allows for low tooling costs, good source utilization and high deposition rates, the necessary film quality is generally only achieved following a controlled post-deposition oxidation step at 250–350 °C.[68]

A post-deposition oxygen anneal process is generally used even when 'reduced' oxide sources are used for mechanical, throughput and/or other production reasons. If high quality is required in as-deposited films, film quality will be a delicate balance between structural quality (*i.e.* the amount of dopant that is properly coordinated into an active donor–defect location) and amount of oxygen incorporated into the film.[69] Depending on the type of source used (*i.e.* encompassing the extent of oxidization of a sputtering target and the reactivity of the dopant with oxygen), most researchers find that a small amount of oxygen must be added to the deposition ambient (∼0.3 vol % oxygen in argon is typical). This suggests that IO-based TCOs form desirable donor defects primarily under oxygen-rich deposition conditions (note that the opposite is observed for fabricating high quality ZnO films). If the ambient contains too much oxygen, the carrier concentration decreases, whereas if the ambient becomes too oxygen deficient, optical transparency decreases. The challenge during ITO deposition is to incorporate enough oxygen to produce films with high transparency and optimal electrical properties. In production coating systems, this critical balance of oxygen must be maintained over a large deposition area and it is typically adjusted for variations in target use, seasonal changes and maintenance activities. As is discussed in a later section, techniques have been identified that allow the target composition to be tailored to reduce the effect of oxygen variation in the deposition ambient.

4.4.2.2   Overview of IO-Based TCO Work

As with most other TCOs studied at NREL, research on IO-based materials has been directed primarily at developing TCO films with desirable properties for specific PV devices of interest. For example, the initial NREL IO studies (*ca.*1980s) were directed at improving the performance of ITO/InP PV devices.[70] At that time, ITO/InP devices demonstrated notably high resistance to the type of high-energy proton radiation experienced by Earth-orbiting satellites in low-Earth orbits. However, these devices also required the ITO to be deposited at near room temperature and with controlled damage to the near-surface layers.[71] Later, after the importance of high-mobility TCOs on the performance of thin film PV was more fully appreciated, IO-based studies were undertaken to better understand the interactions of dopant species on film mobility. This led not only to the development of Mo-doped IO (IMO),[72,73] but also the realization that the real part of the dielectric permittivity can be a much more important and controllable parameter in TCOs than was previously appreciated.[69]

4.4.2.3   Results of ITO Studies

As mentioned above, ITO has long been considered a well-understood TCO material. However, the past 25–30 years of work at NREL has provided additional understanding that can assist with forming good quality ITO using different deposition processes, as well as clues to the interaction of oxygen and Sn in yielding donors in the IO host material that demonstrate low scattering. The first insight into the nature of the defects in ITO resulted from investigating the evolution of the Hall parameters (carrier concentration and mobility) as a function of post-deposition annealing temperature for ion beam deposited ITO films formed at room temperature. Figure 4.5 shows how, for these films, carrier activation increases significantly at post-deposition temperatures between 250 and 350 °C.[68] Subsequent transmission electron microscopy (TEM) studies expanded on these initial results to show that the ITO film underwent complete recrystallization (from amorphous to polycrystalline) at post-deposition temperatures greater than ~250 °C.[74] The implications that ITO with good electrical quality can be produced in either an amorphous or a polycrystalline structure is of great scientific interest, and this continues to be investigated and debated. We further propose that this phenomenon is shared by all IO-based TCOs and can provide an invaluable starting point to explain most IO-based material differences that occur as a function of either post-deposition and/or deposition temperature.[75]

On a more practical note, advantages of the relative moisture insensitivity of the amorphous state ITO continues to be actively exploited for thin film PV solar cells (especially those based on $CuInSe_2$ alloys).[76] X-Ray photoelectron spectroscopy (XPS) and Mössbauer studies revealed that during recrystallization, some of the Sn in the amorphous matrix substitutes for indium in the post-anneal polycrystalline structure to donate one electron. Although the increased carrier concentration provided a noticeable Burstein–Moss shift

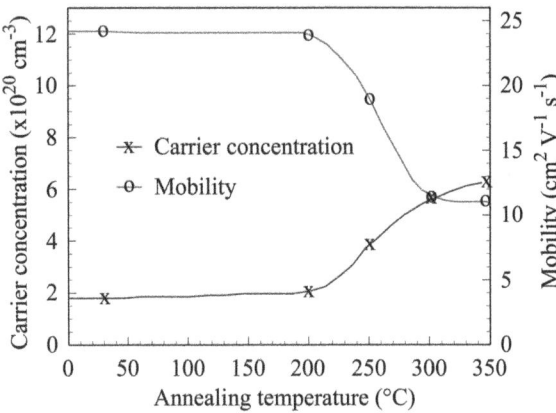

Figure 4.5   Room temperature mobility and carrier concentration *vs.* annealing temperature (5 min in ultra-high purity forming gas at each temperature) for ion-beam sputtered ITO film (91 mol% In_2O_3; 9 mol% SnO_2). The measured data apply to a nearly stoichiometric film (slightly oxygen-deficient) grown with an oxygen partial pressure = 1×10^{-5} torr, rate = 0.04 nm s^{-1} and thickness = 71 nm.[68]

that improves the transparency of the film, it also leads to greater ionized impurity scattering that reduces the mobility.[74] Efforts to improve the mobility eventually led to sputtered ITO films demonstrating mobility of ~ 45 cm^2 V^{-1} s^{-1}, using targets made from 91 wt% In$_2$O$_3$ and 9 wt% SnO$_2$. These were re-oxidized, following hot pressing and machining, and then used to r.f. magnetron sputter films onto Corning 7059 glass, at substrate temperatures of 350–400 °C and a partial pressure of oxygen of ~ 0.3 vol%.[69] It should be noted that, at the time of this writing, ongoing studies to improve the mobility by reducing the Sn concentration from the typical value of 10 wt% are being pursued.

4.4.2.4 Results of IMO Studies

With the ability to produce high-quality ITO at NREL firmly established, it was a logical step to continue to develop IO-based TCOs that may exceed the performance of standard ITO. One idea promoted was to identify an alternative dopant that, unlike Sn, could provide more than a single electron per substitution. It was suggested that in this multi-donor defect scenario, the scattering potential of the multiply ionized defect may remain similar to that of a singly ionized defect, thus allowing higher carrier concentration without greatly increased ionized impurity scattering (*i.e.* multi-donor defects may yield higher mobility at similar carrier concentrations).

Reports at that time suggested that one possible multi-donor dopant was Mo.[77,78] Note that unlike Sn, which has only one oxidation state (+4), Mo has several (+2, +4, +6). Although initial studies using an IO target containing 4 wt% Mo metal suggested that multi-donor substitution was occurring, more detailed investigations revealed that although the Mo could indeed coordinate in one of several oxidation states, the +4 state yielded the highest carrier concentration, similar to Sn, as shown in Figure 4.6.[73] Although this was not the desired result, it was also observed that once the deposition conditions were optimized to coordinate the Mo dopant primarily in the +4 state, electron mobility in excess of 85 cm^2 V^{-1} s^{-1} resulted for films deposited onto glass. At the time, this was the highest mobility ever achieved for any type of sputtered TCO on glass and prompted others to undertake similar studies on IMO using techniques that, although not commercially relevant to PV, were known to yield superior results over sputtering on glass.[79]

4.4.2.5 Results of IZrO Studies

Although the research on IMO discussed above provided positive results that are still being discussed, the results suggested that the idea of incorporating multi-donor defects may not be a straightforward pathway toward developing IO-based TCOs with significantly improved electrical quality. With this in mind, a research effort began that would reconsider all the Group IV elements (both IVa and IVb columns) as potential donors in IO.

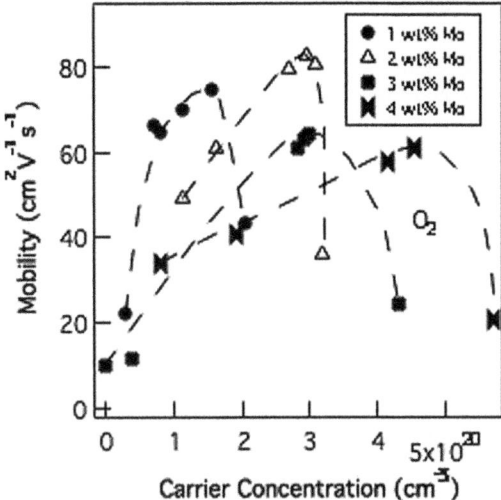

Figure 4.6 Room temperature mobility *vs.* carrier concentration of r.f. sputtered IMO thin films with 0, 1, 2, 3 and 4 wt% Mo content. Direction of arrows indicates increase in oxygen concentration during deposition.[73]

Literature review, including historic literature, suggested that although the IVb-column dopants Zr and Hf had shown promise, research on these dopants had not been reported for decades.[66] Initial efforts at NREL involved simply replacing the 9 wt% SnO_2 in the ITO targets with 9 wt% ZrO_2 (identified as IZrO targets). Electrical analysis of films made from these IZrO targets revealed only modest mobility (maximum electron mobility ~15 cm^2 V^{-1} s^{-1}). However, and more importantly, the optical transmission of the IZrO films did not appear to be nearly as sensitive to the amount of oxygen in the sputtering ambient. Indeed, IZrO films sputtered in pure Ar demonstrated similar visible transmission as ITO or IMO films that were deposited in a carefully controlled partial pressure of oxygen (typically 0.3 vol%).[69]

The insight that the amount of Zr-doped IO could drastically alter the optical properties was next tested to see if co-doping with both Sn and Zr could provide an IO-based TCO film with the electrical properties of ITO but the process latitude of IZrO (*i.e.* tolerance to variation in oxygen ambient). This test was not only successful, but led to an even deeper understanding of the effect of Zr in IO-based TCO (and all TCOs, for that matter).[69] This new insight was based on the knowledge that adding a small amount of high-permittivity (generally refractory-metal) oxide to *dielectric* oxides increases the real part of the dielectric constant of the host oxide beyond the degree that might at first be expected (*i.e.* a little bit of high permittivity addition goes a long way!). Although this had been known in dielectric oxides for many years, the same functionality had not been tested

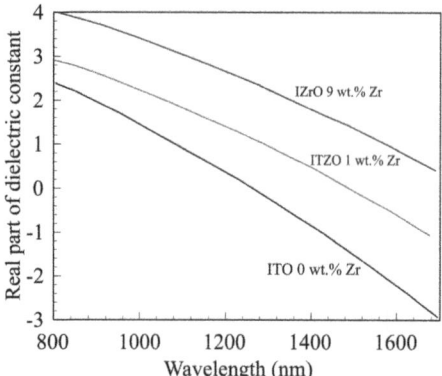

Figure 4.7 Spectroscopic ellipsometry analysis of real part of dielectric permittivity (ε_1) *vs.* wavelength for ITO, IZrO and ITZO films. Data were modeled for the region at the bottom of the film (film–glass interface). Because the plasma wavelength is defined at the point at which ε_1 = zero, figure shows that λ_p increases significantly as Zr is added to the film.[81]

in TCOs.[80] Indeed, as shown in Figure 4.7, by adding even 1 wt% ZrO_2 to an IO or ITO host, the real part of the dielectric constant increased significantly.[81] Because this increase in permittivity occurred without significant change in the scattering potential (*i.e.* both Sn and Zr are Group 4, with only the +4 oxidation state), the increased film permittivity shifted the onset of free-carrier absorption to longer wavelengths (*i.e.* shifted the point at which the real part of the dielectric constant switches from positive to negative).

The idea of engineering TCO dielectric permittivity to improve the films' NIR transmission has since been expanded for application to both ZnO and SnO_2:F films (discussed in the next section) with positive results.[81–83] The application to the benefit of device efficiency—and perhaps long-term stability—remains an outstanding challenge for this group of TCO materials, but it appears to hold substantial promise.

4.4.3 SnO_2

SnO_2 is one of the most popular TCO thin films because of its high electronic conductivity and transparency in the visible spectrum and excellent thermal stability.[67,84,85] SnO_2 is chemically inert and mechanically hard. Because of these properties, SnO_2 is commonly used in gas sensors, electrochromic devices, building glass coatings, solar cells and liquid crystal displays.[86–89] SnO_2 films are often fabricated by metal organic chemical vapor deposition (MOCVD) and their high-temperature stability makes them well suited for the high processing temperatures required for CdTe/CdS solar cells.[90] Typical commercial SnO_2 films are formed by CVD using various tin chloride based cation precursors (*e.g.* tin tetrachloride or TTC), whereas researchers have sometimes used the more toxic tetramethyl tin (TMT).[91-95]

SnO$_2$ is a degenerate semiconductor material. In general, oxygen vacancies (V$_O$) provide n-type doping. The typical resistivity for undoped SnO$_2$ is between 10^{-2} and 10^2 Ω cm, depending on the deposition temperature and precursors. With n-type cationic dopants such as antimony and arsenic, and anionic dopants such as chlorine and fluorine, a SnO$_2$ film with very low resistivity can be obtained.[96-98] Fluorine doping has resulted in the highest conductivity and optical transmission.[67,99,100] Bromotrifluoromethane (CBrF$_3$) has been used in the research laboratory as a fluorine dopant in MOCVD to dope SnO$_2$. However, CBrF$_3$ is a greenhouse gas and has been slowly phased out of the market; thus, an alternative fluorine sources are being investigated.

In our work at NREL, we studied CIF$_3$ and CF$_4$ as possible replacements for CBrF$_3$ and have performed both theoretical calculations and experiments.[101,102] The calculations suggest that CF$_4$ has a very low conversion rate compared with CIF$_3$ and CBrF$_3$; thus, our experiments focused on comparing CIF$_3$ and CBrF$_3$ precursors. We also studied the dependence of electrical and optical properties of SnO$_2$:F films on deposition temperature, as well as the concentrations of dopant and oxygen.

The SnO$_2$ films were prepared by low-pressure chemical vapor deposition (LPCVD) using ultra-high purity TMT and oxygen precursors, and with nitrogen as the carrier gas. The reaction chamber was a cold wall rectangular quartz tube with a high-purity graphite susceptor. The latter was heated using infrared lamps divided into five zones with a maximum temperature of 800 °C. The substrates were either 102 mm diameter silicon wafers or 102 mm × 102 mm 7059 glass squares. No deposition occurred for substrate temperature less than 475 °C, but increased in the range of 500–700 °C. At the higher temperatures, a gas phase reaction occurred and the films became hazy. A wide range of analytical techniques was used to characterize the films.

Undoped films had a carrier concentration of about 10^{18} cm^{-3}, *i.e.* non-degenerate and a mobility of about 1 cm^2 V^{-1} s^{-1}. These properties were essentially independent of the gas composition and substrate temperature, although the mobility improved at the highest deposition temperatures because of improved crystallinity. The fundamental bandgap was determined to be 3.78 eV from measurement of the transmittance and reflectance of films.

Thermodynamic modeling of fluorine incorporation helped design the experiments. The model enabled us to estimate the equilibrium fluorine solubility in SnO$_2$ as a function of temperature and flow rate of the various dopant precursors. First, the saturated solubility of fluorine in the SnO$_2$ film was estimated by the Delta Lattice Parameter (DLP) Model.[103] We calculated the precursor decomposition percentage and equilibrium fluorine concentration in the SnO$_2$ film using Thermo-Calc software based on the interaction energy estimated by the DLP. Thermo-Calc is based on the principle of

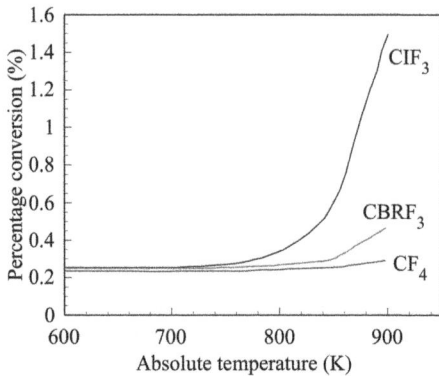

Figure 4.8 Calculated conversion rate of $CBrF_3$ as a function of its initial molar concentration for various temperatures.[101]

minimizing Gibbs energy and was developed for performing various kinds of thermodynamic and phase-equilibria calculations.[104]

The effects of different dopant precursors on fluorine incorporation were compared between the calculated and experimental results. Under chemical equilibrium, different precursors have different conversion rates (ratio of decomposed precursor to initial amount) at the same temperature, which is expected to lead to different fluorine concentrations in SnO_2 films. The calculated conversion rates suggest that CIF_3 decomposes more than $CBrF_3$, especially at SnO_2 growth temperatures greater than 500 °C, and that $CBrF_3$ in turn decomposes more rapidly than CF_4. These results are summarized in Figure 4.8. Experimentally, we determined that SnO_2 grown using CF_4 to provide the fluorine dopant was insulating, in agreement with the low conversion rate predicted by theory. Therefore, we limited our studies to $CBrF_3$ and CIF_3. Although we set out to find a replacement for $CBrF_3$, it was necessary to establish a baseline for this precursor for comparison with potential alternatives such as CIF_3.

4.4.3.1   CBrF₃ as the Fluorine Source

Figure 4.9 shows the calculated precursor conversion fraction of $CBrF_3$ as a function of its initial molar concentration for four deposition temperatures. Although the model predicts that the conversion rate increases with deposition temperature, it is relatively insensitive to the molar concentration for temperatures between 450 and 600 °C. For a wide range of deposition temperatures, the fractional conversion rate is significantly higher at low initial $CBrF_3$ concentration. However, as the initial $CBrF_3$ concentration increases, the fractional conversion rate of the precursor is reduced and saturates at a level that depends on the deposition temperature.

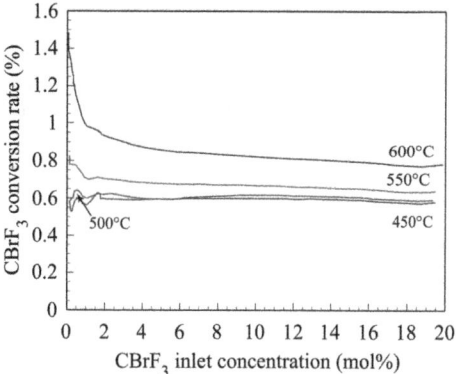

Figure 4.9   Calculated conversion rate of CBrF$_3$ *vs.* its initial molar concentration at the inlet for various temperatures. The total chamber pressure was 40 torr with 44.4 mol% oxygen and 0.59 mol% tetramethyl tin.[101]

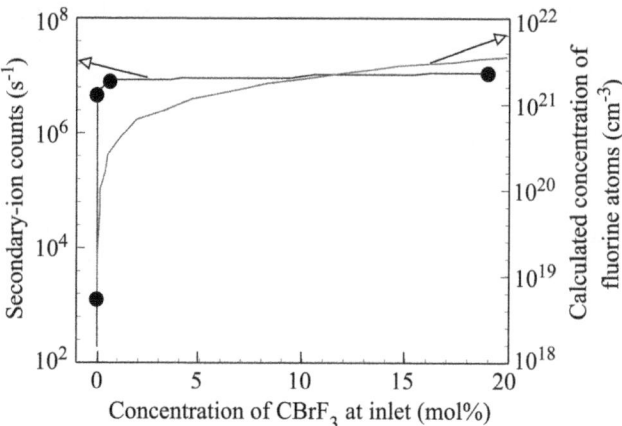

Figure 4.10   Comparison of calculated and measured quantities of fluorine as a function of the molar ratio of the CBrF$_3$ dopant precursor. The arrow on left indicates that the line with solid circle markers represents SIMS-measured ion counts for fluorine, while the arrow on the right indicates that the line without markers represents the calculated concentration of fluorine.

To verify the reliability of the model, we compared the predicted and measured concentrations of fluorine from four different SnO$_2$:F films that were depth-profiled by SIMS. Figure 4.10 shows the average fluorine count measured by SIMS, *i.e.* the symbols and the calculated concentration, both as functions of the inlet CBrF$_3$ molar concentration. Both calculated and experimental data show that the fluorine concentrations increase rapidly for smaller CBrF$_3$ inlet concentrations (0–0.2 mol%). As the inlet CBrF$_3$ concentration increases to above 0.2 mol%, the fluorine concentration in the

SnO$_2$ film saturates and increases by less than a factor of two as the CBrF$_3$ concentration increases to 20 mol%.

When CBrF$_3$ is used as the fluorine dopant source, the conductivity is significantly improved. The carrier concentration increases by about three orders of magnitudes—from 10^{17}–10^{18} cm^{-3} to mid-10^{20} cm^{-3}. The carrier mobility also increases and the resistivity of the films decreases by more than three orders of magnitude—to the mid-10^{-4} Ω cm range. These findings are summarized in Figure 4.11. Generally, as the carrier concentration of a semiconductor increases, the mobility decreases because of ionized impurity scattering. That this does not happen for SnO$_2$ suggests that ionized impurity scattering is not the dominant scattering mechanism. Li's group has argued that grain boundary scattering may be responsible for carrier scattering, although Coutts and co-workers have discounted this mechanism in TCOs on the grounds that the electron mean free path is far smaller than the size of typical grains.[18,81] It seems more likely that neutral impurities and defects within the grains are responsible for scattering carriers. The conversion of the CBrF$_3$ molecules is relatively small and the concentration of neutral fractions may be sufficiently large to scatter electrons. With changing deposition conditions (*e.g.* temperature, pressure, composition of precursor gases), the concentration of neutral species almost certainly changes and modifies the extent of electron scattering.

The role of deposition temperature is less obvious because it may have several roles. Firstly, the conversion of the precursors increases with temperature, but the sticking coefficient of fluorine atoms probably decreases.[105] Hence, the carrier concentration and mobility have uncertain dependences on substrate temperature. SIMS measurements showed that the fluorine count decreased as deposition temperature was increased from 500 to 600 °C, implying that the sticking coefficient decreased by more than the conversion rate increased. Secondly, increasing substrate temperature

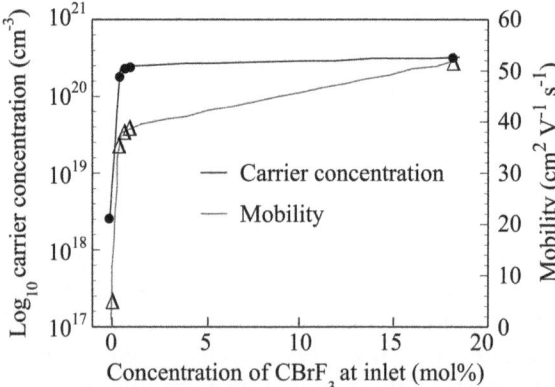

Figure 4.11   Carrier concentration and mobility of SnO$_2$:F films as functions of the molar ratio of the CBrF$_3$ dopant precursor at the inlet. The deposition temperature was 500 °C.[102]

leads to substantial improvement in the crystal quality of the film, with the reduction in defects leading to improved mobility and a small reduction in carrier concentration.

In summary, CBrF$_3$ is a well-understood dopant source and the change to another source would not be necessary if it was not a greenhouse gas.

4.4.3.2 ClF$_3$ as the Fluorine Source

As discussed earlier, because of its greater conversion rate, we believed that ClF$_3$ may be a useful replacement for CBrF$_3$, due to reduced greenhouse gas concerns. To assess this source, we performed various experiments and measurements of film properties. Initially, we fixed the substrate temperature at 550 °C and the oxygen concentration at 20.3 mol%. The ClF$_3$ molar ratio was then varied and a much higher fluorine concentration was found in the films than for a similar quantity of CBrF$_3$, as expected from the thermodynamic calculations. Figure 4.12 shows the carrier concentration and mobility as functions of the inlet concentration of ClF$_3$. Although the former reaches 10^{20} cm^{-3} for very low concentrations of ClF$_3$, the mobility, with a value of only 30 cm^2 V^{-1} s^{-1}, is substantially lower than for CBrF$_3$. Increasing the ClF$_3$ concentration slightly reduced the mobility. While the high conversion rate of ClF$_3$ leads to moderately high carrier concentration for small amounts of the material, it must also lead to a high concentration of neutral impurities, which may again be responsible for the relatively low mobility although to a greater extent than for CBrF$_3$.

With the concentrations of oxygen and ClF$_3$ fixed at 20.3 mol% and 0.03 mol%, we grew a series of films at various temperatures in the range 450–550 °C. At all temperatures, the carrier concentrations were higher and

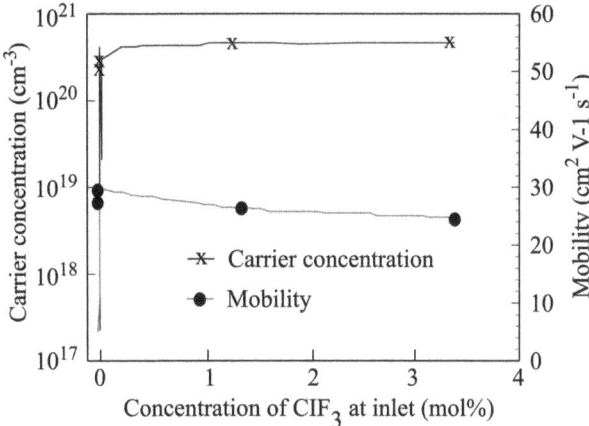

Figure 4.12 Carrier concentration and mobility of SnO$_2$F films as functions of the molar ratio of the ClF$_3$ dopant precursor at the inlet. The deposition temperature was 550 °C and the oxygen content in the chamber was 20.3 mol%.[102]

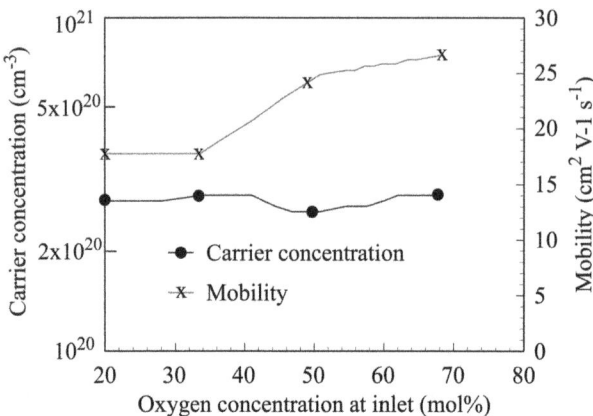

Figure 4.13   Carrier concentration and mobility of SnO$_2$:F films as functions of the oxygen molar ratio.

the mobilities lower than when using CBrF$_3$. The mobility, however, increased systematically with temperature due to substantial improvements in the crystallinity of the films. This supports our earlier postulate that the predominant scattering mechanisms were due to neutral fluorine atoms and crystal defects.

With the temperature fixed at 500 °C and the concentration of ClF$_3$ fixed at 0.03 mol%, we grew a series of films with the oxygen concentration as the variable. These results are shown in Figure 4.13. The results show that the films grown at higher oxygen concentration had higher mobility but slightly lower carrier concentration. X-Ray diffraction (XRD) analysis also showed improved crystallinity with increased oxygen, again linking the mobility to intragrain scattering by defects and perhaps by neutral fluorine atoms.

In summary, although ClF$_3$ as a dopant produces films with slightly inferior properties to CBrF$_3$, this work is at its early stages and it is possible that with further refinement, films of equivalent electrical and optical properties could be achieved. The advantage of ClF$_3$ is that less of it is needed to produce a given carrier concentration, which would make waste management simpler and less expensive. Higher deposition temperatures and oxygen content appear to improve mobility, but it may be necessary to minimize the ClF$_3$ inlet pressure to reduce neutral impurity scattering.

4.4.3.3   *Permittivity-Engineered SnO$_2$:F Films*

As explained in Section 4.2.2, new design options were offered by realizing that an engineered change in the high frequency permittivity could benefit the optical properties of TCOs, especially in the NIR region of the spectrum. Changing the high frequency permittivity had been underappreciated in Drude modeling of TCOs as a means of obtaining improved optical properties, while not having a negative impact on the electrical properties. However, it had been recognized that adding small amounts of high permittivity oxides, such as ZrO_2 and HfO_2, to silicon oxide increased the permittivity by much more than would be expected from a linear extrapolation between SiO_2 and ZrO_2.[80] Adding zirconium to In_2O_3 dopes the material about as effectively as tin although, as will be seen, in the case of tin oxide, adding zirconium can actually form an alloy.[66,81,106]

To test the effects of adding zirconium to SnO_2, we deposited films onto glass by LPCVD with tetramethyl tin, zirconium *tert*-butoxide and $CBrF_3$ precursors. These films are referred to as FTZO. For comparison, fluorine-doped films of SnO_2 without zirconium were also deposited and are referred to as FTO. The films were deposited in an atmosphere of oxygen at a temperature in the range 500–550 °C. The optical properties were determined using a spectroscopic ellipsometer in the range 800–1700 nm.

Figures 4.14 and 4.15 show the optical transmittance of two pairs of FTO and FTZO films. In the former figure, the films each had a carrier concentration of about 1.8×10^{20} cm^{-3} and their thicknesses were about 500 nm. In the latter case, the carrier concentrations were about 4.4×10^{20} cm^{-3} and the thicknesses were again about 500 nm. All four films had mobilities in the range 26–31 cm^2 V^{-1} s^{-1}. The optical properties of the FTZO films were superior in both cases: the transmittances remained higher up to longer wavelengths. Closer inspection of the FTZO films reveals that they have a slightly larger bandgap, presumably due to alloying. The small shift of the band edge to shorter wavelength cannot be accounted for by the Burstein–Moss shift because the carrier concentration is similar for each pair of films.

Figure 4.16 shows the ellipsometric analysis of a series of FTZO films and clearly demonstrates that the real part of the permittivity increases with the quantity of zirconium. Recall that the plasma wavelength occurs when the real part of the permittivity equals zero. Hence, the figure shows that the plasma wavelength increases significantly with zirconium addition and this is responsible for the improved optical properties shown in Figures 4.14 and 4.15.

Although this shift occurs in the NIR, in which region the flux of photons in the solar spectrum is relatively low, it appears that there would be a small benefit to thin film solar cells by modifying the TCO layer as described above. Although demonstrated for only In_2O_3 and SnO_2 to date, we have no reason to suppose that the same technique would fail to work for other TCOs such as ZnO, In_2O_3:ZnO and Cd_2SnO_4. The approach appears to be generally applicable.

In summary, there are clear improvements that can be made to the manufacture of SnO_2 by using CIF_3 as a growth precursor (although improvements in the mobility still need to be achieved) and by incorporating a high permittivity oxide such as ZrO_2. Other oxides may offer even greater benefits.

4.4.4 CdO

Cadmium oxide has a fundamental bandgap of about 2.28 eV and high conductivity, and is typically a degenerate semiconductor, as first established by Koffyberg.[107] It was also shown by Ueda *et al.* that the extent of the

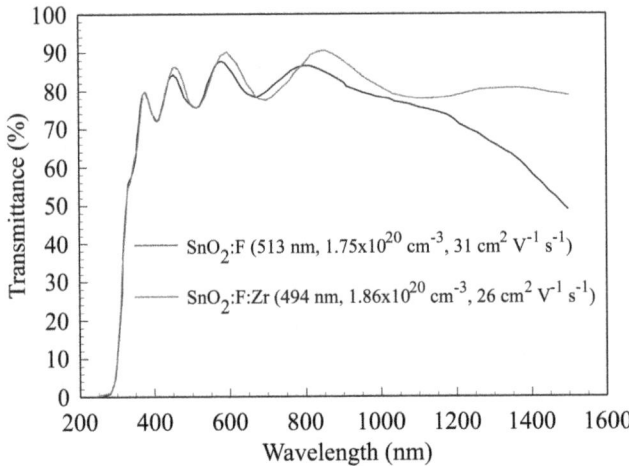

Figure 4.14 Comparison of the optical transmittance of a FTO and a FTZO film. The carrier concentrations were about 1.8×10^{20} cm^{-3} for both films. The thicknesses and mobilities were similar for both films.[81]

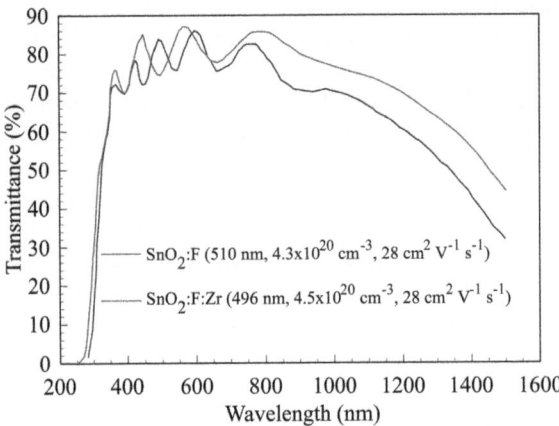

Figure 4.15 Comparison of the optical transmittance of a FTO and a FTZO film. The carrier concentrations were about 4.4×10^{20} cm^{-3} for each film. The thicknesses and mobilities were similar for both films.[81]

degeneracy increased strongly with carrier concentration, implying a conduction band with very small radius of curvature and, therefore, small effective mass.[108] Because of this, it was concluded that the optical bandgap may widen sufficiently with degeneracy to make the material potentially useful as a TCO for PV-related applications.

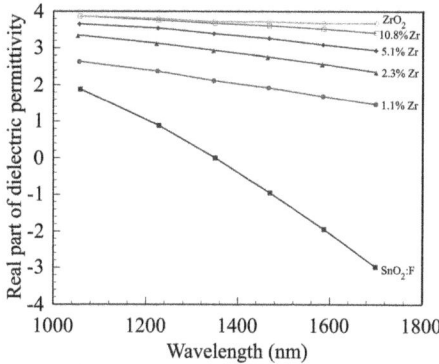

Figure 4.16 Variation of the real part of the dielectric permittivity of films of FTZO with wavelength. The lowermost and uppermost curves represent pure SnO$_2$ and ZrO$_2$, respectively. The intermediate curves show FTZO with varying amounts of Zr.[81]

Li *et al.* studied the properties of CdO films deposited by LPCVD using a quartz reactor with a deposition zone of 10.2 cm × 30.5 cm.[109] Films were deposited onto Corning 7059 substrates, which were heated by a five-zone quartz lamp array used to heat a graphite susceptor, as described above. The reactant gases were dimethyl cadmium and oxygen. Film thicknesses were typically ~100 nm. The electrical and optical properties of the films were a strong function of the deposition temperature. The mobility increased from 5.9 to 216 cm^2 V^{-1} s^{-1}, and the carrier concentration decreased from 2.7–0.24 × 10^{20} cm^{-3} as the deposition temperature increased from 150 to 450 °C. Unfortunately, these changes yielded CdO film resistivity on the order of 10^{-3} Ω cm, which was considered too large for required PV applications at the time. The reduced effective mass of electrons was determined to be 0.11 m_e which, coupled with direct measurement of the density-of-states (DOS) electron effective mass using the method of four coefficients, gave a valence band DOS effective mass of 0.28–0.51 m_e.[110] It was also shown that the optical bandgap increased to almost 3.1 eV at the highest carrier concentration. Hence, for some applications, CdO may indeed be considered a potentially suitable TCO. However, the challenge of obtaining a resistivity of about 10^{-4} Ω cm may be very severe.

4.5 Ternary Compounds and Alloys

4.5.1 Cadmium Stannate

Cd$_2$SnO$_4$ (CTO) combines many beneficial characteristics of both SnO$_2$ and CdO. It is an n-type semiconductor with either orthorhombic or spinel crystal structure. In bulk form, CTO crystallizes as an orthorhombic crystal, but in thin films, the spinel crystal may be observed. Thin film CTO has given electron mobilities up to 65 cm² V⁻¹ s⁻¹, high electrical conductivity and low visible absorption, which make it potentially suitable for several applications.[111] The properties of CTO films were first reported by Nozik, who prepared amorphous films by r.f. sputtering and reported mobilities as high as 100 cm² V⁻¹ s⁻¹ for a carrier concentration of 5×10^{18} cm⁻³. Nozik attributed this unusually high mobility to a low electron effective mass ($m* \sim 0.04\ m_e$).[111] Haacke *et al.* investigated the effects of deposition and annealing parameters on the properties of polycrystalline CTO films prepared by r.f. sputtering and reported excellent transparency and resistivities as low as 1.49×10^{-4} Ω cm.[112] The electrical properties of sputtered CTO films were negatively affected by the presence of secondary phases, namely CdO and $CdSnO_3$. Theoretically, it is not expected that CTO forms in the spinel phase in bulk form, though this is not found to be true in thin films for which the energetics must be distinct. These could be minimized by adjusting the sputtering parameters and by post-deposition annealing. Subsequently, there have been numerous investigations of various deposition methods to prepare CTO thin films, including, but not limited to, dc reactive sputtering, ion-beam sputtering, chemical vapor deposition, spray pyrolysis and electroless deposition.

Wu and co-workers have also studied the preparation of CTO films and used them to fabricate record-performance CdTe solar cells.[6,7,113] The films were sputter deposited in pure oxygen onto an unheated borosilicate (Corning 7059) substrate. The sputtering target was supplied by Cerac, Inc. (presently known as Materion Advanced Materials Group, Milwaukee, WI) and it was made by hot pressing a 2 : 1 molar mixture of CdO and SnO_2. The films were amorphous as deposited. However, they were then annealed in pure argon in close proximity to a film of CdS that had been grown by chemical bath deposition. Annealing temperatures up to 680 °C were used, after which XRD showed the films were single phase spinel without any evidence of amorphous material. Figure 4.17 shows the XRD spectra of a film before and after annealing at 680 °C in argon in close proximity to a film of CdS.

The transport properties improved with progressively higher annealing temperatures, as is shown in Figure 4.18. The highest mobility achieved was 54.5 cm² V⁻¹ s⁻¹ with a carrier concentration of 8.9×10^{20} cm⁻³, corresponding to a resistivity of 1.28×10^{-4} Ω cm.

With increasing annealing temperature, the crystallinity of the films improved and this appeared to relate to the concurrent increase in mobility. The optical properties of an annealed film are shown in Figure 4.19. The striking feature is that the absorbance is extremely small in the visible range of wavelengths, making CTO potentially ideal for use in thin film PV devices. In addition to excellent optical and electrical properties, the films were also very smooth, with a root mean square (r.m.s.) roughness of only ±0.13 nm, were very stable at elevated temperature, and could easily be etched in either HCl or HF. Note that these excellent properties were only found for single phase, well-oriented films—in contrast to the work discussed next.

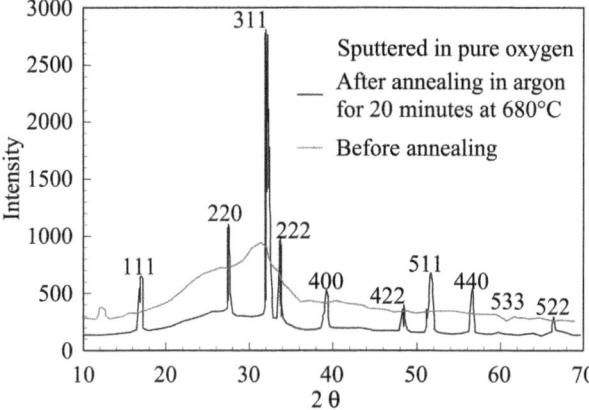

Figure 4.17 XRD spectra of a CTO film before and after annealing. The films were annealed in close proximity to a film of CdS deposited using chemical solution growth. Adapted from Wu *et al.*[7]

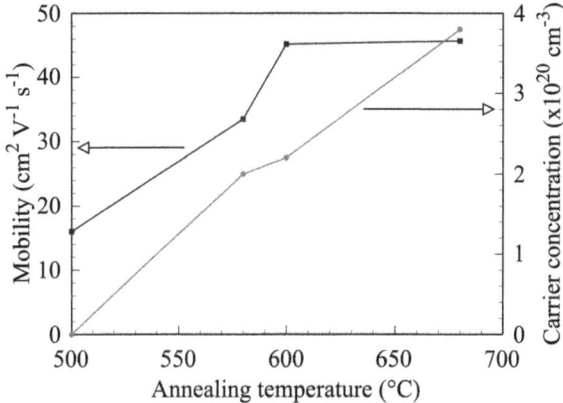

Figure 4.18 Mobility and carrier concentration of CTO as a function of the annealing temperature. The films were typically annealed in an atmosphere of pure argon. The arrow on left indicates that the black line with square markers represents electron mobility, while the arrow on the right indicates that the grey line with circle markers represents electron carrier concentration. Adapted from Wu *et al.*[7]

Unfortunately, the process described above for fabricating high-quality CTO is unsuitable for large-volume manufacture, because it has proved difficult to obtain reproducible targets, requires a high-temperature annealing process incompatible with the low-cost soda lime substrates typically used for the manufacture of thin film solar cells, and the deposition process is rather slow. However, CVD has been used to produce high-quality films, has the advantage of being easily scaled, and is rapid. Consequently, it is widely used by industry and may be compatible with manufacture of CdTe solar cells. Very few data points have been studied along the CdO–SnO$_2$ tie line, but combinatorial synthesis, in which several elements or compounds are deposited non-uniformly, was used by Li *et al.*[114] The latest developments in this work promise to make the use of CTO films for thin film PV practicable.

To exploit the combinatorial approach, thin film libraries of samples are often synthesized using multi-target sputtering or pulsed laser deposition. Li *et al.* established compositionally graded thin film libraries of CTO films using MOCVD.[114] This enabled the dependence of crystal structure, and electrical and optical properties, to be investigated as a function of composition, which varied along the length of the reaction tube.

The CVD reaction chamber used was a cold-walled quartz tube with rectangular cross-section. The reactant gases, dimethyl cadmium (DMCd) and TMT entered at one end of the tube, flowed along the chamber length, and exited after passing through the reaction zone. Three Corning 7059 borosilicate glass substrates, 1 mm thick, and 102 mm × 102 mm in area were coated in each deposition run.

Because TMT and DMCd have different decomposition temperatures, their decomposition rates are different at a given growth temperature. As the precursors flow along the *x*-axis, those with a low decomposition temperature decompose more rapidly and the composition of the Cd–Sn–O film varied along the *x*-axis. The precise compositional, electronic and optical properties were controlled *via* the precursor ratio and the substrate temperature.

Here, we discuss two of the libraries. The first, CTO53, was deposited at 550 °C with a DMCd-to-TMT ratio of 2 : 1, whereas the second, CTO50, was deposited at 500 °C with a DMCd-to-TMT ratio of 0.2 : 1. The former temperature is compatible with the industrially preferred soda lime

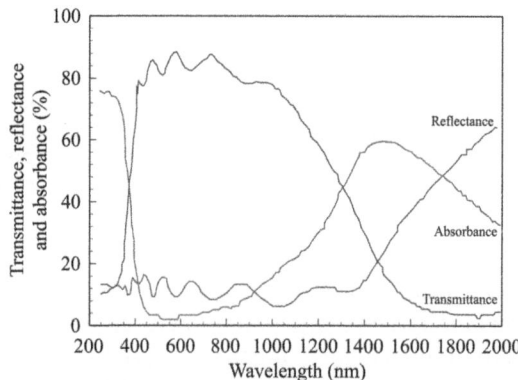

Figure 4.19 Optical properties of an annealed CTO film. The absorption is minimal in the portion of the spectrum of greatest relevance to a CdTe solar cell. Adapted from Wu *et al.*[7]

substrates. Each library consisted of three 102 mm × 102 mm borosilicate glass substrates labeled as A, B and C from the leading edge to the trailing edge of the deposition zone. At these deposition temperatures, the decomposition rate of DMCd was much higher than that of TMT, which led to a Cd-rich film at the leading edge of the substrate and a Sn-rich film at the trailing edge. To characterize these compositional gradient libraries, each piece of glass was cut into six small pieces and numbered from 1 to 6 along the *x*-axis for reference. These 18 samples enabled us to map out the properties of the material along the entire deposition zone.

Figure 4.20 shows an electron probe microanalyzer (EPMA) composition analysis taken from library CTO53, the sampling area being in the millimeter range. At the top of the figure is an illustration of the library and how it is labeled. The results indicate that the highest Cd : Sn atomic ratio in this library was slightly higher than 3 : 1. The composition varied rapidly at the leading edge of the substrate. On the first two glass substrates, the Cd : Sn ratio decreased along the length from 3 : 1 to about 1 : 1. However, on the third substrate, the composition was almost constant with a ratio of about 1 : 1.

The crystal structure of the thin film libraries was investigated using XRD with Cu–K$_\alpha$ radiation ($\lambda = 0.5416$ nm). The lattice constants for the various phases observed were calculated from the XRD peaks. These results are discussed in the following sections.

XRD indicated, as expected, that the first zone of the CTO53 library was a mixed phase of CdO and Cd$_2$SnO$_4$. Farther along the *x*-axis, near the end of substrate A, the intensities of the diffraction peaks associated with CdO are

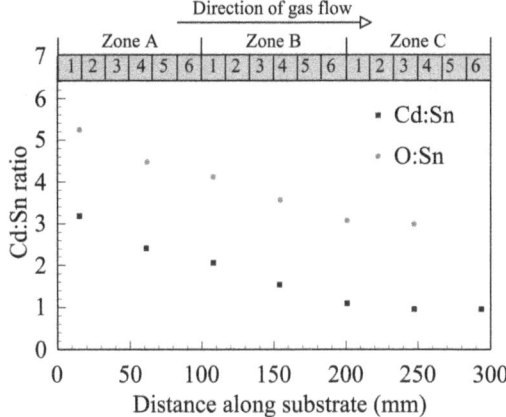

Figure 4.20 EPMA of library CTO53. The classification of the 18 individual samples cut from the three substrates (A, B and C) is shown in the schematic above the EPMA spectrum. The numbers (1–6) refer to the samples cut from each of the main substrates. The library was prepared using a substrate temperature of 550 °C. The analysis indicated that at the leading edge of sample, the ratio of Cd : Sn was 3.2 : 1 and at the trailing edge was about 1 : 1.

low, suggesting that the film was mainly Cd_2SnO_4. In the mixed phase region, it was possible to calculate the lattice constants of the CdO and the Cd_2SnO_4, which showed that the CdO lattice was under compression whereas the Cd_2SnO_4 was under tension.

Farther downstream (substrate B), the Cd : Sn ratio continued to decrease, the cubic Cd_2SnO_4 phase disappeared, and the film became amorphous. In zone B, where the ratio of Cd : Sn varied from about 2 : 1 to 1 : 1, an additional broad peak was observed that was most likely due to $CdSnO_3$. Along zone C where the Cd : Sn ratio was almost constant and equal to 1 : 1, the remaining peaks corresponded mostly to amorphous Cd_2SnO_4 and $CdSnO_3$. Thus, the CTO53 library started at the leading edge with a Cd-rich composition and mixed phase structure of cubic CdO and spinel Cd_2SnO_4. The library ended with amorphous material at the trailing edge.

Library CTO50 was fabricated with a DMCd : TMT ratio of 0.2 : 1 and a deposition temperature of 500 °C, meaning that a Sn-rich film was expected. The XRD pattern of library CTO50 began at a composition similar to the trailing edge of library CTO53 (deposited at 550 °C) and a broad peak indicative of amorphous Cd_2SnO_4 with some short-range order remaining.

Farther downstream (sample B-2), where the Cd : Sn ratio was less than 1 : 1, the broad amorphous peak disappeared and no crystal structure could be detected. Even farther downstream, a new peak appeared that corresponded to the (200) peak of tetragonal SnO_2. The intensity of this peak increased along the length of the reactor, as did the intensity of the other peaks associated with tetragonal SnO_2. We concluded that a pure amorphous film was formed as the film began to change from a cubic to tetragonal structure. Hence, these libraries enabled us to span a broad range of compositions and crystal structures and, as described below, to correlate these with the optical and electrical properties of the materials.

Figure 4.21 shows the variation of the carrier concentration and mobility with the Cd : Sn compositional ratio. In the zone consisting of CdO and Cd_2SnO_4, the carrier concentration was about 5×10^{20} cm^{-3} and the Hall mobility was about 20 cm^2 V^{-1} s^{-1}, both of which are similar to earlier results for CdO film. As the cadmium content decreases along the x-axis, the carrier concentration also decreased but the Hall mobility increased. In the composition range (Cd : Sn ratio) between 2 : 1 and 1 : 1—corresponding to the mixed phases of spinel Cd_2SnO_4 or amorphous Cd_2SnO_4, SnO_2 and orthorhombic $CdSnO_3$—the carrier concentration is about $1–2 \times 10^{20}$ cm^{-3} and the Hall mobility is about 50–60 cm^2 V^{-1} s^{-1}. This value is similar to those discussed earlier for sputter-deposited poly-crystalline Cd_2SnO_4.[113] As the Cd content decreased and the film finally became tetragonal SnO_2, the carrier concentration and mobility decreased to 4×10^{18} cm^{-3} and ~ 1 cm^2 V^{-1} s^{-1}, respectively, similar to undoped SnO_2. Hence, the resistivity of the films varied from the lowest value of 5×10^{-4} Ω cm in the zone consisting of mixed CdO plus Cd_2SnO_4, to the highest value of 5 Ω cm in the SnO_2 zone.

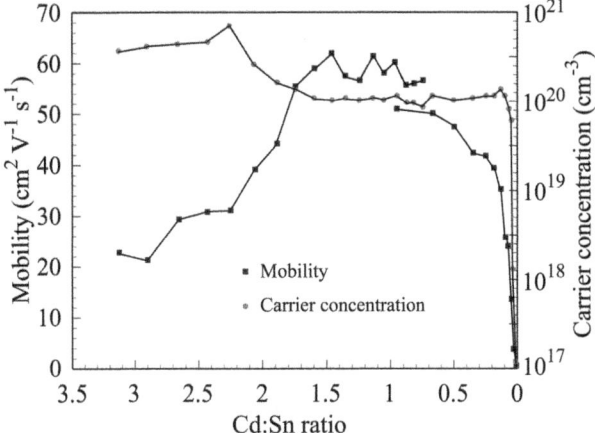

Figure 4.21 Variation of the electrical properties of samples taken from the combinatorial libraries CTO53 and CTO50.

Note that the highest electron mobility was observed in the region corresponding to the least perfect crystallinity. The values of electron mobility reached 61 cm^2 V^{-1} s^{-1}, similar to the value for single-phase polycrystalline material.[113] The electron mobility of amorphous Si is about 10^{-3} cm^2 V^{-1} s^{-1}, which is about three orders of magnitude smaller than for its polycrystalline form. Hosono *et al.*[115] and Kawazoe and Ueda[116] suggested that, for the spinel structure, the orbitals of the heavy metals overlap to form an extended conduction band. In a metal oxide composed of a heavy metal, the relevant orbital may have sufficient overlap, even in the amorphous structure, to enable high electron mobility. This would not be expected for amorphous Si, for which the orbitals are not extended.

Bandgaps calculated from transmittance and reflectance data were obtained for libraries CTO53 and CTO50. Figure 4.22 shows the narrowest bandgap of 2.7 eV (characteristic of Cd$_2$SnO$_4$), which was observed at the leading edge of library CTO53. After reaching its highest value, the bandgap decreased as the quantity of cubic spinel Cd$_2$SnO$_4$ structure decreased, back to 2.7 eV. Within the amorphous régime, the value of the bandgap increased with increasing Sn composition. At the trailing edge of library CTO53, the bandgap is 3 eV with a Cd : Sn ratio of 1 : 1. The bandgap increased progressively as the Sn concentration increased. Library CTO50 starts with a bandgap of 2.9 eV, whereas on its trailing edge, the widest bandgap of 3.65 eV (characteristic of SnO$_2$) was observed.

The single greatest challenge for the future application of CTO is the development of a method of synthesis compatible with large-scale manufacture. The above work has demonstrated that the high temperature annealing step in close proximity to a film of CdS is not essential for fabricating high-quality material. However, at this stage, the electrical and optical

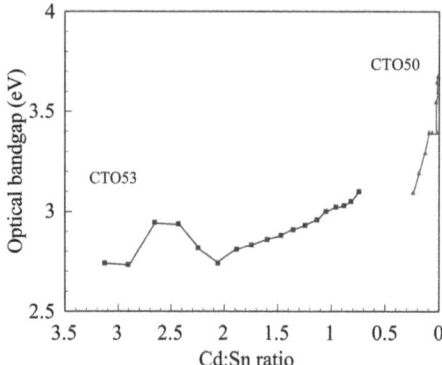

Figure 4.22 Variation of the optical bandgap for samples from libraries CTO53 and CTO50.

properties of CTO films made using LPCVD are not quite equivalent to those made using r.f. sputtering; further optimization may be required, although significantly lower temperatures have been used. In addition, some companies may be unwilling to manufacture CTO because of the presence of cadmium—even though large markets may exist because of the need to increase the efficiency of CdTe modules—in the face of rapidly increasing competition from materials such as polycrystalline silicon manufactured in lower cost regions of the world.

4.5.2 Zinc Stannate

The original motivation for studying zinc stannate (Zn_2SnO_4 or ZTO) was that it does not contain either the toxic element cadmium or the expensive element indium. In thin film form, ZTO, like CTO, has the spinel crystal habit. It also has the inverse spinel form in which half the tetrahedral sites are occupied by zinc and half by tin atoms while the octahedral sites are fully occupied by tin atoms. ZTO has a variety of applications such as flame retardants,[117] photoacoustic devices,[118] dye-sensitized solar cells[119] and CdTe/CdS solar cells.[6,120] Bulk ZTO crystallizes in the orthorhombic form, similar to CTO, but the spinel form is usually found in thin films. As discussed below, the effective mass at the bottom of the conduction band is small, which means that there is a rather large Burstein–Moss shift as the material becomes degenerate.

Young studied the basic materials properties of ZTO with a view to increasing its conductivity,[121] which had been found by other authors to be low.[122,123] The objective was to make TCO films of ZTO to eliminate cadmium, but rather limited success was achieved because of very low carrier concentrations and low mobilities. The approach used by Young and co-workers was to understand the limiting conduction mechanisms.[124]

The ZTO films were r.f. magnetron sputter-deposited from a phase-pure ceramic target in pure oxygen onto Corning 7059 substrates, which could be heated. The films were characterized using XRD, atomic force microscopy (AFM), TEM, Mössbauer spectroscopy, ultraviolet/visible (UV/vis) spectrophotometry, and an advanced charge-transport technique known as the method of four coefficients,[110] which involves measuring the conductivity, Hall coefficient, Seebeck coefficient and Nernst coefficient. The technique provides valuable information about the effective mass of the carriers, scattering time and mechanism of scattering, and position of the Fermi energy, independent of whether or not the conduction band (or valence) is parabolic.

Films deposited at room temperature were found to be amorphous but became randomly oriented after annealing at 600 °C in argon. In the presence of oxygen in the sputtering gas, the rates of deposition were approximately halved. At a deposition temperature of 550 °C in purer argon, the films were uniaxially oriented about the (400) peak. AFM showed that the films were exceptionally smooth, with an r.m.s. roughness of only ±3.3 nm and grain size of about 100 nm. This smoothness facilitated the use of ZTO as a buffer layer between the CTO transparent conductor and the CdS window layer. In the high-efficiency CdTe solar cells, the ZTO smoothness and relatively high resistivity meant that an exceptionally thin layer of CdS could be used without risking shorting of the junction by conduction paths between the CTO and the CdTe. In turn, this enabled larger currents to be generated, leading to record-efficiency devices.[6]

High-resolution TEM of a single grain of ZTO indicated incomplete crystallization and an absence of well-ordered lattice planes, in contrast to CTO, for which well-defined lattice planes were found in the grains. Further evidence of disorder was provided by Mössbauer spectroscopy. This showed that the degree of inversion of the structure was near unity, meaning that the tin atoms were completely in the octahedral sites. In the normal spinel, tin would be half in the tetrahedral and half in the octahedral sites. The octahedral site does not have cubic symmetry that causes quadrupole splitting in the ^{119}Sn Mössbauer data. If perfect atomic arrangements were present in the inverse spinel structure, no splitting would have been observed. Our data revealed two quadrupole splittings, indicating that tin atoms occupied two non-equivalent octahedral sites. Hence, the combination of the atomic disorder (revealed by Mössbauer) and crystallographic disorder (revealed by XRD) grossly disrupts perfect periodicity in the lattice and, consequently, the mobility of electrons in ZTO films was substantially less than that in CTO.

The optical data indicated that the ZTO films were highly transparent because of the relatively low concentration of electrons. The Burstein–Moss shift was rather large, implying a small effective mass of electrons, and the fundamental bandgap was found to be 3.35 eV.[15,16] However, because of the small effective mass, the degeneracy increased the optical gap to as large as 3.89 eV.

The method of four coefficients also provided fundamental information about this material. The negative Hall and Seebeck coefficients confirmed that the films were n-type, while positive Nernst coefficients implied ionized scattering was responsible for electrons scattering. However, the transport theory shows that band non-parabolicity can make interpretation of the Nernst data difficult and its sign alone cannot be taken as unambiguous proof of ionized scattering.[125]

The effective mass increased strongly with carrier concentration showing that the conduction band is non-parabolic. At the bottom of the band, the effective mass is 0.15 m_e, substantially lower than other TCO materials and responsible for the large Burstein–Moss shift.[123]

The value of relaxation time was typically about 1–2 fs, which is consistent with the observed disorder at the atomic and microscopic levels, and is the cause of the low mobilities ($\sim$20 cm^2 V^{-1} s^{-1}) typically observed for this material. The method of four coefficients also provides a 'scattering coefficient' that gives a good indication about the dominant electron scattering mechanism. For ZTO, this coefficient implied that ionized impurities were primarily responsible for scattering electrons though, because of the high electron concentration, screening of the charged impurities was evident. The temperature dependence of mobility also suggested that ionized impurities were responsible. Of course, the background disorder adds to the magnitude of the scattering, further reducing the mobility.

In summary, ZTO is an excellent material as a buffer layer in CdTe device, but it remains a very difficult challenge to improve its properties, particularly mobility, to the extent that it could compete with either CTO or any other high-quality TCO.

4.5.3 $Zn_xMg_{1-x}O$

ZnO has a fundamental bandgap of about 3.2 eV, which corresponds to a wavelength of 0.39 μm. At shorter wavelengths, the ZnO strongly absorbs photons, preventing them from being useful to the junction. The bandgap of MgO is about 7.2 eV, which means that the addition of a small amount of Mg to ZnO increases the bandgap of the alloy significantly. The bandgap of the MgO differs from that of ZnO primarily because the energy of the conduction band edge is much greater than that of ZnO; that is, it has a much smaller electron affinity, whereas the valence band edge is somewhat lower than that of ZnO.[2] The use of $Zn_xMg_{1-x}O$ as a window layer in the CIGS device has two apparent attractions: the first is the wider bandgap (thereby transmitting more photons to the absorber); and the second is the potential improvement in the lineup of the conduction bands of the $Zn_xMg_{1-x}O$ and CdS.

The variation of the bandgap of undoped $Zn_xMg_{1-x}O$ with alloy composition is shown in Figure 4.23. Most of the larger bandgap is due to the shift in the conduction band edge. The two sets of points correspond to different beam energies.

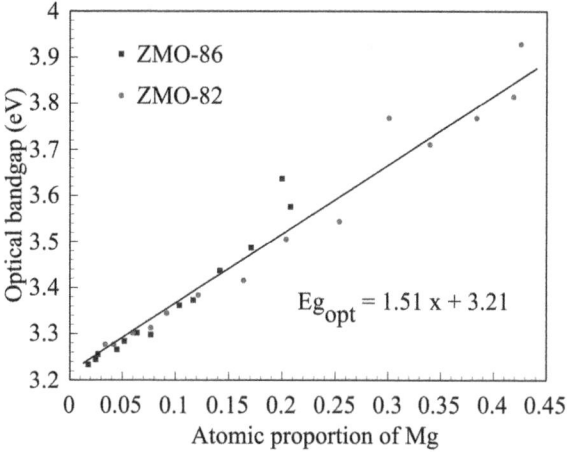

Figure 4.23 Variation of the optical bandgap of undoped $Zn_xMg_{1-x}O$ with the atomic proportion, x, of Mg.[128]

For Al-doped alloys, we found a very different variation of the optical bandgap, as shown in Figure 4.24. In this case, the optical bandgap actually decreases as Mg increases. Without Mg, the ZnO:Al is degenerate and the carrier concentration is greater than 10^{20} cm^{-3}. As Mg is added and the alloy is formed, the conduction band edge moves toward the vacuum level, the ionization energy of the Al dopant increases (implying that the aluminum level does not follow the conduction band but perhaps remains relatively fixed relative to the vacuum level), and less of it is ionized. Consequently, the free carrier concentration decreases severely, causing the Fermi energy to move away from the vacuum level and the optical bandgap to decrease. With further increases of the Mg, the degeneracy decreases further until the Fermi level coincides with the conduction band edge; at this point, adding further Mg causes the fundamental bandgap to increase and to be equal to the optical bandgap.

Figure 4.25 shows the carrier concentration and mobility of $Zn_xMg_{1-x}O$:Al as a function of the atomic proportion of magnesium.

Although modeling showed that the mismatch of the conduction band edges of the ZnO alloy and the CdS in a CIGS device should be reduced by alloying with Mg, the efficiency of the devices was also reduced because of the large increase in resistivity of the alloy. In addition, no significant increase in the open-circuit voltage was observed, as may be expected for improved alignment of the conduction band edges. The use of CdS virtually ensures that the current cannot increase because its bandgap is less than that of the TCO, meaning that very few of the higher energy photons can be transmitted to the CIGS.

Despite the rather negative device results, this approach could lead to significant improvements if a more appropriate TCO could be developed. The remaining challenge is to incorporate a dopant other than Al that

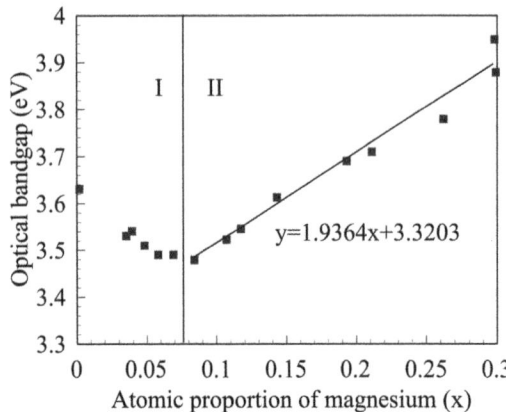

Figure 4.24 Variation of the optical bandgap of the Al-doped $Zn_xMg_{1-x}O$ with the atomic proportion, x, of Mg.[129]

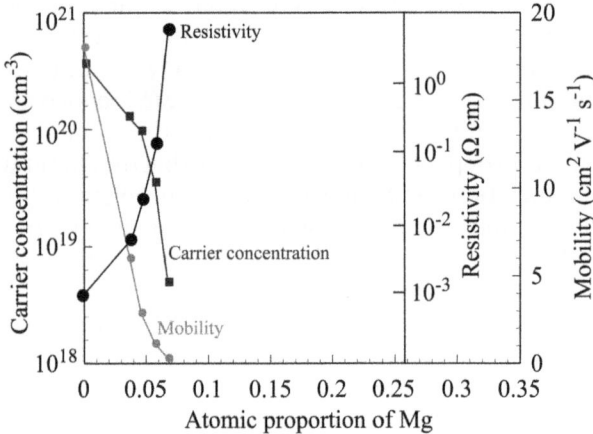

Figure 4.25 Variation in carrier concentration, mobility and resistivity as a function of the atomic proportion of Mg in $Zn_xMg_{1-x}O$:Al.

remains fully ionized, even for larger bandgaps, so that there is not: a severe decrease in carrier concentration, a corresponding increase in the series resistance and a deterioration of fill factor. Perhaps there exists a dopant that follows the conduction band rather than the vacuum level. Additionally, the concept of improved alignment of the conduction band edges between the TCO and the n-CdS window layer warrants further investigation.

4.6 Summary

In this chapter, we have reviewed our work at NREL on some of the TCOs we have studied during the course, for at least two of us (T.J.C. and T.A.G.) of the past 25–30 years. Initially, we described some of the requirements of TCOs for use in thin film solar cells. For each material discussed, we have described the method of deposition along with some of the deposition parameters, provided illustrative experimental data, identified remaining challenges and suggested further work. This has enabled us to highlight some of the more significant results achieved, which we repeat below.

For zinc oxide, our work has shown that far less of the aluminum dopant may be used than is typically used in devices. This has the potential to benefit the performance of CIGS devices. In addition, our work showed that there are beneficial effects of adding a small, controlled amount of hydrogen to the sputtering atmosphere.[126,127]

Our work on indium tin oxide and indium oxide began at NREL with the formation of the Device Development Group in 1984. In this research, we established the importance of controlling the deposition atmosphere and the changes that can occur with deposition and annealing temperature.[74] Also, this work eventually led to the incorporation of novel dopants such as molybdenum and zirconium.[73] These high permittivity materials provided significant benefits in material properties that remain to be fully exploited in devices. It remains to be seen whether or not these alternative dopants enhance the corrosion stability of materials such as zinc oxide.

Our discussion of tin oxide focused on the low-pressure chemical vapor deposition technique and our ability to make good quality material using the fluorine precursor $CBrF_3$. Because this is a greenhouse gas, we discussed the properties of films made using the alternative precursor ClF_3 and pointed out the remaining challenges.[101] In this section, we also discussed our work on the properties of fluorine-doped tin oxide films with zirconium as a secondary dopant. This showed the same shift of the free carrier absorption band away from the region of the solar spectrum, *i.e.* to longer wavelengths, that we had previously observed for indium tin oxide and indium oxide. With further optimization, we can confidently expect these developments to be incorporated into commercially manufactured solar modules.

Cadmium oxide is the least likely of the materials that we have researched to be applied in solar cells because its bandgap is simply too low. However, the low effective mass results in a large Burstein–Moss shift of the optical gap, which may negate this criticism.[17] In addition, the minimum resistivity achieved for material without an extrinsic dopant was too large for the material to be relevant. However, extraordinarily high mobilities were achieved and, if this could be maintained together with extrinsically doped material with higher carrier concentration, the material may have applicability in solar cells. This has yet to be studied.

We discussed ternary materials next, with particular emphasis on cadmium stannate because it was an important part of the world record CdTe solar cell and still appears to have considerable potential to help increase the efficiency of CdTe modules.[120] First, we briefly revisited the work by Wu *et al.*[113] on the synthesis of films using r.f. sputtering and then discussed the work by Kerr *et al.*[101] using LPCVD, much of which has not previously been published. Unpublished work by Li and co-workers used two cation precursor gases, which enabled us to perform linear combinatorial synthesis depositions. This work showed, rather remarkably, that the highest mobilities were found in what seemed to be the least-organized region of the compositionally graded film. It made the point that a single phase film is unnecessary. The greatest challenge facing the use of cadmium stannate is integrating it into the manufacture of CdTe modules. For safety reasons, glass coating companies are not currently interested in manufacturing coated glass with integral films of cadmium stannate.

Next, we reviewed some of our work on the fundamentals of zinc stannate and, in particular, those characterized using the method of four coefficients.[110,123] Young was able to establish that the conduction band was strongly non-parabolic and that the carriers typically had a very short relaxation time leading to a very low mobility. The carrier concentrations achieved were typically at least an order of magnitude less than those observed for cadmium stannate. In addition, he showed that the scattering mechanism changed from low to high carrier concentrations. Using Mössbauer spectroscopy and high-resolution electron microscopy, he also showed that zinc stannate typically has far less order at the atomic and crystallographic scales. This reduced order is intrinsic to the material and appears to limit the mobility to values far lower than for cadmium stannate. The main challenge for zinc stannate is therefore to realize much lower resistivities. If this could be achieved, there would be considerable attraction in using this material as both a buffer layer and a TCO in CdTe solar cells.

Finally, we discussed our work on zinc magnesium oxide. So far, our results for this material are not promising and it appears to have fundamental issues that may make it difficult to achieve its potential benefits as a TCO. One approach is to establish a dopant that remains near the conduction band minimum as magnesium is added to widen the bandgap. This is an interesting opportunity for both theoreticians and experimentalists.

Acknowledgements

The authors wish to thank their colleagues in the NREL Measurements and Characterization Group and the NREL Theoretical Sciences Group for long-term collaboration in analyzing various materials and fundamental attributes of TCO films. The authors also wish to acknowledge the Colorado School of Mines and Vanderbilt University for collaborations involving students. This work was largely supported by US Department of Energy under various contracts and subcontracts, most recently involving Contract No. DE-AC36-08-GO28308 to NREL.

References

1. T. J. Coutts, T. O. Mason, J. D. Perkins and D. S. Ginley, in *Photovoltaics for the 21st Century, Proceedings of the Electrochemical Society International Symposium*, ed. V. K. Kapur, R. D. McConnell, D. Carlson, G. P. Ceasar and A. Rohatgi, Electrochemical Society Proceedings, The Electrochemical Society Pennington, NJ, 1999, vol. 99-11, pp. 103–109.
2. S. Calnan and A. N. Tiwari, *Thin Solid Films*, 2010, **518**, 1839.
3. B. O'Neill, presented at the Minor Metals Conference, London, 2004.
4. N. Blair, C. Christensen, M. Mehos, S. Janzou, P. Gilman, C. Cameron, F. Burkholder and G. Glatzmaier, *The Solar Advisor Model (SAM) and its Usage for the Solar America Initiative*, http://www.nrel.gov/analysis/sam.
5. K. Zweibel, J. Mason and V. Fthenakis, *Sci. Am.*, December 2008, **298**, 48.
6. X. Wu, R. G. Dhere, D. S. Albin, T. A. Gessert, C. DeHart, J. C. Keane, A. Duda, T. J. Coutts, S. Asher, D. H. Levi, H. R. Moutinho, Y. Yan, T. Moriarty, S. Johnston, K. Emery and P. Sheldon, in *Proceedings of the 2001 NCPV Program Review Meeting*, NREL/CP-520-31025, National Renewable Energy Laboratory, Golden, CO, 2001.
7. X. Wu, W. P. Mulligan and T. J. Coutts, *Thin Solid Films*, 1996, **286**, 274.
8. V. M. Fthenakis, H. C. Kim and E. Alsema, *Environ. Sci. Technol.*, 2008, **42**, 2168.
9. E. J. Luna-Arredondo, A. Maldonado, R. Asomoza, D. R. Acosta, M. A. Meléndez-Lira and M. d. l. L. Oivera, *Thin Solid Films*, 2005, **490**, 132.
10. J. Wennerberg, J. Kessler, M. Bodegård and L. Stolt, presented at 2nd World Conference on Photovoltaic Solar Energy Conversion, Vienna, 1998.
11. F. J. Pern, S. H. Glick, X. Li, C. DeHart, T. Gennet, M. A. Contreras and T. A. Gessert, in *Reliability of Photovoltaic Cells, Modules, Components, and Systems II*, Proceeedings SPIE, ed. N. G. Dhere, J. H. Wohlgemuth and D. T. Ton, SPIE, Bellingham, WA, 2009, vol. 7412.
12. F. J. Pern, S. H. Glick, R. Sundaramoorthy, B. To, X. Li, C. DeHart, S. Glynn, T. Gennett, R. Noufi and T. Gessert, presented 35th IEEE Photovoltaic Specialists Conference, Honolulu, 2010.
13. ASTM, *Reference Solar Spectral Irradiance: Air Mass 1.5*, American Society for Testing and Materials, West Conshohocken, PA, 2003.
14. P. Drude, *Ann. Phys.*, 1900, **3**, 566.
15. E. Burstein, *Phys. Rev.*, 1954, **93**, 632.
16. T. S. Moss, *Proc. Phys. Soc. London*, 1954, **382**, 775.
17. X. Li, Y. Yan, A. Mason, T. A. Gessert and T. J. Coutts, *Electrochem. Solid-State Lett.*, 2001, **4**, C66.
18. T. J. Coutts, D. L. Young and X. Li, *MRS Bull.*, 2000, **25**, 58.
19. T. Minami, *Mater. Res. Soc. Bull.*, 2000, **25**, 38.
20. B. G. Choi, I. H. Kim, D. H. Kim, K. S. Lee, T. S. Lee, B. Cheong, Y.-J. Baik and W. M. Kim, *J. Eur. Ceram. Soc.*, 2005, **25**, 2161.
21. M. de la L. Olivera, A. Maldonado and R. Asomoza, *Sol. Energy Mater. Sol. Cells*, 2002, **73**, 425.

22. M. de la L. Olivera, A. Maldonado, R. Asomoza, O. Solorza and D. R. Acosta, *Thin Solid Films*, 2001, **394**, 242.
23. I. Kim, K.-S. Lee, T. S. Lee, J.-H. Jeong, B.-K. Cheong, Y.-J. Baik and W. M. Kim, *J. Appl. Phys.*, 2006, **100**, 063701.
24. T. Minami, S. Ida, T. Miyata and Y. Minamoto, *Thin Solid Films*, 2003, **445**, 268.
25. T. Miyata, S. Ida and T. Minami, *J. Vac. Sci. Technol., A,* 2003, **21**, 1404.
26. T. Minami, *Semicond. Sci. Technol.,* 2005, **20**, S35.
27. R. G. Gordon, *Mater. Res. Bull.,* 2000, **25**, 52.
28. K. Ellmer, *J. Phys. D: Appl. Phys.,* 2001, **34**, 3097.
29. K. Ellmer, *J. Phys. D: Appl. Phys.,* 2000, **33**, R17.
30. Z.-C. Jin, I. Hamberg and C. G. Granqvist, *J. Appl. Phys.,* 1988, **64**, 5117.
31. Ü. Özgür, Y. I. Alivov, C. Liu, A. Teke, M. A. Reshchikov, S. Doğan, V. Avrutin, S.-J. Cho and H. Morkoç, *J. Appl. Phys.,* 2005, **98**, 041301.
32. T. Makino, Y. Segawa, S. Yoshida, A. Tsukazaki, A. Ohtomo and M. Kawasaki, *Appl. Phys. Lett.,* 2004, **85**, 759.
33. V. Bhosle and J. Narayan, *J. Appl. Phys.,* 2006, **100**, 093519.
34. E. Fortunato, V. Assunção, A. Gonçalves, A. Marques, H. Águas, L. Pereira, I. Ferreira, P. Vilarinho and R. Martins, *Thin Solid Films,* 2004, **451–452**, 443.
35. H. Fujiwara and M. Kondo, *Phys. Rev. B,* 2005, **71**, 075109.
36. T. Minami, S. Ida and T. Miyata, *Thin Solid Films,* 2002, **416**, 92.
37. B. Szyszka, *Thin Solid Films,* 1999, **351**, 164.
38. I. Ivanov and J. Pollmann, *Phys. Rev. B,* 1981, **24**, 7275.
39. P. Schröer, P. Krüger and J. Pollmann, *Phys. Rev. B: Condens. Matter Mater. Phys.,* 1993, **47**, 6971.
40. H. Q. Ni, Y. F. Lu and Z. M. Ren, *J. Appl. Phys.,* 2002, **91**, 1339.
41. M. Usuda, N. Hamada, T. Kotani and M. van Schilfgaarde, *Phys. Rev. B: Condens. Matter Mater. Phys.,* 2002, **66**, 125101.
42. A. F. Kohan, G. Ceder, D. Morgan and C. G. Van de Walle, *Phys. Rev. B,* 2000, **61**, 15019.
43. G. D. Mahan, *J. Appl. Phys.,* 1983, **54**, 3825.
44. E. Ziegler, A. Heinrich, H. Opperman and G. Stover, *Phys. Status Solidi A,* 1981, **66**, 635.
45. D. C. Look, J. W. Hemsky and J. R. Sizelove, *Phys. Rev. Lett.,* 1999, **82**, 2552.
46. A. Janotti and C. G. Van de Walle, *Appl. Phys. Lett.,* 2005, **87**, 122102.
47. A. Janotti and G. G. Van de Walle, *J. Cryst. Growth,* 2006, **287**, 58.
48. F. Oba, S. R. Nishitani, S. Isotani and H. Adachi, *J. Appl. Phys.,* 2001, **90**, 824.
49. J.-L. Zhao, W. Zhang, X.-M. Li, J.-W. Feng and X. Shi, *J. Phys.: Condens. Matter,* 2006, **18**, 1495.
50. C. G. Van de Walle and J. Neugebauer, *Nature,* 2003, **423**, 626.
51. C. G. Van de Walle, *Phys. Status Solidi B,* 2003, **235**, 89.
52. Ç. Kiliç and A. Zunger, *Appl. Phys. Lett.,* 2002, **81**, 73.
53. A. Janotti and C. G. Van de Walle, *Nat. Mater.,* 2007, **6**, 44.

54. C. G. Van de Walle, *Phys. Rev. Lett.,* 2000, **85**, 1012.
55. E. V. Lavrov, J. Weber, F. Börrnert, C. G. Van de Walle and R. Helbig, *Phys. Rev. B,* 2002, **66**, 165205.
56. C. A. Wolden, T. M. Barnes, J. B. Baxter and E. S. Aydil, *J. Appl. Phys.,* 2005, **97**, 043522.
57. T. Miyata, S. Suzuki, M. Ishii and T. Minami, *Thin Solid Films,* 2002, **411**, 76.
58. M. Lv, X. Xiu, Z. Pang, Y. Dai and S. Han, *Appl. Surf. Sci.,* 2005, **252**, 2006.
59. J. N. Duenow, T. A. Gessert, D. M. Wood, T. M. Barnes, M. Young, B. To and T. J. Coutts, *J. Vac. Sci. Technol., A,* 2007, **25**, 955.
60. N. J. Dayan, S. R. Sainkar, R. N. Karekar and R. C. Aiyer, *Thin Solid Films,* 1998, **325**, 254.
61. A. M. Gas'kov and M. N. Rumyantseva, *Russ. J. Appl. Chem.,* 2001, **74**, 430.
62. K. Ip, M. E. Overberg, Y. W. Heo, D. P. Norton, S. J. Pearton, C. E. Stutz, B. Luo, F. Ren, D. C. Look and J. M. Zavada, *Appl. Phys. Lett.,* 2003, **82**, 385.
63. J. N. Duenow, T. G. Gessert, D. M. Wood, A. C. Dillon and T. J. Coutts, *J. Vac. Sci. Technol., A,* 2008, **26**, 692.
64. J. N. Duenow, *Dopants and Transport Properties of Transparent Conducting ZnO Thin Films,* PhD thesis, Colorado School of Mines, 2008.
65. T. Tohsophon, J. Hüpkes, S. Calnan, W. Reetz, B. Rech, W. Beyer and N. Sirikulrat, *Thin Solid Films,* 2006, **511–512**, 673.
66. R. Groth, *Phys. Status Solidi A,* 1966, **14**, 69.
67. K. L. Chopra, S. Major and D. K. Pandya, *Thin Solid Films,* 1983, **102**, 1.
68. T. A. Gessert, D. L. Williamson, T. J. Coutts, A. J. Nelson, K. M. Jones, R. G. Dhere, H. Aharoni and P. Zurcher, *J. Vac. Sci. Technol., A,* 1987, **5**, 1314.
69. T. A. Gessert, Y. Yoshida, C. Fesenmaier and T. J. Coutts, *J. Appl. Phys.,* 2009, **105**, 083547.1.
70. T. J. Coutts, X. Wu, T. A. Gessert and X. Li, *J. Vac. Sci. Technol., A,* 1988, **6**, 1722.
71. T. A. Gessert, X. Li, M. W. Wanlass, A. J. Nelson and T. J. Coutts, *J. Vac. Sci. Technol., A,* 1990, **8**, 1912.
72. Y. Yoshida, T. A. Gessert, C. L. Perkins and T. J. Coutts, *J. Vac. Sci. Technol., A,* 2003, **21**, 1092.
73. Y. Yoshida, T. A. Gessert and T. J. Coutts, *Appl. Phys. Lett.,* 2004, **84**, 2097.
74. T. J. Coutts, T. A. Gessert, R. G. Dhere, A. J. Nelson and H. Aharoni, *Rev. Bras. Apl. Vacuo,* 1986, **6**, 289.
75. A. J. Leenheer, J. D. Perkins, M. F. A. M. van Hest, J. J. Berry, R. P. O'Hayre and D. S. Ginley, *Phys. Rev. B: Condens. Matter,* 2008, **77**, 115215.
76. R. Sundaramoorthy, F. J. Pern, C. DeHart, T. Gennett, F. Y. Meng, M. Contreras and T. A. Gessert, in *Reliability of Photovoltaic Cells, Modules, Components, and Systems II,* Proceeedings SPIE, ed. N. G. Dhere, J. H. Wohlgemuth and D.T. Ton, SPIE, Bellingham, WA, 2009, vol. 7412.
77. Y. Meng, X. Yang, H. Chen, J. Shen, Y. Jiang, Z. Zhang and Z. Hua, *Thin Solid Films,* 2001, **394**, 219.

78. Y. Meng, X. Yang, H. Chen, J. Shen, Y. Jiang, Z. Zhang and Z. Hua, *J. Vac. Sci. Technol., A,* 2002, **20**, 288.
79. C. Warmsingh, Y. Yoshida, D. W. Ready, C. W. Teplin, J. D. Perkins, P. A. Parilla, L. Gedvilas, B. Keyes and D. S. Ginley, *J. Appl. Phys.,* 2004, **95**, 3831.
80. G. Lucovsky and G. B. Rayner, *Appl. Phys. Lett.,* 2000, **77**, 2912.
81. T. A. Gessert, J. Burst, X. Li, M. Scott and T. J. Coutts, presented at Spring Meeting of the European Materials Research Society, Strasbourg, France, 2010.
82. T. A. Gessert, Y. Yoshida and T. J. Coutts, *Patent App. No.,* 11/718,628, 2007.
83. T. A. Gessert, J. N. Duenow, T. M. Barnes and T. J. Coutts, *Patent App. No.,* PCT 07-42, 2007.
84. G. Haacke, *Ann. Rev. Mater. Sci.,* 1977, **7**, 73.
85. P. S. Patil, *Mater. Chem. Phys.,* 1999, **59**, 185.
86. I. Stambolova, K. Konstantinov, S. Vassilev, P. Peshev and T. Tsacheva, *Mater. Chem. Phys.,* 2000, **63**, 104.
87. R. J. Hill and S. J. Nadel, *Coated Glass: Applications and Markets,* BOC Coating Technology, Fairfield, CA, 1999.
88. R. G. Gordon, J. Proscia, F. B. Ellis and A. E. Delahoy, *Sol. Energy Mater.,* 1989, **18**, 263.
89. R. H. Kyoung, C.-S. Kim, T. K. Keon, J. K. He, I. K. Deok and H. Jingwen, *J. Electroceram.,* 2002, **10**, 69.
90. X. Li, R. Ribelin, Y. Mahathongdy, D. Albin, R. G. Dhere, D. Roswe, S. Asher, H. Moutinho and P. Sheldon, in *Proceedings of the 15th NCPV Photovoltaics Program Review,* ed. M. Al-Jassim, J. P. Thornton and J. M. Gee, Conference Proceedings 462, American Institute of Physics, Woodbury, NY, 1998, pp. 230–235.
91. K. Ishiguro, T. Sasaki, T. Arai and I. Imai, *J. Phys. Soc. Jpn.,* 1958, **13**, 296.
92. R. E. Aitchison, *Aust. J. Appl. Sci.,* 1954, **5**, 10.
93. J. Proscia and R. G. Gordon, *Thin Solid Films,* 1992, **214**, 175.
94. C. G. Borman and R. G. Goldon, *J. Electrochem. Soc.,* 1989, **136**, 3820.
95. V. K. Miloslavskii, *Opt. Spectrosc.,* 1959, **7**, 154.
96. B. Thangaraju, *Thin Solid Films,* 2002, **402**, 71.
97. E. Shanthi, V. Dutta, A. Banerjee and K. L. Chopra, *J. Appl. Phys.,* 1980, **51**, 6243.
98. S. R. Vishwakarma, J. P. Upadhyay and H. C. Prasad, *Thin Solid Films,* 1989, **176**, 99.
99. E. Shanthi, A. Banerjee and K. L. Chopra, *Thin Solid Films,* 1982, **88**, 93.
100. F. Simonis, M. van der Leij and C. J. Hoogendoorn, *Sol. Energy Mater.,* 1979, **1**, 221.
101. L. L. Kerr, T. J. Anderson, O. D. Crisalle, S. Li, X. Li, R. Noufi, T. J. Coutts, M. Bai, S. Asher and T. A. Gessert, *Chem. Eng.,* 2003, 303.
102. W. Lin, R. Ma, J. Xue and B. Kang, *Sol. Energy Mater. Sol. Cells,* 2007, **91**, 1902.

103. G. B. Stringfellow, *Organometallic Vapor-Phase Epitaxy*, Academic Press, London, 1989.
104. N. Saunders and A. P. Miodownir, *CALPHAD (Calculation of Phase Diagrams): A Comprehensive Guide*, Pergamon Press, Oxford and New York, 1998.
105. R. J. Bennett and C. Parish, *J. Phys. D: Appl. Phys.*, 1976, **9**, 2555.
106. T. Koida and M. Mondo, *Appl. Phys. Lett.*, 2006, **89**, 082104.1.
107. F. Koffyberg, *Phys. Rev. B: Condens. Matter*, 1976, **13**, 4470.
108. K. Ueda, H. Maeda, H. Hosono and H. Kawazoe, *J. Appl. Phys.*, 1998, **11**, 6174.
109. X. Li, D. L. Young, H. R. Moutinho, Y. Yan, C. Narayanswamy, T. A. Gessert and T. J. Coutts, *Electrochem. Solid-State Lett.*, 2001, **4**, C43.
110. D. L. Young, T. J. Coutts, V. I. Kaydanov, A. S. Gilmore and W. P. Mulligan, *J. Vac. Sci. Technol., A*, 2000, **18**, 2978.
111. A. J. Nozik, *Phys. Rev. B: Condens. Matter*, 1972, **6**, 453.
112. G. Haacke, W. E. Mealmaker and L. A. Siegel, *Thin Solid Films*, 1978, **55**, 67.
113. X. Wu, W. P. Mulligan and T. J. Coutts, in *Proceedings of 39th Annual Technical Conference of the Society of Vacuum Coaters*, Society of Vacuum Coaters, Albuquerque, NM, 1996, pp. 217–221.
114. X. Li, T. A. Gessert and T. J. Coutts, *Appl. Surf. Sci.*, 2004, **223**, 138.
115. H. Hosono, M. Yasukawa and H. Kawazoe, *J. Non-Cryst. Solids*, 1996, **203**, 334.
116. H. Kawazoe and K. Ueda, *J. Am. Ceram. Soc.*, 1999, **82**, 3330.
117. A. Petsom, S. Roengsumran, A. Ariyaphattanakul and P. Sangvanich, *Polym. Degrad. Stab.*, 2003, **80**, 17.
118. T. Ivetić, M. V. Nikolić, D. L. Young, D. Vasilijević-Radović and D. Urošević, *Mater. Sci. Forum*, 2006, **518**, 465.
119. B. Tan, E. Toman and Y. Li, *J. Am. Chem. Soc.*, 2007, **129**, 4162.
120. X. Wu, P. Sheldon, Y. Mahathongdy, R. Ribelin, A. Mason, H. Moutinho and T. J. Coutts, presented at the National Center for Photovoltaics Program Review Meeting, Denver, CO, 1998.
121. D. L. Young, *Electron Transport in Zinc Stannate (Zn₂SnO₄) Thin Films*, PhD thesis, Colorado School of Mines, 2000.
122. H. Enoki, T. Nakayama and J. Echigoya, *Phys. Status Solidi A*, 1992, **129**, 181.
123. X. Wu, T. J. Coutts and W. P. Mulligan, *J. Vac. Sci. Technol., A*, 1997, **15**, 1057.
124. D. L. Young, T. J. Coutts and D. L. Williamson, in *Proceedings of Materials Research Society Symposium on Transport and Microstructural Phenomena in Oxide Electronics*, Conference Proceedings 666, Materials Research Society, Warrendale, PA, 2001, pp. F3.8.1–F3.8.6.
125. J. Kolodziejczak and L. Sosnowski, *Acta Phys. Pol.*, 1962, **21**, 399.
126. J. N. Duenow, T. A. Gessert, D. M. Wood, D. L. Young and T. J. Coutts, *J. Non-Cryst. Solids*, 2008, **354**, 2787.

127. J. N. Duenow, T. A. Gessert, D. M. Wood, B. Egaas, R. Noufi and T. J. Coutts, presented at 33rd IEEE Photovoltaic Specialists Conference, IEEE, San Diego, CA, 2008.

128. T. A. Gennett, Y. Yoshida, C. Fesenmaier and T. J. Coutts, *J. Appl. Phys.*, 2009, **105**, 083547.1.

129. X. Li, H. Ray, C. L. Perkins and R. Noufi, in *Proceedings of 2008 MRS Spring Meeting, Symposium KK, Light Management in Photovoltaic Devices–Theory and Practice*, ed. C. Ballif, R. Ellingson, M. Topic and M. Zeman, MRS Proceedings, Materials Research Society, Warrendale, PA, 2008, vol. 1101, 1101-KK05-15.

CHAPTER 5

Thin Film Cadmium Telluride Solar Cells

ANDREW J. CLAYTON* AND VINCENT BARRIOZ

Centre for Solar Energy Research, OpTIC, Glyndŵr University,
St Asaph Business Park, St Asaph LL17 0JD, UK
*E-mail: a.clayton@glyndwr.ac.uk

5.1   Introduction

Research into the cadmium telluride (CdTe) thin film solar cell has been carried out for more than 40 years,[1] with the p–n junction typically consisting of p-type CdTe and n-type cadmium sulfide (CdS). Both layers are poly-crystalline and generally used in a 'superstrate' configuration, consisting of a glass/TCO/CdS/CdTe structure. As discussed in Chapter 1, CdTe has a direct bandgap (E_g) of 1.45 eV with a high optical absorption coefficient $>10^{-4}$ cm^{-1} at visible wavelengths, making it a good photovoltaic (PV) solar absorber in a single junction device. The n-type CdS has a direct E_g of 2.42 eV and, as well as being the n-type of the p–n junction, it acts as a window layer. The thin film CdS/CdTe solar cell is active in the solar spectrum between ~514 nm and ~850 nm, with carrier generation occurring in the CdTe absorber layer.

A number of deposition techniques can be used to produce polycrystalline CdS/CdTe solar cells including chemical bath deposition (CBD), vapour transport deposition (VTD), close space sublimation (CSS), metal organic chemical vapour deposition (MOCVD), radio-frequency (r.f.) sputtering and electrochemical deposition. Typically a number of deposition techniques, such as the ones described above, are used to produce the complete PV structure and generally depend on the quality achieved for each layer.

RSC Energy and Environment Series No. 12
Materials Challenges: Inorganic Photovoltaic Solar Energy
Edited by Stuart J C Irvine
© The Royal Society of Chemistry 2015
Published by the Royal Society of Chemistry, www.rsc.org

CdTe may be doped n-type or p-type, but also self-compensates when dopant concentrations are high.[2-8] Unwanted defect levels acting as carrier traps are introduced into CdTe due to its self-compensation nature and require passivation using post-growth treatment with $CdCl_2$ deposition and annealing, as discussed further in this chapter. The conductivity of CdTe can be induced by depositing under certain conditions. Nouhi *et al.*[9] employed Te-rich conditions using MOCVD to intrinsically dope CdTe p-type onto commercially available glass/SnO_2/CdS achieving a device efficiency of 9.4%. This approach was adopted by Chou and Rohatgi,[10] who managed to obtain a cell efficiency of 11.5%. CSS became the favourable technology for depositing CdTe with its fast growth rates and large grain sizes as a consequence of high process temperatures. In 1993 Britt and Ferekides[11] reported a maximum efficiency of 15.8%, improved on by Aramoto *et al.*[12] with a 16.0% efficiency cell. The 15.8% efficiency cell was produced[11] with sputtered CdS, whereas the 16.0% efficiency cell had[12] a MOCVD CdS layer, while in both cases CdTe was deposited by CSS. In 2001 the record for cell efficiency was increased to 16.5%,[13] which stood for a decade. This was achieved by engineering the device structure to improve the transmittance and conductivity of the transparent conducting oxide (TCO), including incorporation of a high resistive transparent (HRT) layer to improve the cell shunt resistances. Inter-diffusion between these layers played a part in the observed improvements to the device characteristics of which extended details are given in the following section.

Commercial production of CdS/CdTe PV modules using the CBD and CSS technologies started towards the turn of the century with First Solar[2] and Antec GmbH.[14] Transfer of thin film deposition from laboratory scale solar cells to large area modules introduces challenges such as lateral homogeneity across the device area, which becomes more difficult to control as the area increases. The device engineering used[13,15-18] for the 16.5% efficiency record cell was not suitable for industrial scale-up. This was partly due to the high post-growth annealing temperatures required after sputtering the TCO and HRT layers, requiring the more costly bariumsilicate glass[18] to be used replacing the cheaper soda lime glass (SLG), but also due to the added complexity of the device structure. The commercial drive within the thin film industry has led to a number of world record CdTe solar cells and modules in recent years using the SLG substrate. GE Global Research have now achieved a 19.6% cell efficiency, whilst First Solar have produced a thin film CdTe solar module with conversion efficiency of 16.1%.[19] The recent progress has been encouraging and the industry is optimistic of further improvements to thin film CdTe solar cell performances. In 2011 First Solar reported[20] a cumulated production capacity of 5 GW, substantiating the role of CdTe and thin film PV within the commercially viable technologies for producing low cost solar energy.

This chapter looks at the current material aspects and challenges that still present themselves with thin film CdTe PV solar cells. The focus is on MOCVD as an upcoming technology for producing these cells with consideration for prospects as a large-scale commercial production process and with more versatility than the current production methods.

5.2 CdS n-type Window Layer

CdS was determined[1] to be a suitable material to create a heterojunction with CdTe, with n-type conductivity occurring by intrinsic doping through the formation of sulfur vacancies (V_S).[21,22] It has an E_g of 2.42 eV allowing photons greater than wavelengths of ~514 nm to transmit through to the p-type CdTe layer for carrier generation. Generation of carriers from high energy photons absorbed in the CdS window layer are lost and cannot contribute the photocurrent generation as illustrated in Figure 5.1. This loss of current from absorption in the CdS window accounts for 5–7 mA cm^{-2} of the short-circuit current density (J_{sc}).[23,24] Therefore, the approach is generally to reduce the CdS window layer thickness.[21,25]

Whilst a gain in J_{sc} is observed using thinner CdS it can be accompanied by a drop in open circuit voltage (V_{oc}) and fill factor (FF)[23] due to formation of localised TCO/CdTe microjunctions. These microjunctions resulting from pinholes or areas of insufficient window layer coverage can cause regions of shorts and can significantly reduce the shunt resistance.[16,23,25] However, the CdS layer grown by MOCVD used in the 16.0% device produced by Aramoto *et al.*[12] in 1997 had a thickness below 100 nm. A later study[27] on the quality of MOCVD-deposited CdS using temperatures ranging from 430 to 470 °C found that CdS/CdTe solar cells produced using the lower temperature CdS gave the best PV performances, obtaining 13% efficient cells for a contact area of 1 cm.[2] The lower temperature CdS layers were confirmed[27] by scanning electron microscopy (SEM) and atomic force microscopy (AFM) to have smaller grains than those deposited at higher temperature as well as smoother surfaces. These structural properties were implicated in reducing the density of voids that may be present between grains and improving the quality of the CdS/CdTe interface without the presence of pinholes.

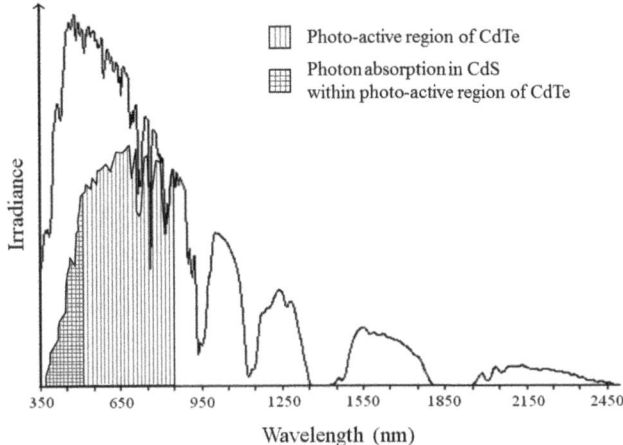

Figure 5.1 Solar spectrum at AM1.5 showing the photo-active region available for CdTe and its Shockley-Queisser related limit,[26] the portion of photons within this area that are absorbed by CdS is also highlighted.

5.2.1 Doped CdS

The introduction of O_2 into the ambient during CSS has been found to reduce the growth rate and grain size of the CdS,[24,28,29] producing a denser window layer and thus helping to reduce the density of pinholes present in the layer. The introduction of oxygen as an impurity is considered to occur naturally in CBD–CdS.[28,29] Depositing CdS in the presence of O_2 can lead to an insulating layer forming at the CdS–CdTe interface, thought to be $CdSO_3$,[24] which can be removed by carrying out a post-growth anneal in an O_2-free environment. Wu reported[15] the use of sputtering at room temperature in an Ar/O_2 atmosphere mixture, forming nanocrystalline CdS:O as the n-type window layer, with tuning of the E_g from 2.5 to 3.1 eV by altering the O_2 content in the ambient. The grains were found to be much smaller than conventional CdS, hence being ascribed nanocrystalline, with grain sizes of only a few nanometres. Bosio and co-workers[24,30] reported improvements to CdS films with superior optical and structural properties using sputtering at a temperature of ~220 °C in an Ar/CHF atmosphere. It was thought that the F^- ions bombarded the weakly bonded Cd and S atoms sufficiently to sputter them back. The resulting CdS:F films did not require any post-growth annealing and the reverse saturation current in the resulting CdTe devices was found to decrease relative to standard CdS giving better PV performances.

5.2.2 High Resistive Transparent Layer

Regardless of what technology is used to deposit the CdS window layer it has been suggested by Bonnet[31] that the lower limit of window layer thickness with the current technology is 50–60 nm. Therefore, to avoid the issue of shunts due to formation of TCO/CdTe micro-junctions, HRT (high resistive transparent) buffer layers have been investigated[13,15–17,23] acting as an insulating layer between the TCO and CdS. The buffer layers are typically a HRT oxide, forming a bilayer with the TCO to give a low-resistivity (ρ)/high-ρ structure on the glass substrate before the CdS is deposited.[16]

Wu and co-workers[13,15,17] used r.f. magnetron sputtering to deposit cadmium stannate (Cd_2SnO_4 or CTO) as the TCO, which was found to have higher transmittance and lower resistivity, as well as smoother surfaces relative to the standard F-doped SnO_2 for low-ρ high conductivity TCO layers. The high-ρ TCO was based on zinc stannate (Zn_2SnO_4 or ZTO),[13,15,17] which was later modified to $ZnSnO_x$,[15] also produced by sputtering. Both the low-ρ/high-ρ films were deposited at room temperature in a pure O_2 ambient with a post-growth anneal. These two ternary compounds are discussed in detail in Chapter 4.

5.2.3 Wide Bandgap $Cd_{1-x}Zn_xS$ Alloy Window Layer

After the $CdCl_2$ treatment and anneal, inter-diffusion between both the CdS–CdTe and ZTO–CdS interfaces were observed to occur,[15] resulting in consumption of CdS from both sides. Not only did the CdS thickness reduce but Zn diffusing into the CdS from the ZTO buffer layer resulted in

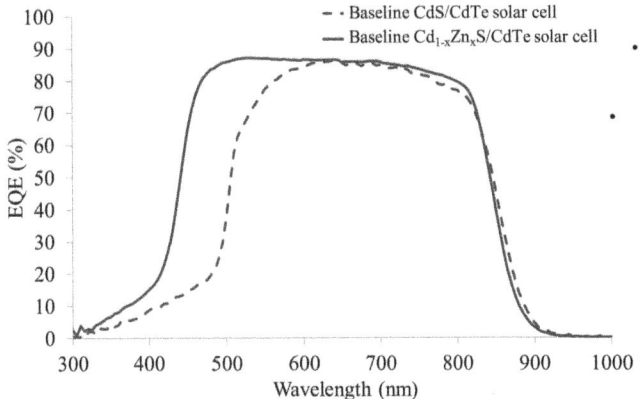

Figure 5.2 External quantum efficiency of two baseline solar cells produced by MOCVD at CSER showing the improved photo-response in the blue region of the solar spectrum for the cell using a $Cd_{1-x}Zn_xS$ window layer with wider E_g.

a $Cd_{1-x}Zn_xS$ alloy, which has a higher E_g than CdS[15,32-36] resulting in a greater transmission of photons in the blue region and thus enhancing the J_{sc} in the solar cells.

Introduction of Zn into the CdS window layer to form the $Cd_{1-x}Zn_xS$ alloy can be achieved by direct deposition onto the TCO using MOCVD.[32,34-36] The amount of Zn incorporated into the CdS is easily controlled by varying the partial pressure of the Zn precursor. Irvine and co-workers[34,36] demonstrated this by optimising the level of Zn to give a $Cd_{1-x}Zn_xS$ E_g around 2.7–2.9 eV. The wider E_g improved solar cell performance significantly, which is visible in the external quantum efficiency (EQE) measurements (Figure 5.2). This improved blue response resulting in a conversion efficiency of 13.3%.[34] Jones et al.[35] followed the same method[34] utilising $Cd_{1-x}Zn_xS$ window layers for solar cell devices with CdTe thicknesses of 1 µm. This also resulted in improved cell efficiencies relative to equivalent CdS/CdTe solar cells with a significant increase in J_{sc}.

5.3 CdTe p-type Absorber Layer

Deposition of the CdTe layer by CSS to produce the CdS/CdTe thin film solar cell has led to the highest reported efficiencies.[11-13,15,19] This is partly due to the high growth temperatures, up to ~600 °C, used in the process which leads to large grain formation during deposition. The advantage of this is to limit the density of grain boundaries within the CdTe layer that can contain many defects acting as recombination centres for the charge carriers.[24,37,38]

Deposition of CdTe using MOCVD is typically carried out at temperatures <400 °C,[34,39-42] and therefore results in smaller grains relative to CSS, which may introduce a large density of grain boundaries. However, this also minimises the formation of pinholes that can form between large grains[37] and allows thinner CdTe with thickness of ∼2 μm to be used.[34,39–41] Typically, 5–10 μm thicknesses for CSS–CdTe are required[3,43–47] in order to avoid the issue of pinholes that may propagate through the CdTe layer to the back contact. Due to the high absorption coefficient, a CdTe layer thickness of 2 μm is sufficient[48–52] to absorb the majority of photons in the visible and near infrared region, in which CdTe solar cells are active. Carriers that are generated far into the CdTe absorber layer have more chance of recombination as they diffuse towards the junction. Therefore controlling the grain size and the defect level within the CdTe layer is essential to utilise charge carriers produced by longer wavelength photons for contribution to the generated photocurrent.[37]

5.3.1 Doping CdTe

5.3.1.1 Intrinsic

The growth conditions are important in order to produce p-type CdTe, particularly when no external dopant is used to increase the acceptor concentration within the layer. Deposition under Te-rich conditions at the high temperatures ($\geq$500 °C), typically employed in CSS,[11–13,53] induces the p-type character of the growing layer due to intrinsic doping by the formation of cadmium vacancy (V_{Cd}^-) acceptor centres, as well as interstitial Te (Te_i).[24,51,54] Te-rich conditions favouring formation of V_{Cd}^- and Te_i have also been explored[9,10] using the MOCVD process and found at the time to result in optimum solar cell performances relative to more stoichiometric, or Cd-rich conditions. Other growth conditions, such as annealing ambient, can affect the intrinsic doping characteristics.

5.3.1.2 Extrinsic

In processes such as MOCVD, dopants can be introduced as the CdTe layer is growing in order to obtain p-type material, offering a good method of control on dopant concentration levels. This may be done by adding an element that is electron deficient relative to either Cd or Te increasing the hole concentration of the CdTe film. Group I elements, as well as Cu or Ag, may be used to occupy Cd sites,[24] or group V elements may be used[4,55,56] to occupy Te sites. Too much dopant material can cause issues by segregating in the grain boundaries along which they diffuse to the junction producing shunting pathways.[4,24,56] Diffusion of Sb has been observed[46,57] to be slower in CdTe than Cu after stability tests were carried out with assessment of PV degradation in solar cells with different back contact materials.

Addition of As to CdTe will induce p-type character in a similar way to Sb, acting as a shallow acceptor by substituting into a Te site (As_{Te}), which is enhanced under Cd-rich conditions.[55,56] The As_{Te} shallow acceptor is also an important p-type dopant for CdTe as it only requires a low formation energy of 1.68 eV ensuring an effective doping.[58] A study[56] using capacitance–voltage (C–V) measurements found that As dopant concentration in CdTe of PV solar cells grown by MOCVD was several orders of magnitude greater than the acceptor concentration. This was attributed to segregation of As at the grain boundaries, interfaces, back surface or at defects within the CdTe layer.[10,56] This correlated[56] with a decrease in shunt resistance as As concentration was increased from 1×10^{17} to 1.5×10^{19} atoms cm^{-3}. Nevertheless, acceptor concentration levels achieved $\sim 10^{15}$ cm^{-3} and, if doping concentrations are optimised, then group V elements such as As and Sb can be used as effective dopants to form p-type CdTe.[4,55,56]

5.3.1.3 Associated Defects in CdTe

CdTe grown under Te-rich conditions is likely to introduce Te interstitials into the layer.[51,54] Such deep acceptor centres can form without the presence of extrinsic dopant precursors and may act as minority carrier recombination sites.[51,59] Donor levels such as tellurium vacancies (V_{Te}^{2+}) or cadmium interstitials (Cd_i^{2+}) may also exist,[51,56] but are less likely with Te-rich growth conditions. Cadmium vacancies (V_{Cd}^{-}) are thought[54,60] to be the predominant defects in most Cd-based compounds and introduce deep acceptor states in CdTe. However, positron annihilation lifetime spectroscopy (PALS) measurements on CdTe deposited by CSS have indicated that lifetimes associated with V_{Cd}^{-} revealed an associated longer lifetime component,[59] which may be associated to Cd/Te divacancy ($V_{Cd}^{-}V_{Te}^{2+}$) complexes that may also act as deep level traps.

Under Cd-rich conditions, there is a strong likelihood that Cd_i^{2+} or V_{Te}^{2+} will form introducing donor levels into the CdTe,[56] which will compensate for any extrinsic dopants used to obtain p-type character. These defect states will introduce energy levels within the forbidden energy gap of CdTe that can affect minority carrier lifetimes in particular.[8,61–63] Self-compensation of CdTe limits the acceptor level that can be achieved;[2–8,64] if this effect can be reduced to increase p-type doping in CdTe, overall PV cell efficiencies can improved.[2,4,8]

5.4 CdCl$_2$ Activation Treatment

Due to their polycrystalline nature, as grown CdTe solar cells have too many recombination centres within the grain boundaries, at the CdS–CdTe interface and at the CdTe–metal interface, resulting in poor device performance. Treatment of the CdTe layer and the CdS–CdTe interface is typically carried out using a CdCl$_2$ deposition followed by anneal to allow Cl to diffuse into the CdTe *via* the grain boundaries and towards the interface with CdS. This leads to a marked improvement in PV cell performances, which in some cases can increase efficiency by an order of magnitude.[43,44] This treatment process is widely accepted to bring about a number of changes to the CdTe, as described below:

i Re-crystallisation of the CdTe resulting in grain growth
ii Promotion of inter-diffusion between the CdS–CdTe interface to form CdS$_{1-x}$Te$_x$, relieving some strain caused by the 9.7% lattice mismatch between CdS and CdTe
iii Passivation of the grain boundary defects.

5.4.1 Recrystallisation of CdTe Grains

Activation using $CdCl_2$ deposition and annealing treatment results in recrystallisation of the CdTe grains, causing growth[16,25,38] and hence reduction in the density of grain boundaries. Figure 5.3 shows a comparison between two SEM images of polycrystalline CdTe deposited by MOVCD: (a) as grown; and (b) after $CdCl_2$ treatment. High temperature processes such as CSS result in large grain sizes for the as-grown layers and little or no grain growth is observed[16,25,38] after $CdCl_2$ treatment, or removal of the few smaller grains present in the CdTe layer occurs.[65] The effect of grain growth is more notable for CdTe layers with small grains (<1 μm).[38,64] Taking this into consideration it could be construed that improvement in PV cell performance may be more significant for cells with small grained CdTe after $CdCl_2$ treatment due to a more significant reduction in the number of grain boundaries in which defect centres are concentrated.

5.4.2 Inter-diffusion at the CdS–CdTe Interface

The promotion of inter-diffusion between the CdS–CdTe interface is important for improving the quality of the junction by removal of defects related to structural effects between the CdS and CdTe layers caused by a lattice mismatch of 9.7%.[15,66,67] An abrupt CdS–CdTe interface is considered to result in a poor junction.[67,68] Diffusion of S into CdTe and Te into CdS during the $CdCl_2$ annealing treatment forms the alloy $CdS_{1-x}Te_x$ at the CdS–CdTe interface and is considered to reduce the density of recombination centres in the region of the p–n junction.[16,43,69]

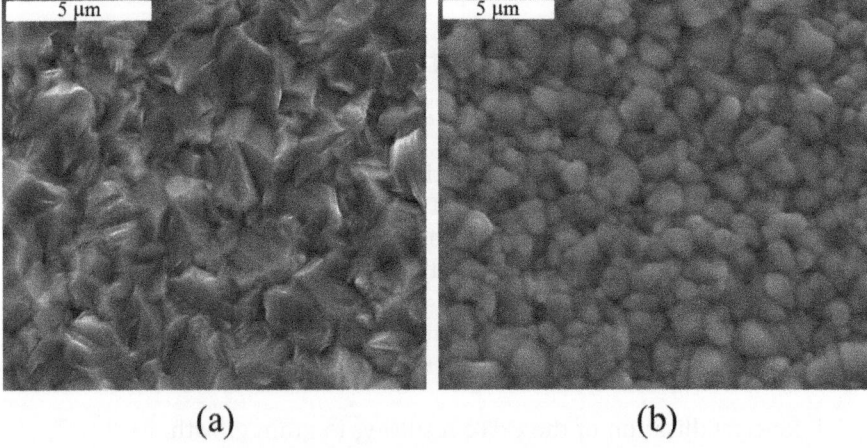

(a)   (b)

Figure 5.3 SEM images of polycrystalline CdTe grown by MOCVD at 390 °C: (a) as grown; and (b) after $CdCl_2$ treatment. The scale for each micrograph is at 5 μm.

The procedure for the $CdCl_2$ treatment is strongly dependant on the technique and conditions used to grow the device structure. The high growth temperatures employed in CSS can also induce inter-diffusion,[16,28,70] but not to the same extent as during annealing with $CdCl_2$.[66] However, too much inter-diffusion of S into CdTe is also thought to degrade the p–n junction,[15,71] which may be due to excessive consumption of CdS leading to shunting issues or by introduction of new defect centres into the interface region.[28,29] Therefore, limiting the inter-diffusion process may be desirable and was reported to be possible by introducing O_2 into the ambient during CdS growth.[15,24,28,29] This resulted in a reduced growth rate and a decrease in grain size to form a more compact layer. It was suggested that incorporating O_2 into the CdS layer helped to also relieve some of the strain at the CdS/CdTe junction and reduce associated defects.[15,28,29] Similar results were observed by annealing the CdS layer post-growth in the presence of O_2,[28,29] leading to improvements to PV solar cell performances.

5.4.3 Passivation of Grain Boundary Defects within CdTe

After $CdCl_2$ deposition and anneal, Cl is incorporated into the bulk CdTe as either a donor, substituting a Te site (Cl_{Te}^+), or by forming a complex with V_{Cd}^{2-}. The V_{Cd}/Cl complexes can have a neutrally charged state with two Cl atoms,[54] or the A-centre ($V_{Cd}^{2-}Cl_{Te}^+$) arrangement with one Cl, which acts as a shallow acceptor state.[38,54,65] The latter is considered to be the compensating effect, reducing the concentration of V_{Cd}^{2-}.[72] Formation of the $V_{Cd}^{2-}Cl_{Te}^+$ A-centre was studied using photo-induced current transient spectroscopy (PICTS) by observing energy band intensities associated with the complex.[73] Comparison between Cl-doped and undoped samples determined that the compensation effect occurred readily during $CdCl_2$ treatment.[73] The resulting decrease in deep levels that act as traps for minority carriers leads to improvements in V_{oc} and efficiency.[54]

Figure 5.4 represents a schematic of changes to EQE curves according to changes within a solar cell structure. The importance of defect passivation using $CdCl_2$ treatment is illustrated where incident photon conversion efficiency (IPCE) reduces with absorption of lower energy photons towards the back contact. Also shown in Figure 5.4 is what may be expected from EQE for a CdTe solar cell with a poor CdS–CdTe interface with a large number of defects.

Net acceptor activity may be further enhanced by using O_2 in the CdTe growth ambient,[4] or post-growth annealing at 390 °C for 20 minutes in an O_2/N_2 mixed atmosphere containing some HCl.[44] Photoluminescence (PL) measurements showed an increase in exciton band intensity associated with active shallow centres involving V_{Cd}^{2-} and Cl_{Te}^+ when O_2 was present during CdTe annealing treatment.[44] However, another report[65] using deep level transient spectroscopy (DLTS) concluded that $CdCl_2$ treatment in the presence of O_2 did not influence the defect structure in CdTe compared with $CdCl_2$ activated cells carried out in the absence of O_2. Such observations emphasise the fact that the optimisation of the $CdCl_2$ treatment is strongly

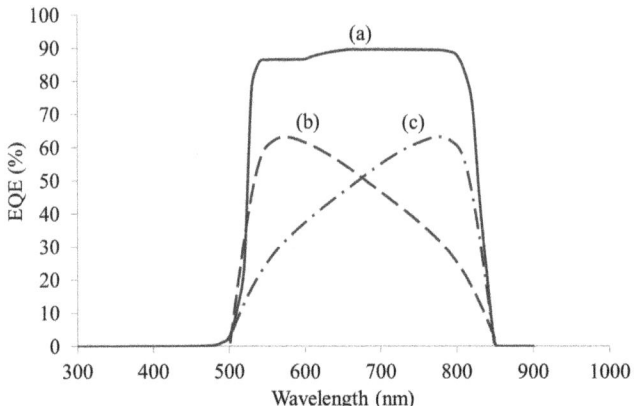

Figure 5.4 A schematic representation of: (a) ideal spectral response considering an idealised standard CdS–CdTe PV solar cell (represented by the solid line); (b) a CdTe solar cell with no CdCl$_2$ treatment (represented by the dashed line) showing poorer incident photon conversion efficiency (IPCE) of lower energy photons due to greater recombination of charge carrier towards the back contact; and (c) a CdTe solar cell with poor CdS–CdTe interface (represented by the dashed line with central dot).

dependent on the condition in which the CdTe device was grown. Even though there is debate as to the role of O$_2$ in the CdCl$_2$ annealing ambient, there are many reports[38,44,74] stating improvements to CdTe solar cell performances using O$_2$ in the CdCl$_2$ treatment process.

5.5 Back Contact Formation

Formation of a back contact to the CdTe layer requires good ohmic properties at the CdTe–metal interface, in contrast to the rectifying p–n junction associated with the CdS–CdTe interface.[5,25] Finding a suitable metal for this purpose is a challenge due to the high electron affinity of CdTe and difficulty in obtaining high p-type conductivity to form a good ohmic contact.[5-7,25] This can lead to the formation of the Schottky barrier (ϕ_b) which would restrict majority (hole) carrier conduction.[5-7] In these conditions, the device can be electrically described[75-77] as a two diode model, with the p–n junction representing one diode and the back contact acting as an opposed diode with associated shunt resistance. Figure 5.5 shows a schematic representing such a model. Capacitance measurements can give the saturation current, I_2, from which the barrier height at the back contact may be derived.[75,76] The presence of a large barrier at the back contact results in 'roll-over' (kink) effects, observed in I–V curves, due to current saturation at high forward voltage.[5,75,76]

Another approach for investigating barrier formation at the back contact is to consider the space charge regions at the p–n junction and back contact using the energy band diagrams.[75,78] When the two space charge regions overlap, depending on the barrier height and CdTe thickness, the effect of

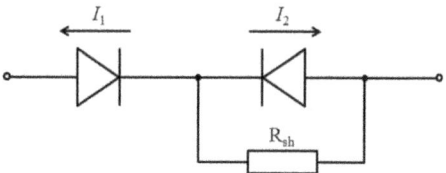

Figure 5.5 Equivalent circuit of a solar cell with back contact barrier represented by a diode at the p–n junction with dark saturation current I_1 and a diode in the reverse direction at the back contact with dark saturation current I_2 and shunt resistance R_{sh}. Figure reproduced from *Thin Solid Films*[75] with permission from Elsevier Publishing.

the barrier can be reduced due to closer CdS–CdTe band alignment.[75] This overlap is more likely for thin CdTe layers (sub-microns) whereas typical absorber layer thicknesses are of the order of several microns.

Attempts are usually made to form a p$^+$-type layer before depositing the metal contact to create tunnelling pathways and/or back surface field (BSF) in order to reduce the effects of ϕ_b at the back contact interface.[5–7]

5.5.1 Cu$_x$Te

The conventional approach[46,61–63] towards achieving this is by using Cu, forming a Cu$_x$Te phase with a Te-rich CdTe surface prepared by a bromine/methanol or nitric/phosphoric acid chemical etching process. The Cu increases the acceptor concentration in CdTe by substituting Cd to form Cu$_{Cd}$ and results in back contacts with a reduced barrier.[61–63,79] Typically, Cu is used as an alloy with Te and mixed with HgTe in conjunction with a form of carbon paste.[61,62] Cu may be deposited as a pure metal or as a Cu$_x$Te alloy.[61] When Cu is deposited as a pure metal the value of x in the Cu$_x$Te phase is dependent on the Cu thickness.[46] Thick Cu deposits are more likely to form the Cu$_2$Te phase, which is electrically superior to the other Cu$_x$Te phases,[61,80] but also releases more Cu into the CdTe layer as well as being more susceptible to oxidation with the formation of Cu$_2$O. Cu-related oxides are suspected to lead to roll-over, the effects of which can be reduced by using less Cu to form Cu$_x$Te phases with $x < 1.5$.[61] Increased Cu concentration in CdTe is associated with a greater acceptor concentration, as Cu$_{Cd}$ is considered as a deep acceptor level; including compensation effects that are caused by interstitial Cu (Cu$_i$) deep donor levels. High Cu concentrations in CdTe have been linked[62] to a reduced minority carrier lifetime (τ) due to an increase in deep level trap density. Control of the Cu content in the Cu$_x$Te alloy is therefore essential.

5.5.2 ZnTe:Cu

Incorporation of Cu with ZnTe deposited onto CdTe has been reported[63,79] to give good ohmic contacts by introducing p$^+$-type conductivity, depending on the thickness of ZnTe:Cu used. After the Cu is deposited onto the ZnTe layer the contact is annealed to promote Cu diffusion to form the ZnTe:Cu layer. Increased ZnTe:Cu thickness whilst maintaining contact annealing temperature correlated with other investigations,[61,62] with degradation of PV cell performance due to excessive Cu diffusion.[63] Corresponding low values of τ were determined by time-resolved photoluminescence (TRPL). An optimised ZnTe:Cu thickness and contact annealing temperature of 280 °C resulted in high PV device performances with increased values of $\tau \sim 8 \times 10^3$ picoseconds, whereas higher annealing temperatures accelerated Cu diffusion towards the CdS layer causing τ to fall.[63]

5.5.3   Ni–P

Some reports suggest[80,81] Ni may be used as a back contact to give good PV performance and stability. Addition of dopant P levels with the Ni improves the contact properties where P acts as a shallow acceptor. Reaction of Ni with a Te-rich CdTe surface at 200 °C can form NiTe or $NiTe_2$, confirmed using X-ray diffraction (XRD).[81] Ni is a slow diffuser[7] and could provide a suitable and stable back contact, which stability tests have confirmed.[80]

5.5.4 Sb_2Te_3

A number of investigations report the use of Sb_2Te_3 and Sb–Te phases as suitable back contact materials for long-term stability.[6,7,57] Back contact annealing of Sb–Te can lead to the formation of Sb_2Te_3 at certain temperatures, which is considered to be more stable.[6] A thin dense layer of Sb_2Te_3 is considered the most effective for use as a diffusion barrier to the back contact metal, such as Ni or Mo.[7] Studies showed that Ni can react with Sb_2Te_3 to form Ni–Sb and Ni–Te alloys,[6,7] liberating some Sb or Te which are more likely to diffuse into the CdTe when isolated. The formation of deep acceptors Sb_{Te} and Te_i will increase CdTe p-type conductivity, but also diffuse towards the junction, if in excess, deteriorating PV cell performance. Reaction of Sb_2Te_3 with Mo was found to be less likely,[7] with accelerated stability testing showing no PV cell degradation.[6,46,57] This is dependent on the Sb_2Te_3 layer preventing diffusion of Mo to the $CdTe/Sb_2Te_3$ interface, which would form a high density of traps in the bulk absorber material.[7]

5.5.5 $CdTe:As^+$

A back contact layer (BCL) creating a p^+ layer suitable for metallisation of the back contact, giving an effective npp^+ cell structure has been reported.[34,40,82] This has been used in conjunction with As doping for bulk CdTe p-type conductivity.[55,56] A $CdTe:As^+$ BCL with increased As concentration of 1×10^{19} atoms cm^{-3} is deposited,[40] creating a BSF lowering the contact barrier.[40,64] This process also removed the necessity of any wet chemical processing steps which would require transfer of the solar cell device from the chamber between steps.[34,40,82] A suitable back contact can then be deposited to complete the solar cell device.

5.6 MOCVD CdTe Cells

Although less dominant within the CdTe research community, thin film CdTe solar devices can be produced by MOCVD. A reported baseline process utilises a horizontal reactor configuration with vapour species mixing at the inlet of the chamber with laminar flow over the substrate.[34,83] The cells are deposited in a superstrate configuration where the substrates used are either NSG Pilkington SLG coated with fluorine doped tin oxide (FTO) or Corning boroaluminosilicate coated with indium tin oxide (ITO). The coated sample area for the baseline process is typically 50×50 mm^2.

5.6.1 MOCVD Cd$_{1-x}$Zn$_x$S vs. CdS Window Layer

Solar cell performance is optimised by controlling the concentration of Zn in the Cd$_{1-x}$Zn$_x$S alloy to give an E_g up to 2.9 eV.[34,35,84] Addition of too much Zn causes resistance on the n-type side of the junction to increase,[32,33] reducing the benefit of the improved solar cell response in the blue region. Careful preparation of the substrate before deposition is required, particularly with the smoother boroaluminosilicate/ITO substrates to achieve good adherence of the Cd$_{1-x}$Zn$_x$S to give complete window layer coverage. This is essential for avoiding areas of no growth which leads to regions of TCO/CdTe microjunctions that are inferior to the Cd$_{1-x}$Zn$_x$S/CdTe junction.[41,85,86] A thicker window layer thickness of 240 nm can be used without significantly limiting solar cell performance from blue absorption,[34,87] because of the increased transmittance of the Cd$_{1-x}$Zn$_x$S alloy relative to CdS. However, consideration towards reducing this window layer thickness will be necessary for further improvement to J_{sc} to be realised. The film is deposited up to a substrate temperature of 360 °C using dimethylcadmium (DMCd), di-*tert*-butylsulfide (DtBS) and diethylzinc (DEZn) as precursors in one growth step,[34,35,84,87] regulating gas phase partial pressure to control the stoichiometry of the growing layer. Figure 5.6 illustrates the widening E_g of the Cd$_{1-x}$Zn$_x$S alloy window layer as the Zn content (x) is increased.

The quality of the Cd$_{1-x}$Zn$_x$S deposition can affect the whole device structure. Laser beam induced current (LBIC) can be used as a diagnostic tool for assessing the quality of the complete cell and can reveal the effect that pinholes and other defects may have on a solar cell device. The principle of the technique is the same as optical beam induced current (OBIC), but with utilisation of specific wavelengths to probe the cell being studied. Spatially resolved assessment of photocurrent generation can be achieved with a triple laser set up[85] using wavelengths, for example, at 405, 658 and 810 nm each having different penetration depths through the cell structure. A three-dimensional map is represented in Figure 5.7 for a CdTe PV cell produced on a plasma-cleaned boroaluminosilicate/ITO substrate. The induced current can be generated within different regions of the Cd$_{1-x}$Zn$_x$S/CdTe cell scanning over the whole contact area, typically 5×5 mm^2 or with higher resolution over a smaller area. The resolution of the LBIC instrument beam can

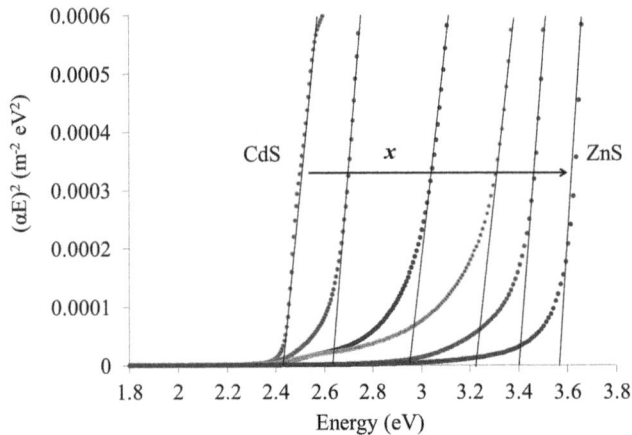

Figure 5.6   Bandgap (E_g) determination of $Cd_{1-x}Zn_xS$ alloy window layer with increasing Zn content (x) using α^2E^2 *vs.* E curves. Figure reproduced from *Progress in Photovoltaics: Research and Applications* with permission from Wiley Publishing.

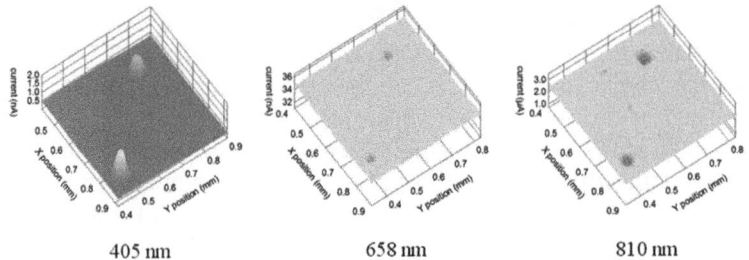

Figure 5.7   LBIC three-dimensional map showing current (nA) against X–Y position (mm) of a $Cd_{1-x}Zn_xS$/CdTe solar cell produced by MOCVD for three different laser wavelengths. Regions of pinholes are identified by peaks at 405 nm due to high transmittance through localised areas with no $Cd_{1-x}Zn_xS$ coverage and corresponding dips at 658 and 810 nm signifying poorly generated photocurrent within the CdTe absorber layer. Figure reproduced from *Energy Procedia*[87] with permission from Elsevier Publishing.

be reduced down to 10 μm, allowing smaller features to be mapped into localised photo-active regions. Contribution from each photo-active region can then be separated and quantified. A high photocurrent response at the shortest laser wavelength (405 nm) corresponds to regions of thin $Cd_{1-x}Zn_xS$ coverage and results in a low photocurrent response at the longer wavelengths, associated with poor minority carrier generation in the CdTe absorber layer. Likewise, a poor response observed in the LBIC measurements at 405 nm correlates with a thick region of $Cd_{1-x}Zn_xS$, giving good photocurrent generation in the CdTe layer at the two longer wavelengths and described as a strong photo-active region.

Variation in cell efficiency has shown,[85] from the LBIC measurements, to have strong dependence on V_{oc}. Cells with majority weak photo-active regions corresponded with an overall low V_{oc} and cells having a majority of strong photo-active regions with high overall V_{oc}, the latter having the highest conversion efficiencies. Shunt resistance (R_{sh}) also had an effect on overall cell efficiency, although independent of $Cd_{1-x}Zn_xS$ thickness distribution, which led to poor PV performance; attributed to micro-shunts within the solar cell device.

5.6.2   MOCVD CdTe:As Absorber and Contact Layer

Deposition of the CdTe layer occurs sequentially to the $Cd_{1-x}Zn_xS$ window layer, using the same MOCVD growth chamber. Relatively low growth temperatures are used (390 °C) in comparison with other techniques such as CSS, which leads to small grain sizes ($\sim$1 µm). An as-grown CdTe thickness of 2 µm is sufficient to capture the majority of photons above the E_g of CdTe for photocurrent generation.

Extrinsic doping with arsenic is carried out *in situ* using tris(dimethyla-mino)arsine (tDMAAs) during the CdTe growth step, along with precursors dimethylcadmium (DMCd) and diisopropyltelluride (DiPTe). Cd-rich conditions[55,56] are used to promote As incorporation into Te vacancies and excessive deep donor V_{Te} must be avoided. An As-dopant concentration of 2×10^{18} atoms cm^{-3} was found[55,56] to give the optimum dopant levels in order to obtain the p-type conduction, while increasing the As-dopant concentrations [As] above this value resulted in reduced solar cell performance. The likely reason for the drop in efficiency with increased [As] beyond a value of 2×10^{18} atoms cm^{-3} is increased segregation of As into the grain boundaries, which may then diffuse towards the junction creating localised areas of high electrical conduction leading to micro-shorts and increased shunting.[4,24,56]

Following deposition of the bulk CdTe:As absorber layer, a second CdTe:As$^+$ layer with high As-dopant concentration of 1×10^{19} atoms cm^{-3} is deposited as the BCL to create a low contact barrier.[40] *In situ* CdCl$_2$ deposition can be carried out followed by anneal at 420 °C for 10 minutes for passivation of the CdTe grain boundaries.[34,88] Apart from the metallisation process, the complete device structure is carried out in a single growth chamber, at atmospheric pressure, reducing processing time and simplifying the process to produce the solar cell device. Outside of the MOCVD chamber, any excess CdCl$_2$ is rinsed off using deionised water followed by drying under high pressure N_2. The final metallisation, using gold, is carried out using a thermal evaporator to form the back contacts with areas of 0.25–1.0 cm^2 leading to a conversion efficiency of 15.3%.[83] The *J–V* curve for this MOCVD-grown $Cd_{1-x}Zn_xS$/CdTe cell, 0.25 cm^2 in area, is represented in Figure 5.8. *J–V* parameters are shown in Table 5.1 for the MOCVD-CdTe PV cell with comparison with the 16.5% efficient CTO/ZTO/CdS/CdTe PV cell[13,15]

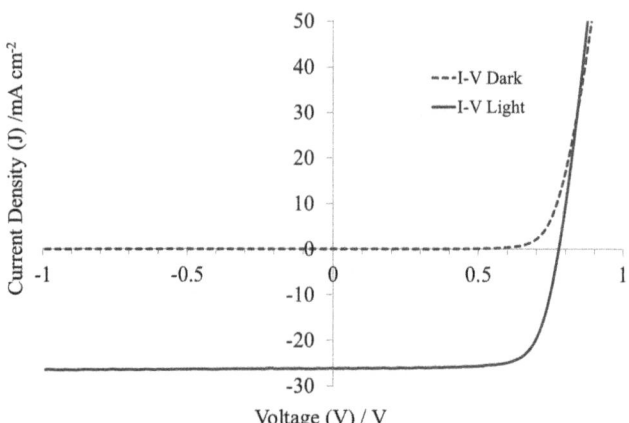

Figure 5.8 *J–V* curve of MOCVD-CdTe PV cell with 15.3% conversion efficiency.

Table 5.1   Comparison of *J–V* parameters of a MOCVD-grown CdTe PV cell[83] relative to the 16.5% efficient CTO/ZTO/CdS/CdTe cell produced at NREL (2001).[13,15] The data in the table were obtained from: proceedings of the 8th Photovoltaic Science Applications and Technology conference, C94, April 2012, Newcastle upon Tyne, UK;[83] *Solar Energy*[15] with permission from Elsevier Publishing; and *Progress in Photovoltaics: Research and Applications*[19] with permission from Wiley & Sons

Cell structure	η (%)	J_{sc} (mA cm^{-2})	V_{oc} (mV)	FF (%)	Area (cm^2)	Ref.
FTO/ZnO/CdS/CdTe	19.6	28.6	857	80	1.01	19
CTO/ZTO/CdS/CdTe	16.5	25.9	845	75.5	1.03	13, 15
ITO/CdZnS/CdTe	15.3	26.2	767	76.2	0.25	83

produced by the National Renewable Energy Laboratory (NREL) in the USA and the current world record 19.6% CdTe PV cell.[19] Table 5.1 shows that the 19.6% CdTe PV cell has a J_{sc} close to the theoretical limit[24] of 30.5 mA cm^{-2}. Improvements to V_{oc} and FF have also been achieved. The MOCVD-CdTe solar cell is currently limited by a lower V_{oc} compared with the CSS–CdTe solar cells. In order to achieve further improvements to the best CdTe PV cells, efforts will most likely be directed towards greater p-type doping to obtain higher V_{oc}.

5.6.2.1   Reducing the CdTe Absorber Layer Thickness

With high volume manufacture of CdTe solar modules and limited global availability of Te,[48,49,89] the use of ultra-thin CdTe becomes an attractive prospect. This would include a decrease in overall material consumption along with lower module production costs. It has been shown in a theoretical study that most of the carrier generation occurs close to the CdS/CdTe junction within the PV cell.[50] The carrier generation was calculated to reduce by two orders of magnitude within the first 1 μm of the CdTe absorber layer.

Investigation into reduction of the CdTe absorber layer thickness has been carried out using MOCVD and assessed in relation to solar cell performance.[35,90] Little effect on the series resistance was observed between varying CdTe absorber thicknesses, demonstrating that the close proximity to the p–n junction does not affect the back contact barrier in these devices. As the CdTe layer thickness is reduced, however, reduction of pinhole formation becomes more crucial[41] to preserve the photocurrent generation and PV cell performance within the optical absorption limits. V_{oc} and FF have been observed to drop as a result of thinning the CdTe absorber.[35,90] This is confirmed by EQE characterisation (Figure 5.9) where loss of photocurrent generation occurred over the whole photo-active region for CdTe, with a corresponding decrease in shunt resistance for thinner absorber layer thicknesses. Also, recombination at the back contact may become more prominent, as identified by Plotnikov *et al.*[49] from current–density–voltage *vs.* temperature (*J–V–T*) characterisation. This particular study established that bifacial illumination would be important when considering photon capture in ultra-thin CdTe PV cells. Reports have discussed large losses in EQE curves at longer wavelengths for ultra-thin PV cells due to a decrease in photon absorption towards the CdTe band edge.[48,50]

It has also been found that $CdCl_2$ treatment can have an adverse effect on the ultra-thin $Cd_{1-x}Zn_xS$/CdTe PV cells, such that the improved response in the blue region deteriorates with reduction of absorber layer thickness.[35,90] The window layer band edge shifted towards the red region of the solar

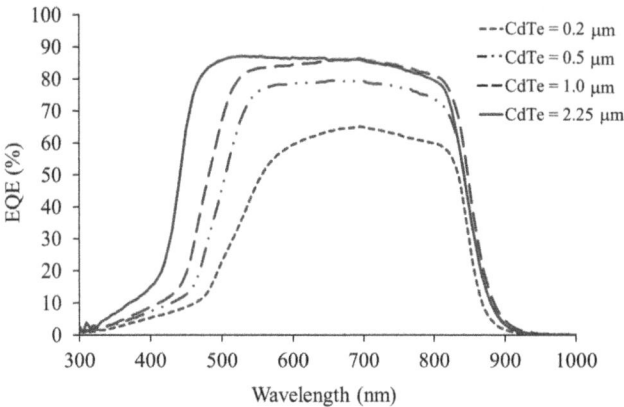

Figure 5.9   External quantum efficiency (EQE) curves for ultra-thin CdZnS/CdTe PV cell deposited by MOCVD using various CdTe absorber thicknesses showing red-shift of the $Cd_{1-x}Zn_xS$ absorption edge as effective Cl concentration increased for thinner CdTe. Figure reproduced from *Solar Energy Materials and Solar Cells*[90] with permission from Elsevier Publishing.

spectrum (Figure 5.9) showing closer characteristic to the absorption edge of CdS, particularly for the thinnest CdTe absorber layer thicknesses. This was attributed to Cl leaching Zn out of the $Cd_{1-x}Zn_xS$ alloy during $CdCl_2$ treatment,[90] becoming more severe for thinner CdTe layers where the Cl concentration effectively increased. Reducing the $CdCl_2$ deposition and anneal time proportionally, for the ultra-thin CdTe thicknesses of 500 nm relative to the baseline CdTe thickness, recovered PV cell response in the blue region of the solar spectrum. The overall PV performances of the cells with recovered blue response did not improve after optimisation of the $CdCl_2$ treatment. A further study[91] showed that the V_{oc} reduced with $CdCl_2$ layer thickness indicating that activation treatment needed to be preserved. Optimisation of the window alloy composition was found to be the necessary approach towards improving device blue response and overall ultra-thin $Cd_{1-x}Zn_xS$/CdTe PV solar cell performances.

5.7 Prospects for Large-scale Manufacture using MOCVD

In order for a new technology—albeit materials and/or deposition techniques—to have prospects for large-scale manufacturing, its module production cost, *i.e.* cost per Watt peak (W_p), should be as low as possible, which in the current market targets ~$0.5 per W_p. Such a scale of manufacture towards cost reduction has been demonstrated by First Solar, through economy of scale, achieving $0.75 per W_p at the end of 2010[92] and aiming towards $0.5 per W_p by 2015. Although the gap in production cost is narrowing due to increased production of crystalline silicon PV modules in China, thin film PV modules still offer advantages and flexibility from the materials viewpoint. Furthermore, an improvement in conversion efficiency is paramount in order to increase W_p per m.^{2}

Currently, most commercial CdTe PV modules use physical vapour deposition techniques, namely, the vapour transport deposition (VTD) technique or its alternative and more widely reported CSS. Both techniques rely on high temperatures for the source material and the substrate reaching 500–900 °C to achieve deposition rates of 1–10 $\mu m\ min^{-1}$.[93] It was reported that a CdTe PV module takes 2.5 hours from a TCO coated glass to a finished product ready for shipment.[92] The annual maximum production capacity of one of these lines is reported to be 70 MW per line, which extrapolates to a single module produced every minute. Therefore, in such production lines multiple deposition arrangements and/or parallel processes must be used to maximise throughput and 'overcome' slower processes, which could include the substrate heating and cooling steps. MOCVD can be a scalable and low energy process alternative, offering controllability of materials and utilisation, with tuneability of materials' properties. These factors have been detailed throughout this chapter so far for CdTe, and emphasis will now be placed on its suitability as an alternative high throughput process.

The first advantage of the MOCVD process is that atmospheric pressure (AP) is used, reducing running and maintenance costs as well as removing some of the complexity in chamber designs and therefore reducing the capital cost of each line. Secondly, the temperatures employed during the process are in the range of −10 to 50 °C for the source materials, while the substrate temperature range is within 200–450 °C. Finally, as was introduced in Section 5.6, the full structure (except for the metal contacts) can be deposited by AP-MOCVD, therefore simplifying the duplication and scaling of the whole process. Barrioz *et al.*[94] assessed the feasibility of using AP-MOCVD by considering the molar supply limits during the pyrolysis of CdTe and found that 1 μm could be deposited at a process line speed of 60 cm min^{-1}. In terms of material utilisation, Hanket *et al.*[93] reported values of up to ∼50% with dynamic deposition rates of up to 0.8 μm min^{-1} for CdTe deposited by VTD on a substrate moving at 1.25 cm min^{-1}. By comparison, with an inline AP-MOCVD and a substrate moving at 1.13 cm min^{-1}, it was reported that a material utilisation of more than 40% was achieved with a dynamic deposition rate of 0.3 μm min^{-1} (Figure 5.10).[95] Therefore, both VTD and AP-MOCVD appear to offer similar performance both from the throughput and material utilisation point of view.

In terms of scalability and with potential scarcity of some materials,[96] both high material utilisation and minimal amount of materials being used per module should be a priority although not at the detriment of conversion efficiency. Ultra-thin absorbers are readily suited for the MOCVD process, but as mentioned in Section 5.6.2.1, due to the optical absorption limits this approach would have to be combined with photon trapping solutions in order to improve the low energy photon path length within the structure to maintain or improve conversion efficiency.

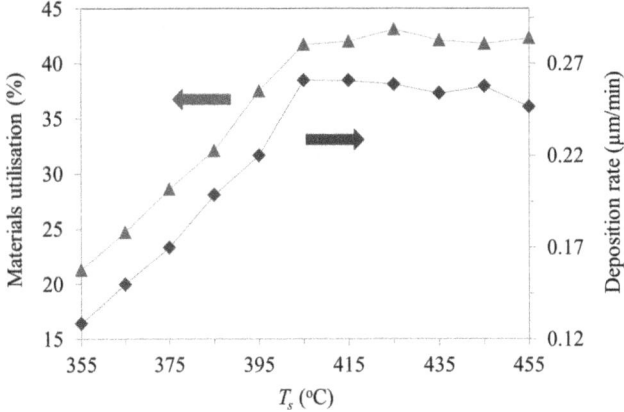

Figure 5.10   Materials utilisation and growth rate as a function of substrate temperature for an inline MOCVD-CdTe PV process. Figure reproduced from the *Journal of Crystal Growth*[95] with permission from Elsevier Publishing.

Finally, it is essential to be able to scale up a deposition process to enable production of thin film modules with desired production cost and conversion efficiency. The deposition process must also be adaptable for any future improvements to be made to the structure to enhance conversion efficiencies. As described in Section 5.6, not only can MOCVD be used for depositing all of the structure layers, but it has also been shown to be tuneable where the E_g of the window layer can be accurately and repeatedly alloyed to gain photocurrent by up to ~ 4 mA cm^{-2}. Extrinsic As *in situ* doping of p-type CdTe was also demonstrated by MOCVD with controllable levels to dope both the bulk p-CdTe and the BSF p$^+$-CdTe with acceptor carrier concentration of up to $\sim 10^{15}$ cm^{-3}. The increased As doping to create the BSF enabled ohmic behaviour to be observed at room temperature with low series resistance of ~ 2 $\Omega \cdot$cm^2. CdTe devices deposited using an inline AP-MOCVD process in a preliminary study showed promising results with 8% conversion efficiency with a dynamic deposition rate of ~ 0.6 μm min^{-1}.[97] These results highlight AP-MOCVD as a potential alternative to the usual VTD process for large scale manufacture of thin film PV on rigid as well as flexible substrates.

5.8 Conclusions

This chapter provides a brief discussion of CdTe solar cell progress in recent years, introducing a number of deposition techniques employed for producing the complete photovoltaic structure. A relationship between selected process methods for deposition and the properties of the layers forming the CdS–CdTe p–n junction has been given. Focus is on the CdTe absorber and the effects that impurities have on its p-type character, which is dependent on the process conditions used, but also due to the nature of the material itself with self-compensation influencing the acceptor levels in the layer. Some of the defects associated with the compensation have been identified and their suspected origin in relation to the deposition conditions. These impurities can introduce deep donor/acceptor levels that can act as traps for both majority and minority carriers. This leads to greater recombination and reduced carrier lifetimes, causing a loss in the level of generated photocurrent and overall performance of the PV solar cell. The impurities are concentrated at the grain boundaries of the polycrystalline CdTe making grain size an important parameter in controlling the density of defect states, which is influenced by the deposition process employed. However, post-growth treatment using CdCl$_2$ and annealing results in recrystallisation and grain growth. It also causes inter-diffusion at the CdS–CdTe interface, removing some of the defects related to the lattice mismatch between the two layers including passivation of deep acceptor states through complex formation with the Cl$_{Te}^+$ shallow donor. Increase in the p-type character of CdTe whilst avoiding the compensation effects is required, coupled with control of defect density, in order to improve on current PV solar cell performances. High p-type doping is necessary for the formation of a back contact with good ohmic properties without the formation of a Schottky barrier that can restrict conduction of majority carriers. Due to the high work function of CdTe there is no suitable metal available without the use of a p$^+$-type layer to create tunnelling pathways for reducing the effects of the barrier at the back contact interface. Diffusion from the back contact metal and p$^+$-layer into the bulk CdTe needs to be prevented to avoid formation of deep level states. Stable back contacts during illumination and temperature cycling are required, with the Sb$_2$Te$_3$–Mo system possibly offering the best solution. Finally in this chapter, MOCVD is offered as a prospective technique for large-scale commercial production of CdTe solar modules, with discussion of the beneficial impact in reducing CdTe absorber thickness and the processing challenges associated with it.

References

1. D. Bonnet and H. Rabenhorst, presented at 9th IEEE Photovoltaic Specialist Conference, Silver Springs, 1972.
2. R. W. Birkmire and B. E. McCandless, *Curr. Opin. Solid State Mater. Sci.*, 2010, **14**, 139–142.
3. D. L. Bätzner, G. Agostinelli, M. Campo, A. Romeo, J. Beier, H. Zogg and A. N. Tiwari, *Thin Solid Films*, 2003, **431–432**, 421–425.
4. H. Zhao, A. Farah, D. Morel and C. S. Ferekides, *Thin Solid Films*, 2009, **517**, 2365–2369.
5. P. Nollet, M. Burgelman and S. Degrave, *Thin Solid Films*, 2000, **361–362**, 293–297.
6. A. E. Abken and O. J. Bartelt, *Thin Solid Films*, 2002, **403–404**, 216–222.
7. A. E. Abken, *Sol. Energy Mater. Sol. Cells*, 2002, **73**, 391–409.
8. J. Sites and J. Pan, *Thin Solid Films*, 2007, **515**, 6099–6102.
9. A. Nouhi, O. V. Meyers, R. J. Stirn and C. H. Lui, *J. Vac. Sci. Technol.*, 1989, **7**, 833.
10. H. C. Chou and A. Rohatgi, *J. Electron. Mater.*, 1994, **23**, 31.
11. J. Britt and C. S. Ferekides, *Appl. Phys. Lett.*, 1993, **62**, 2851.
12. T. Aramoto, S. Kumusawa, H. Higuchi, T. Aritta, S. Shibutani, T. Nishio, J. Nakajima, M. Tsuji, A. Hanafusa, T. Hibino, K. Omura, H. Ohyama and M. Murozono, *Jpn. J. Appl. Phys.*, 1997, **36**, 6304.
13. X. Wu, R. G. Dhere, D. S. Albin, T. A. Gessert, C. DeHart, J. C. Keane, A. Duda, T. J. Coutts, S. Asher, D. H. Levi, H. R. Moutinho, Y. Yan, T. Moriarty, S. Johnston, K. Emery and P. Sheldon, presented at NCPV Program Review Meeting, NREL/CP-520-31025, Lakewood, Colorado, 2001.
14. D. P. Halliday, J. M. Eggleston and K. Durose, *J. Cryst. Growth*, 1998, **186**, 543–549.
15. X. Wu, *Sol. Energy*, 2004, **77**, 803–814.
16. C. S. Ferekides, U. Balasubramanian, R. Mamazza, V. Viswanathan, H. Zhao and D. L. Morel, *Sol. Energy*, 2004, **77**, 823–830.
17. X. Wu, P. Sheldon, Y. Mahathongdy, R. Ribelin, A. Mason, H. R. Moutinho and T. Coutts, presented at NCPV Program Review Meeting, NREL/CP-520-25656, Denver, Colorado, 1998.

18. A. Bosio, D. Menossi, S. Mazzamuto and N. Romeo, *Thin Solid Films,* 2011, **519**, 7522–7525.
19. M. A. Green, K. Emery, Y. Hishikawa, W. Warta and E. D. Dunlop, *Prog. Photovoltaic,* 2013, **21**, 827–837.
20. First Solar, First Solar achieves 5 GW photovoltaic production milestone [online], www.firstsolar.com [accessed 14 December, 2011].
21. D. Bonnet, *Thin Solid Films,* 2000, **361–362**, 547–552.
22. J. Aguilar-Hernández, J. Sastre-Hernández, N. Ximello-Quiebras, R. Mendoza-Pérez, O. Vigil-Galán, G. Contreras-Puente and M. Cárdenas-García, *Sol. Energy Mater. Sol. Cells,* 2006, **90**, 2305–2311.
23. C. S. Ferekides, R. Mamazza, U. Balasubramanian and D. L. Morel, *Thin Solid Films,* 2005, **480–481**, 224–229.
24. A. Bosio, N. Romeo, S. Mazzamuto and V. Canevari, *Prog. Cryst. Growth Charact. Mater.,* 2006, **52**, 247–279.
25. K. Durose, P. R. Edwards and D. P. Halliday, *J. Cryst. Growth,* 1999, **197**, 733–742.
26. W. Shockley and H. J. Queisser, *J. Appl. Phys.,* 1961, **32**, 510.
27. M. Tsuji, T. Aramoto, H. Ohyama, T. Hibino and K. Omura, *J. Cryst. Growth,* 2000, **214–215**, 1142–1147.
28. Y. Yan and M. M. Al-Jassim, *Curr. Opin. Solid State Mater. Sci.,* 2012, **16**, 39–44.
29. Y. Yan, K. M. Jones, M. M. Al-Jassim, R. Dhere and X. Wu, *Thin Solid Films,* 2011, **519**, 7168–7172.
30. A. Podestà, N. Armani, G. Salviati, N. Romeo, A. Bosio and M. Prato, *Thin Solid Films,* 2006, **511–512**, 448–452.
31. D. Bonnet, in *Clean Energy from Photovoltaics*, ed. M. D. Archer and R. Hill, Imperial College Press, New York, 2001.
32. T. L. Chu, S. S. Chu, J. Britt, C. Ferekides and C. Q. Wu, *J. Appl. Phys.,* 1991, **70**, 2688.
33. J.-H. Lee, W.-C. Song, J.-S. Yi, K.-J. Yang, W.-D. Han and J. Hwang, *Thin Solid Films,* 2003, **431–432**, 349–353.
34. S. J. C. Irvine, V. Barrioz, D. Lamb, E. W. Jones and R. L. Rowlands-Jones, *J. Cryst. Growth,* 2008, **310**, 5198–5203.
35. E. W. Jones, V. Barrioz, S. J. C. Irvine and D. Lamb, *Thin Solid Films,* 2009, **517**, 2226–2230.
36. G. Kartopu, A. J. Clayton, W. S. M. Brooks, S. D. Hodgson, V. Barrioz, A. Maertens, D. A. Lamb and S. J. C. Irvine, *Prog. Photovoltaic,* 2014, **22**, 18–23.
37. J. D. Major, Y. Y. Proskuryakov, K. Durose, G. Zoppi and I. Forbes, *Sol. Energy Mater. Sol. Cells,* 2010, **94**, 1107–1112.
38. K. Durose, M. A. Cousins, D. S. Boyle, J. Beier and D. Bonnet, *Thin Solid Films,* 2002, **403–404**, 396–404.
39. V. Barrioz, D. A. Lamb, E. W. Jones, Y. Y. Proskuryakov, S. J. C. Irvine and K. Durose, presented at 23rd European Photovoltaic Solar Energy Conference, Valencia, Spain, 2008.

40. V. Barrioz, Y. Y. Proskuryakov, E. W. Jones, J. D. Major, S. J. C. Irvines, K. Durose and D. A. Lamb, presented at Materials Research Society Symposium, 2007, **1012**, Y12–08.

41. S. J. C. Irvine, V. Barrioz, A. Stafford and K. Durose, *Thin Solid Films*, 2005, **480–481**, 76–81.

42. R. Sudharsanan and A. Rohatgi, *Sol. Cells*, 1991, **31**, 143–150.

43. M. D. G. Potter, D. P. Halliday, M. Cousins and K. Durose, *Thin Solid Films*, 2000, **361–362**, 248–252.

44. M. A. Hernández-Fenollosa, D. P. Halliday, K. Durose, M. D. Campo and J. Beier, *Thin Solid Films*, 2003, **431–432**, 176–180.

45. M. Emziane, K. Durose, N. Romeo, A. Bosio and D. P. Halliday, *Thin Solid Films*, 2005, **480–481**, 377–381.

46. D. L. Bätzner, A. Romeo, M. Terheggen, M. Döbeli, H. Zogg and A. N. Tiwari, *Thin Solid Films*, 2004, **451–452**, 536–543.

47. M.-A. Arturo, *Sol. Energy Mater. Sol. Cells*, 2006, **90**, 678–685.

48. A. Gupta, V. Parikh and A. D. Compaan, *Sol. Energy Mater. Sol. Cells*, 2006, **90**, 2263–2271.

49. V. Plotnikov, X. Liu, N. Paudel, D. Kwon, K. A. Wieland and A. D. Compaan, *Thin Solid Films*, 2011, **519**, 7134–7137.

50. N. Amin, K. Sopian and M. Konagai, *Sol. Energy Mater. Sol. Cells*, 2007, **91**, 1202–1208.

51. T. M. Razykov, G. Contreras-Puente, G. C. Chornokur, M. Dybjec, Y. Emirov, B. Ergashev, C. S. Ferekides, A. Hubbimov, B. Ikramov, K. M. Kouchkarov, X. Mathew, D. Morel, S. Ostapenko, E. Sanchez-Meza, E. Stefanakos, H. M. Upadhyaya, O. Vigil-Galan and Y. V. Vorobiev, *Sol. Energy*, 2009, **83**, 90–93.

52. M. Hädrich, C. Kraft, C. Löffler, H. Metzner, U. Reislöhner and W. Witthuhn, *Thin Solid Films*, 2009, **517**, 2282–2285.

53. B. Dieter, *Thin Solid Films*, 2000, **361–362**, 547–552.

54. M. Xavier, *Sol. Energy Mater. Sol. Cells*, 2003, **76**, 225–242.

55. R. L. Rowlands, S. J. C. Irvine, V. Barrioz, E. W. Jones and D. A. Lamb, *Semicond. Sci. Technol.*, 2008, **23**, 15017.

56. Y. Y. Proskuryakov, K. Durose, J. D. Major, M. K. Al Turkestani, V. Barrioz, S. J. C. Irvine and E. W. Jones, *Sol. Energy Mater. Sol. Cells*, 2009, **93**, 1572–1581.

57. D. L. Bätzner, A. Romeo, H. Zogg, R. Wendt and A. N. Tiwari, *Thin Solid Films*, 2001, **387**, 151–154.

58. S.-H. Wei and S. B. Zhang, *Phys. Rev. B*, 2002, **66**, 155211.

59. D. J. Keeble, J. D. Major, L. Ravelli, W. Egger and K. Durose, *Phys. Rev.*, 2011, **84**, 174122.

60. B. T. Ahn, J. H. Yun, E. S. Cha and K. C. Park, *Curr. Appl. Phys.*, 2012, **12**, 174–178.

61. X. Wu, J. Zhou, A. Duda, Y. Yan, G. Teeter, S. Asher, W. K. Metzger, S. Demtsu, S.-H. Wei and R. Noufi, *Thin Solid Films*, 2007, **515**, 5798–5803.

62. S. H. Demtsu, D. S. Albin, J. R. Sites, W. K. Metzger and A. Duda, *Thin Solid Films*, 2008, **516**, 2251–2254.

63. T. A. Gessert, W. K. Metzger, P. Dippo, S. E. Asher, R. G. Dhere and M. R. Young, *Thin Solid Films,* 2009, **517**, 2370–2373.
64. M. Burgelman, J. Verschraegen, S. Degrave and P. Nollet, *Thin Solid Films,* 2005, **480–481**, 392–398.
65. V. Komin, B. Tetali, V. Viswanathan, S. Yu, D. L. Morel and C. S. Ferekides, *Thin Solid Films,* 2003, **431–432**, 143–147.
66. M. Terheggen, H. Heinrich, G. Kostorz, A. Romeo, D. Baetzner, A. N. Tiwari, A. Bosio and N. Romeo, *Thin Solid Films,* 2003, **431–432**, 262–266.
67. W. Jaegermann, A. Klein, J. Fritsche, D. Kraft and B. Späth, presented at Materials Research Society Symposium, 2005, **865**, F6.1.
68. Z. C. Feng, H. C. Chou, A. Rohatgi, G. K. Lim, A. T. S. Wee and K. L. Tan, *J. Appl. Phys.,* 1996, **79**, 2151.
69. J. Fritsche, T. Schulmeyer, A. Thißen, A. Klein and W. Jaegermann, *Thin Solid Films,* 2003, **431–432**, 267–271.
70. C. S. Ferekides, D. Marinskiy, V. Viswanathan, B. Tetali, V. Palekis, P. Selvaraj and D. L. Morel, *Thin Solid Films,* 2000, **361–362**, 520–526.
71. B. E. McCandless, I. Youm and R. W. Birkmire, *Prog. Photovoltaic,* 1999, **7**, 520.
72. A. Castaldini, A. Cavallin and B. Fraboni, *J. Appl. Phys.,* 1998, **83**, 2121.
73. A. Castaldini, A. Cavallini, B. Fraboni, L. Polenta, P. Fernandez and J. Piqueras, *Mater. Sci. Eng. B,* 1996, **42**, 302–305.
74. S. Vatavu, H. Zhao, V. Padma, R. Rudaraju, D. L. Morel, P. Gaşin, I. Caraman and C. S. Ferekides, *Thin Solid Films,* 2007, **515**, 6107–6111.
75. M. Hädrich, C. Heisler, U. Reislöhner, C. Kraft and H. Metzner, *Thin Solid Films,* 2011, **519**, 7156–7159.
76. D. L. Bätzner, M. E. Öszan, D. Bonnet and K. Bücher, *Thin Solid Films,* 2000, **361–362**, 288–292.
77. S. H. Demtsu and J. R. Sites, *Thin Solid Films,* 2006, **510**, 320–324.
78. Y. Roussillon, V. G. Karpov, D. Shvydka, J. Drayton and A. D. Compaan, *J. Appl. Phys.,* 2004, **96**, 7283.
79. T. A. Gessert, S. Asher, S. Johnston, M. Young, P. Dippo and C. Corwine, *Thin Solid Films,* 2007, **515**, 6103–6106.
80. B. Ghosh, *Microelectron. Eng.,* 2009, **86**, 2187.
81. B. Ghosh, S. Purakayastha, P. K. Datta, R. W. Miles, M. J. Carter and R. Hill, *Semicond. Sci. Technol.,* 1995, **10**, 71.
82. V. Barrioz, D. A. Lamb, E. W. Jones, Y. Y. Proskuryakov, S. J. C. Irvine and K. Durose, presented at 23rd European PV Solar Energy Conference, Valencia, Spain, 2008.
83. A. J. Clayton, S. L. Rugen-Hankey, W. S. M. Brooks, G. Kartopu, V. Barrioz, D. A. Lamb, S. D. Hodgson and S. J. C. Irvine, presented at 8th Conference on Photovoltaic Science Applications and Technology, Newcastle upon Tyne, UK, 2012.
84. G. Kartopu, A. J. Clayton, W. S. M. Brooks, S. D. Hodgson, V. Barrioz, A. Maertens, D. A. Lamb and S. J. C. Irvine, *Prog. Photovoltaic,* 2014, **22**, 18–23.

85. W. S. M. Brooks, S. J. C. Irvine, V. Barrioz and A. J. Clayton, *Sol. Energy Mater. Sol. Cells,* 2012, **101**, 26–31.
86. S. J. C. Irvine, D. A. Lamb, V. Barrioz, A. J. Clayton, W. S. M. Brooks, S. Rugen-Hankey and G. Kartopu, *Thin Solid Films,* 2011, **520**, 1167–1173.
87. W. S. M. Brooks, S. J. C. Irvine and V. Barrioz, *Energy Procedia,* 2011, **10**, 232–237.
88. V. Barrioz, S. J. C. Irvine, E. W. Jones, R. L. Rowlands and D. A. Lamb, *Thin Solid Films,* 2007, **515**, 5808–5813.
89. W. Xia, H. Lin, H. N. Wu and C. W. Tang, *Thin Solid Films,* 2011, **520**, 563–568.
90. A. J. Clayton, S. J. C. Irvine, E. W. Jones, G. Kartopu, V. Barrioz and W. S. M. Brooks, *Sol. Energy Mater. Sol. Cells,* 2012, **101**, 68–72.
91. A. J. Clayton, S. Babar, M. A. Baker, G. Kartopu, D. A. Lamb, V. Barrioz and S. J. C. Irvine, presented at 28th European Photovoltaic Solar Energy Conference, Paris, 2013.
92. First Solar, *Annual report [online],* www.firstsolar.com [accessed 12 April, 2012].
93. G. M. Hanket, B. E. McCandless, W. A. Buchanan, S. Fields and R. W. Birkmire, *J. Vac. Sci. Technol.,* 2006, **A24**, 1695.
94. V. Barrioz, E. W. Jones, D. Lamb and S. J. C. Irvine, presented at Materials Research Society Symposium Conference, 2009, **1165**, M07–03.
95. V. Barrioz, G. Kartopu, S. J. C. Irvine, S. Monir and X. Yang, *J. Cryst. Growth,* 2012, **354**, 81–85.
96. C. Candelise, M. Winskel and R. Gross, presented at 26th European Photovoltaic Solar Energy Conference and Exhibition, Valencia, Spain, 2011.
97. V. Barrioz, A. Clayton, W. S. M. Brooks, X. Yang, S. J. C. Irvine and S. Rugen-Hankey, presented at 25th European Photovoltaic Solar Energy Conference/5th World Conference on Photovoltaic Energy Conversion, Valencia, Spain, 2010.

CHAPTER 6

New Chalcogenide Materials for Thin Film Solar Cells

DAVID W. LANE[*a], KYLE J. HUTCHINGS[a],
ROBERT McCRACKEN[a], AND IAN FORBES[b]

[a]Cranfield University, Defence Academy of the United Kingdom, Shrivenham, SN6 8LA, UK; [b]NPAC, School of Computing Engineering and Information Sciences, Northumbria University, Ellison Place, Newcastle upon Tyne NE1 8ST, UK
*E-mail: D.W.Lane@cranfield.ac.uk

6.1 Introduction and Background

Photovoltaic (PV) technologies were developed for terrestrial energy generation during the 1970s as a result of the oil crisis. The dependence of the western economies on fossil fuels was highlighted and this vulnerability demonstrated a need to address the security of energy supply. During this period the potential threat to the world was beginning to become apparent, especially to two of the major energy dependent economies that had few fossil fuel resources, Japan and West Germany. The following three decades saw the development of silicon terrestrial PV and the research into thin film amorphous silicon, cadmium telluride and copper indium diselenide based technologies move towards commercialisation.

One of the main barriers to photovoltaics being implemented as a major energy generation technology is its relatively high cost. Thin film inorganic photovoltaics is a proven technology that offers significant reductions in cost compared with crystalline silicon which is the current dominant market PV

RSC Energy and Environment Series No. 12
Materials Challenges: Inorganic Photovoltaic Solar Energy
Edited by Stuart J C Irvine
© The Royal Society of Chemistry 2015
Published by the Royal Society of Chemistry, www.rsc.org

technology. Of these, copper indium diselenide (CIS) based photovoltaics and in particular its variant, copper indium gallium diselenide (CuInGaSe$_2$ or CIGS), has yielded the highest laboratory efficiency of all the thin film technologies. Currently both cadmium telluride (CdTe) and CIGS based photovoltaics are now produced at the gigawatts peak (GWp) scale and their cost reductions and potential for thin film photovoltaics is starting to be achieved.

During the period of these developments, the threats of climate change and security of energy supply have become increasingly apparent and the need to find alternative sources of energy ever more urgent. The growth of developing economies such as China, India and Brazil and the future need to meet the aspirations of countries in Africa, Central and South America, and Asia have resulted in a future need for the primary energy of tens of terawatts by the end of the 21st century. As the photovoltaic market develops the current CdTe and CIGS technologies are making a significant contribution and this is expected to continue. However, the scale of energy supply needs makes it necessary to consider sustainability of any future energy source.

The solar resource is by far the largest source of energy available and is also the most widely geographically distributed. However, the resources that are required for photovoltaics to convert the solar resource at the required scale must be considered and therefore emphasis must be placed on sustainability concerns relating to both the materials and the processes used. By definition the thin film technologies use relatively small amounts of material. Various studies have examined components of PV modules and their costs and abundance.[1-5] Substrate costs represent a major cost factor, though various routes to reduce this are being considered. These include the use of foils and other materials, and can be expected to be common to any thin film technology.

In general, therefore, it is the active materials that represent the most critical components where abundance and cost are concerned. Typically, thin films absorber layers are a few micrometres thick which corresponds to a volume of one or two cubic centimetres for each square meter of photovoltaic panel. The following simple calculation shows the mass of CIGS that is needed for each GWp of installed capacity:

$$\text{Mass of CIGS for 1 GW installed capacity} =$$
$$\text{Area in m}^2 \text{ equivalent to the installed 1 GW capacity} \times \text{Mass per m}^2 \quad (6.1)$$

If a 2 μm thick CIGS absorber layer is used within the module, then the volume, in cm^3, of CIGS per m^2 would be 2 cm^3, and using a density value of 5.77 g cm^{-3}, the mass per m^2 would be 11.54 g. Assuming a module conversion efficiency of 15% and calculating for standard measurement conditions (1000 Wm^{-2}, 25 °C, AM1.5), each m^2 yields 150 W.

Assuming 1000 Wm^{-2} insolation, 1 GW of installed capacity $= 10^9$ W.

Area in m^2 required $= 10^9/150 \cong 6.7 \times 10^6$ m^2 (6.7 km^2).

Mass of CIGS in 1 GW installed capacity $\cong (6.7 \times 10^6 \times 11.4)$ g $\cong 77.3$ tonnes.

The mass of each element can be estimated using the number of that element's atoms per molecule together with the ratio of atomic to molecular weight and the mass of the absorber.

A general approach that can be used to estimate the potential for PV materials is to consider the minerals resource and translate this into an equivalent scale of deployment or production in the context of estimated future energy demand. Estimates of the abundance of the particular element in the Earth's crust could be used as an indicator of the amount of that element available. However, the crustal abundance is an estimate that does not provide any indication of the distribution in the Earth's crust. To be economically viable for exploitation the elements need to be available in sufficient concentration. The term 'resources' is used to provide a better estimate and takes account of the concentration of the particular elements within exploitable minerals either at the present time or at some time in the future depending on the economic drivers. Other terms that attempt to define the availability of a mineral to be exploited includes the 'reserves base' and mineral 'reserves'. A definition of these terms is given in by the US Geological Survey (USGS).[6]

Where they are available, the reserves base or resources can be used as a very crude upper limit to the amount of a material that is available. However, these are not definitive and there are other factors, such as a given mineral being a byproduct of extracting a more abundant mineral, that become important in determining the supply of such minerals. There are numerous factors that determine the amount of a particular mineral that can be economically exploited and these will all have an impact on the availability of photovoltaic materials. Assuming such resources are available for economic exploitation, then a more reliable metric would be to consider the current world production levels of a material. Where the proposed use, in this case PV thin films, would be expected to take up a significant fraction of this production, then it would be reasonable to assume that the need for additional mining and refining would act as an obstacle to development that would be likely to result in disruption to the market and variations in price. Such an issue was demonstrated during the growth of photovoltaic production in the first decade of the 21st century when a shortage of silicon wafers resulted in increased cost and limited the supply of crystalline silicon based modules.

A more reliable metric to assess the potential of a material to satisfy a terawatt (TW) photovoltaic market and benefit from the cost reductions associated with large-scale production would be to consider production facilities of 1 GWp annual output in the context of annual mineral production for the component material. The area of PV material produced will vary depending on the performance of the technology and the quantity of material used will depend on this and technology choice. For CIGS and CdTe the global production of metals and chalcogens, together with an estimate of the mass needed for 1 GWp production, is illustrated in Figure 6.1.

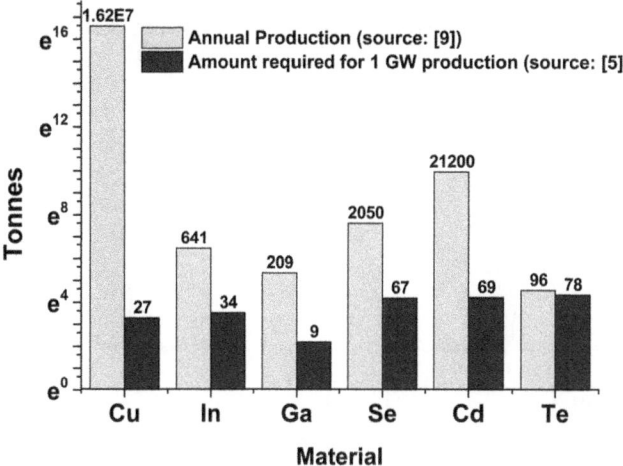

Figure 6.1 Comparison between annual production of selected materials with consumption of those materials at a production level of 1 GWp per annum (*For sources [5] and [9], please see references of the same numbers in this chapter, respectively*).

Figure 6.1 illustrates a number of features including the incomplete nature of the available data. Considering the case for tellurium, the USGS figures imply that more than the total annual production is used for the production of 1 GWp of CdTe. First Solar produced more than 1.8 GW in 2012[7] and claimed over 1.5 GWp production in its 2010 annual report. The figures for global tellurium production do not include figures that were either withheld or not available, and the figures published indicate an annual production of more than 90 tonnes according to the USGS[8] or approximately 100 tonnes according to the British Geological Society (BGS).[9] There is no indication of 'missing data' for the other elements and their figures can be considered more reliable than for tellurium.

From the figures available, if the entire annual production of indium, gallium and selenium were used for CIGS devices then between approximately 100 GWp and 1000 GWp could be produced. However, indium and gallium are both byproducts of refining other metals. Indium is most commonly found in zinc sulfide (Spalerite ore) and is extracted when in concentrations of between 1 and 100 ppm. Gallium can be produced from processing zinc and bauxite, and is extracted when in concentrations >50 parts per million (ppm). Whilst there is estimated to be significantly more gallium than indium that can be extracted, it is unlikely that this will occur independently from the exploitation of bauxite.[6] These materials are also used for other products such as transparent conducting oxides (TCOs) in the case of indium (indium tin oxide, ITO) and light-emitting diodes (LEDs) and other semiconductor devices based on GaAs, in the case of gallium. In addition, these refinery figures do not take into account the purification needed for using the materials in solar

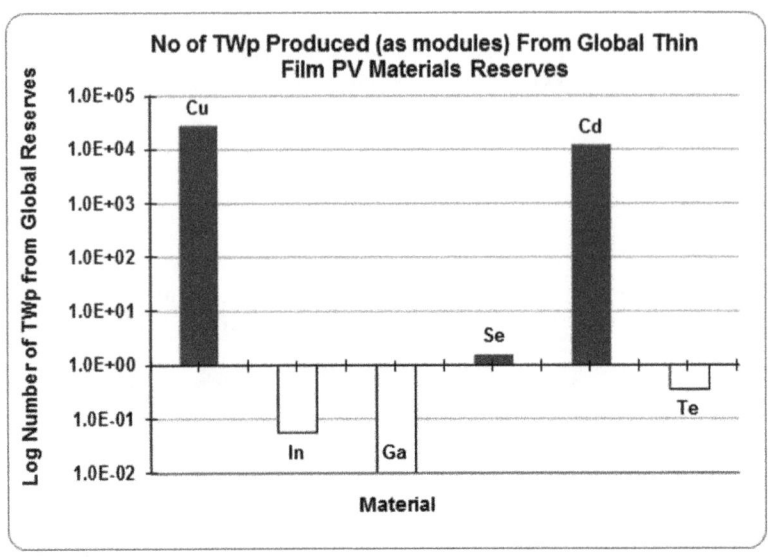

Figure 6.2 Number of TWp produced as modules from global thin film PV materials reserves (filled – TWp production, greater than global material reserves; white – TWp production, less than global material reserves).

cells. Therefore significant breakthroughs in materials extraction and increase in production facilities would be needed if these materials alone are to be able to satisfy a multi-TW photovoltaic market. If the material's reserves are considered—as known resources that can be exploited economically at present—then these can be compared with the requirements for a multi-TWp market. Figure 6.2 shows the number of TWp that could be produced from the reserves as indicated by the USGS.

Figure 6.2 is for illustration purposes and the Indium Corporation considers that the amount of indium available through all sources, including recycling, is in the region of 50 000 tonnes—nearly three times the last published data from the USGS.[10] However, assuming the Indium Corporation's figures can be achieved then indium appears to remain a limiting factor for CIGS-based photovoltaics. These figures imply that the resources will limit CdTe and CIGS photovoltaic technologies to multi-GWp production levels. Assuming that there is no change in production capacity for gallium then this is considered to limit annual maximum production of CIGS to ~26 GWp.[11] However, the estimates used in this study were based on conservative efficiency values for CIGS that are currently being achieved or exceeded in production. In all inorganic PV technologies, especially those based on inorganic thin films, the efficiency values would be expected to increase during the 21st century. However, even with these increases, it is clear that the current generation of materials will struggle to satisfy the predicted demand for solar power.

The figures for reserves are those that are often used in calculations and are based on USGS reports that include estimates for reserves and resources. Figures 6.1 and 6.2 indicate that the most sensitive materials as indicated by their production and reserves are indium and tellurium, and to a lesser extent selenium. Gallium is a high-cost, low-production volume material and the figures for reserves are not given. Estimates for reserves are not equivalent to those for other relatively scarce materials. There are very large reserves of bauxite known, but as byproduct of the exploitation of these, gallium while not likely to be limited in quantity is likely to be limited by the extraction rate. It is not economically viable to exploit bauxite more rapidly than the demand for aluminium dictates. The current distribution of indium indicates that the production is overwhelmingly concentrated within China. The Indium Corporation, however, considers that at least half of its estimated ~50 000 tonnes is available outside China.

This discussion has focused on the abundance and production of these materials. A further factor that is becoming increasingly important for many materials is the security of supply. A number of studies have considered materials that have economic significance to the UK, European Union and the USA,[12] and both from USGS and BGS data on the mining and refining of metals, it is clear that the source of these materials should be considered when projecting their use into future large-scale manufacturing. In particular, the proportion of minerals that is dominated by production based in China needs to be considered. As an emerging super economic power with a rapidly increasing need for its own energy generation, materials originating in China will be increasingly used within that country. These strategic factors are likely to be an increasingly important factor for materials choices and it will be necessary to ensure that vulnerability to fossil fuel supplies is not replaced by a new vulnerability to materials supply.

CIGS-based devices, at over 20% conversion efficiency, have demonstrated the highest performance of any thin film technology. The technology is being developed using a range of processing techniques for commercial production. Solar Frontier has a two-stage vacuum based fabrication facility in Japan which will have a total capacity of about 1 GWp and a mini-module efficiency record of over 17%. The technology has a significant potential to contribute to the cost reduction and development of terrestrial PV solar electricity generation for many decades to come. However, the potential global PV market is predicted to reach the multi-TWp scale by 2050, and CIGS and CdTe are unlikely to be able to meet more than a few hundreds of GWp.

Various technologies have been proposed as capable of meeting a future multi-TWp photovoltaic market, including dye-sensitized technologies and organic photovoltaics (OPV). The former is, at present, limited by ruthenium and the latter is considered as a practical solution based on the abundance of carbon. However, both technologies will need to overcome long-term stability issues if they are to contribute to the power generation market. OPV is considered to be likely to be confined to the small power consumer market for the next 15 years or more. All current OPV require inorganic materials such as indium tin oxide (ITO) transparent conducting oxide contact layers. It has been suggested by proponents of OPV that short lifetimes are overcome by the very low potential cost of the materials. Assuming the cost of the materials is reduced as the scale of production increases, the cost of the photovoltaics represents only part of the system cost and frequent replacement of the panels will incur additional dismantling and installation costs. At present there are significant stability challenges to overcome and such materials will need to demonstrate 20 year lifetimes to be taken seriously as power generating technologies because approximately 30% of the system cost is related to the installation.

The application of crystalline silicon photovoltaics is not considered to be limited by the active absorber layer material because silicon is one of the most abundant materials on Earth. Most studies consider silicon to be limited by the silver used in the contacts.

The success of CIGS technologies, together with the concerns about the dependency on materials with low availability and abundance, has stimulated interest in alternative materials in the same family of inorganic compounds. These include $Cu_2ZnSn(Se,S)_4$ Sn–S based materials, together with Cu–Sb and Cu–Bi based compounds, with some researchers revisiting CuO and PbO based absorbers that were investigated during the 1960s and 1970s. Of these materials the most promising to date are the $Cu_2ZnSn(S,Se)_4$ (CZTS) kesterite materials that have demonstrated laboratory conversion efficiencies of over 12%.[13] Intuitively, CZTS-based devices utilise materials with significantly lower cost and greater abundance than those based on CIGS. Copper pipes and electrical cables are found in most buildings, while zinc is used on as galvanic protection for steel streetlights, fencing and corrugated iron roofing. Tin is used in tinplate and is produced in volume for cans. Considering the materials that form CZTS in the context of PV and future TWp markets, however, it becomes clear that these materials have great potential for a sustainable stable thin film technology. This illustrated in Figure 6.3 which shows the percentage of annual materials production of each component materials consumed in the production of 1 GWp PV, with the constituents of CZTS PV presenting a far lower demand than those for existing inorganic thin film technologies.

In this chapter we explore alternative inorganic materials for thin film solar cells. These are predominantly chalcogenides, or materials based on the group 16 elements O, S, Se and Te. This family of materials has been considered for many years with early studies including the oxides† (*e.g.* Cu_2O and PbO)[14-18] and the sulfides (*e.g.* CdS and FeS_2).[19,20] There have, of course, been some very notable successes such as CdTe and CIGS. However, the sustainability and toxicity of both of these materials presents potential obstacles to their continued and expanded use, with long-term planning subject to geopolitical uncertainties such as the availability of indium

---

†Neglecting the extensive work on oxides that act as transparent conductors for electrical contacts to solar cells and other optoelectronic devices.

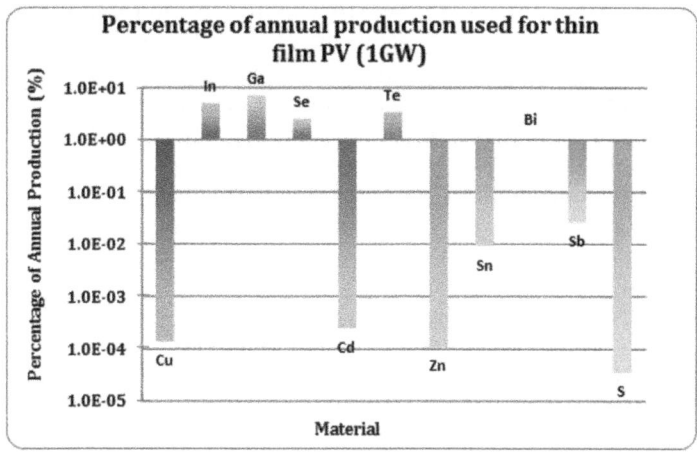

Figure 6.3   Percentage of annual production used for 1 GW of thin film PV (for each material).

(discussed above) or wide-ranging legislation against the use of cadmium on the grounds of its toxicity.[21]

On the most basic level identifying a suitable alternative material is simply a matter of finding those with an appropriate bandgap, around 1.5 eV for a single junction device.[22] However, growing a material in a form suitable for devices presents many challenges. For example, while FeS_2 has been identified as the most promising candidate material for large-scale photovoltaics based on cost and availability,[23] this was based on achieving its theoretical conversion efficiency of ~32% as indicated by its bandgap (0.95 eV). However, the current record efficiency for solar conversion based on a FeS_2 device is only 2.8% for a photoelectrochemical cell,[24] with solid state thin film devices not performing as well. Therefore, while FeS_2 is theoretically a strong candidate for future thin film PV, in reality there is much work remaining to be done.

As we have seen in the previous chapters of this book, increasing the efficiency of a particular device technology generally takes several decades of intense research on a global scale, with sudden step changes rare. Therefore moving to a relatively unexplored material presents a significant risk as far as meeting our immediate demands for low cost renewable energy, even if the prospects for low toxicity and sustainably are good. Potentially there would be far less risk associated with modifying an existing high performing technology, such as CIGS, to improve its credentials in these areas. In principle this would involve substituting the indium and gallium with other more desirable elements, such as zinc and tin. The resulting material's similarity to CIGS would hopefully mean that some of the lessons learnt during CIGS development could be transferred over, speeding up development of the new material. However, this would only work up to a point, so to increase the rate of development of the new material more efficient ways of working would have to be developed, such as the high throughput or combinatorial approach that effectively removes time-consuming process stages from a materials investigation and potentially improves experimental reproducibility.

The following sections introduce the concept of combinatorial synthesis of experimental samples and the resulting high throughput exploration of new materials for absorber layers in thin film solar cells. In the sections that follow new chalcogenide materials, loosely based upon CIGS, are described. These include $Cu_2ZnSn(Se,S)_4$, $CuSb(S,Se)_2$ and Cu_3BiS_3.

6.2 Investigating New Materials

6.2.1 Conventional *versus* High Throughput Techniques

Thoroughly investigating a new material, semiconductor or otherwise, is a protracted process. Small changes in composition and the method of growth can have a major impact. Conventional experimental methods where an individual specimen is grown at each change of composition or process are expensive in both time and labour, and the need to create a manageable number of specimens can mean that important changes are either missed or are poorly resolved.

Process stages that can affect the time required to grow an individual specimen include.

- cleaning individual substrates;
- loading and unloading each substrate into the deposition system;
- evacuation of a vacuum chamber;
- heating and cooling times;
- mixing precursor chemicals;
- deposition time at each composition.

The high throughput or combinatorial approach reduces process time by compressing a wide range of (usually) material compositions into a single specimen, effectively eliminating many of the stages listed above by growing specimens in parallel.[14,25] This can be achieved by several different methods, but can be as simple as using multiple deposition sources to deliberately grow a thin film with a compositional gradient across its surface. This film can be interrogated at any location that the experimenter chooses, with each individual location being equivalent to a single convention specimen.

The 'combinatorial specimen' can therefore be considered to contain an almost endless number of 'conventional specimens' and it is therefore often referred to as a 'library'. The experimenter can examine a particular location in the library and then return to that same location at a future time after the library has been processed, for example by annealing, and determine how its properties have been changed. Because the location is arbitrary, they can also move to any alternative location, should they wish to examine a particular feature. In this case, if the conventional approach was followed it would require the growth of a completely new specimen. In addition, all of the 'specimens' within the combinatorial library have been grown under the same conditions, which offers improved process control compared those grown conventionally.

A composition gradient can be created using the inherently non-uniform flux from most types of deposition source (*e.g.* sputter sources, thermal evaporation sources, spray nozzles)[26] and the careful alignment of the deposition geometry. Alternatively, either a translating shutter that gradually exposes the substrate to the deposition source, or removing or immersing the substrate into a deposition bath can be used. While these methods benefit from simplicity and flexibility, a composition gradient can present a problem during analysis or in the construction of libraries of discrete PV devices. This problem can been overcome by creating a library that has a mosaic structure with different uniform composition at each location. This has been done in practice by using spray pyrolysis and a moving heated substrate, and by using a complicated process with repeated deposition through different fractal masks.[27,28]

6.2.2 One- and Two-dimensional Libraries

The application of thin film technology to the growth of individual layers over the large areas required for PV panels usually demands films with a highly uniform thickness and composition. This can present a major challenge, especially when growing thin films from multiple sources, and is made particularly difficult by the non-uniform flux of material emitted by most types of deposition source. Sources have to be very carefully aligned to ensure a uniform composition and ultimately, complicated substrate rotation systems[‡] may be needed to create a uniform thickness. This 'problem' can be used to the experimenter's advantage when creating a combinatorial library, as the non-uniform flux produced by a source can be used to create the compositional gradient across the library, with multiple sources deliberately operated out of alignment. And while the ability to design and grow a film of a particular thickness and composition gradient assists with experimental design, it is not essential as the library is fully characterised before each subsequent processing stage.

Let us consider the growth of a library on a 75 mm × 25 mm substrate (such as a glass microscope slide). The growth of a bimetallic alloy (AB_x) requires two sources and would create a one dimensional library along direction x as shown in Figure 6.4(a). Repeated analysis of the library at intervals along its length is used to reveal the compositional gradient, for example, intervals in x of 5 mm would make the library equivalent to 14 separate specimens. This is illustrated in Figure 6.4(b) for the metallic

---

[‡]These are of only practical for experimental systems, not the large areas of commercial thin film solar cells.

(a)

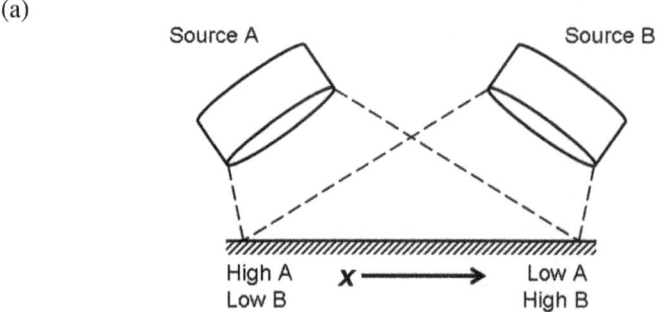

(b)

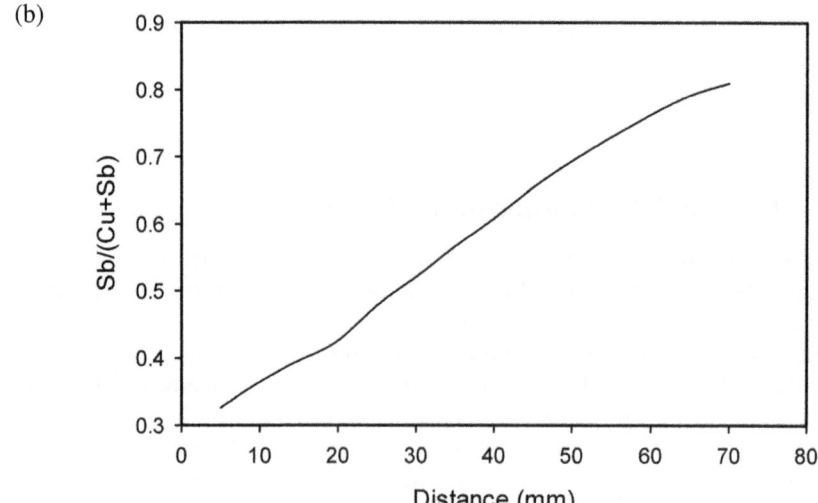

(c)

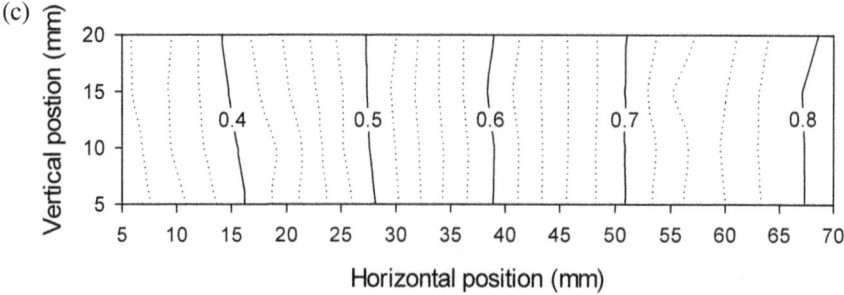

Figure 6.4 (a) Arrangement for producing a one-dimensional library by DC magnetron sputtering. (b) Sb to (Cu+Sb) profile for a $CuSbi_x$ library. (c) Sb to (Cu+Sb) map for the same library.

precursor for the semiconductor $CuSbS_2$, with a $CuSb_x$ alloy library grown by one of the authors (McCracken) by co-deposition by DC magnetron sputtering, with the film composition characterised by the Sb to (Cu+Sb) atomic ratio. Before deposition a quartz crystal microbalance was placed immediately in front of each source and used to measure the deposition rate as a function of power. This was used to establish the power settings for each source to create the desired composition in the centre of the library. The film composition was measured using scanning electron microscopy (SEM) with energy dispersive X-ray analysis (EDXA), giving the profile shown. While the compositional gradient is largely along the direction of the slide, the spread of the flux from the sources does result in a slight gradient in a perpendicular direction [see Figure 6.4(c)].

The range of compositions covered within a library depends on two factors: the compositional gradient and the physical size of the substrate. To a large extent the gradient will be limited by the size of the source and achieving the desired range of compositions can be difficult. In this case the flux from the source can be modified by the inclusion of a shield or baffle. This is illustrated in Figure 6.5(a) where the shield partially blocks the flux from the crucible and produces a diffuse shadow or penumbra across the substrate similar to that produced from an extended light source. Substrate position 1 has direct line of sight to the whole of the source and will receive the maximum flux, whereas position 2 only has direct line of sight to half of the source and consequently the flux will halve. As one proceeds from position 1 through position 2 to position 3, the area of the source masked by the shield increases until it is completely blocked. Figure 6.5(b) presents the composition profile for a library of the candidate intermediate bandgap material Cr doped ZnS, which was deposited using two alumina crucible evaporation sources, one partially cover by a shield. The compositional gradient is far greater than without the shield, allowing a wide range of Cr doping levels to be included on a single substrate.

The growth of a bimetallic precursor layer opens up three possible deposition sequences. Co-deposition where the metals are deposited simultaneously and sequential deposition where they are deposited in turn, either Cu–Bi or Bi–Cu. DC magnetron sputtering is well suited to co-deposition, although other techniques, such as electrodeposition, may only be able to deposit a single element at a time. Electrodeposition is a low temperature, non-vacuum technique in which the metal is deposited from an aqueous precursor.[29] It is inexpensive and capable of producing uniform layers over large areas. However, the electrochemistry of the process means that it can be very difficult to co-deposit alloys. Figure 6.6 shows the experimental arrangement for growing a Cu–Bi library by sequential electrodeposition,[30] where a thickness gradient was created by gradually raising the level of the electrolyte during deposition. The substrate was then rotated through 180° and the process repeated for the second element. It should be noted that

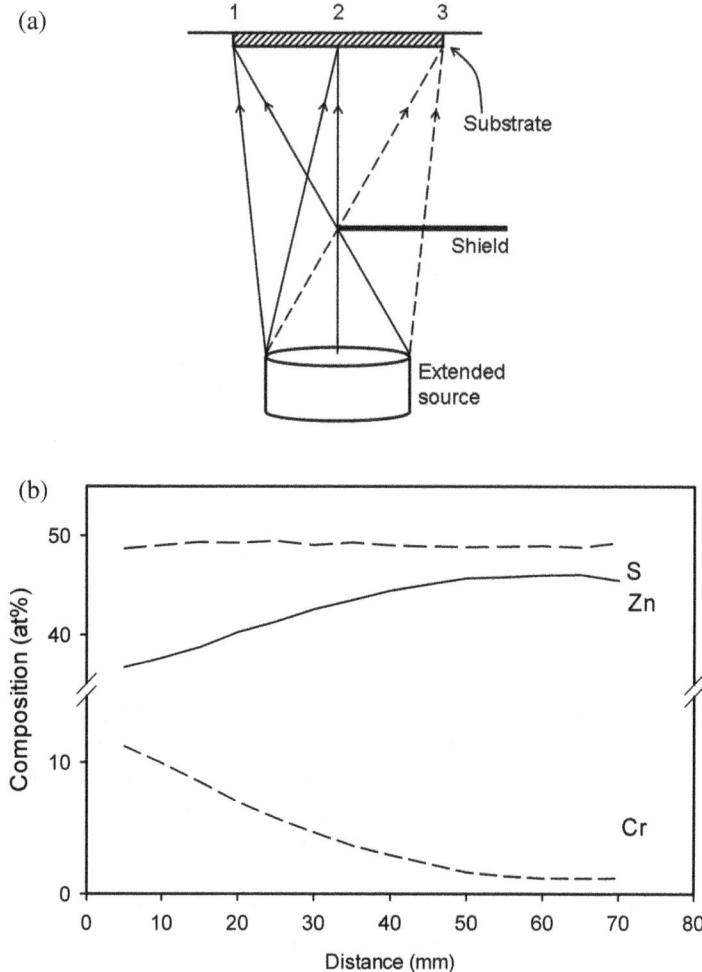

Figure 6.5   (a) Inclusion of a shield in an evaporation system. (b) Cr concentration profile produced for a ZnS(Cr) library.

electrodeposition requires an electrically conductive substrate, in this case sputtered Mo on glass.

   The growth of more complicated materials, such as CZTS, require ternary precursors and generally three sources that can be examined through a two dimensional library of type AB_xC_Y. Following our earlier example this could be achieved using three substrates placed side by side to give a 75 mm × 75 mm library. Repeated analysis of this library over a grid of spacing 5 mm would be equivalent to 196 individual specimens. Figure 6.7(a) shows the experimental arrangement to produce such a library using three Torus sputter sources.[31] This type of source is well suited to combinatorial studies as it can be equipped with a flexible mount that allows the magnetron to be tilted to the

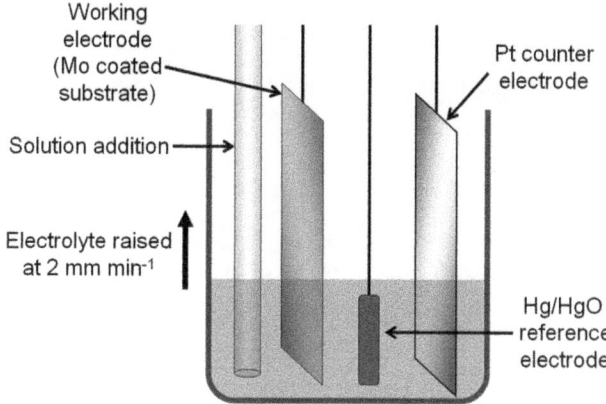

Figure 6.6 Arrangement for producing a one-dimensional library by electrode-position.

angle required for the desired compositional gradient, and it can be rotated on its support tube to achieve the desired alignment. This can be assisted by fixing a small laser to its target and noting where the beam strikes the substrate surface. The deposition process is very similar to that for the one-dimensional library, although many more deposition sequences are now possible. These include the co-deposition of all three metals and sequential deposition in the order A-B-C, A-C-B, B-A-C, B-C-A, C-A-B, C-B-A and sequential deposition of binary alloys by co-deposition. As for binary alloys, the deposition sequence can prove highly influential for subsequence annealing and conversion to the sulfide or selenide. Figure 6.8(b) and (c) shows the range of compositions achieved for a Cu–Zn–Sn precursor library for the investigation of CZTS. In this case the (7.5 cm × 7.5 cm) library was deposited over three unheated microscope slides by co-sputter deposition as described above. The ratios illustrate the targeted stoichiometric regions for Cu_2ZnSnS_4 (Zn : Sn = 1 and Cu : (Zn+Sn) = 1) forming distinct central bands that cross, with a spatial deviation either side spanning the approximate range from near zero to 4.5 across the whole library.

6.2.3 Mapping Libraries

The application of high-throughput techniques to thin film PV materials usually involves correlating the chemical composition at each point within a library with its corresponding structural, optical and electrical properties. These can be measured using a range of different analytical techniques, and the ultimate precision to which the library is investigated will depend on many factors, not least the analysis area required for a particular technique and the time it requires at each location. Mapping is generally done sequentially with each location examined in turn, with the total mapping time being equal to the sum of the analysis times at each point plus the total

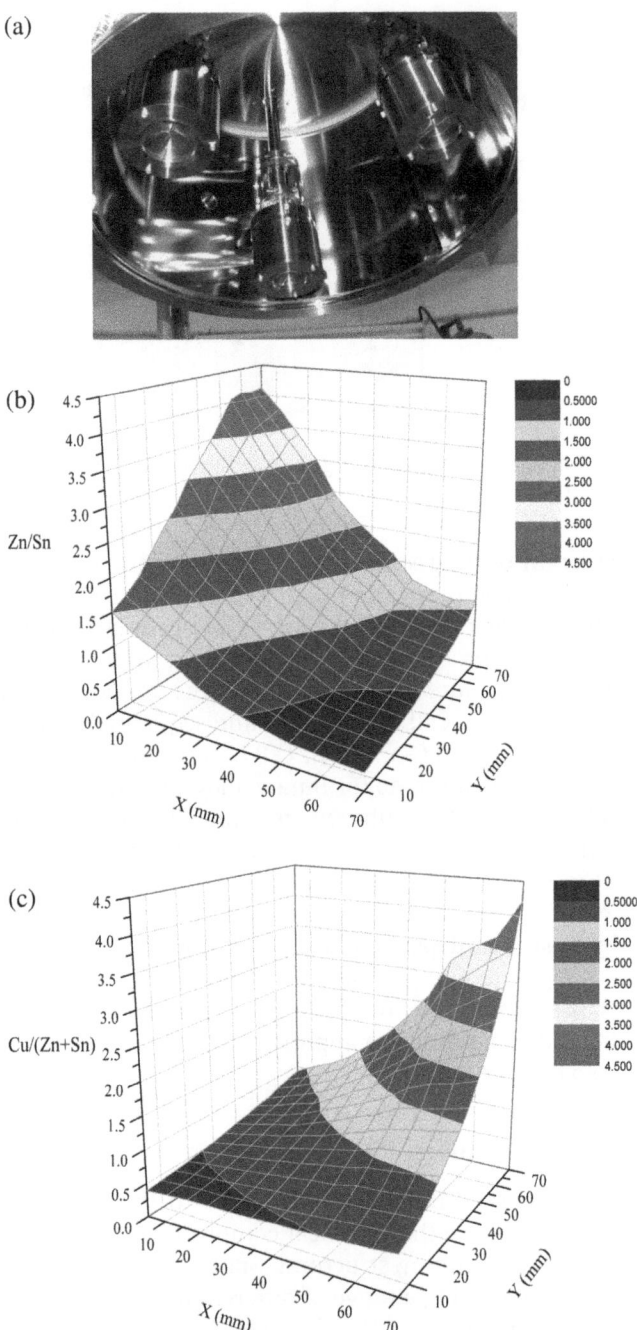

Figure 6.7 (a) Three 2 inch Torus sputter sources supplied by Kurt J. Lesker Ltd for a ternary alloy library. (b) and (c) show elemental ratio contour plots for a single Cu–Zn–Sn library grown by co-sputter deposition for Zn/Sn and Cu/(Zn+Sn), respectively.

time required moving between points. A small probe size is particularly desirable because the properties of the material inherently change over the library's surface and analysis can therefore be considered to be averaged over the probe area. However, a small probe area often means a weaker detected signal, reducing the signal to noise ratio and increasing the required data collection (integration) time if high quality data are to be maintained. This increases the overall mapping time.

The direct comparison of material properties clearly requires the ability for each analytical technique to reliably return to exactly the same location. Therefore careful attention should be paid to accurate library registration and the precision and accuracy to which each location is reached, for example, by ensuring that motors are driven in the same direction on the approach to each point to avoid backlash. The probe areas for each analytical technique should also have similar sizes to ensure averaging effects remain the same.

The analysis of the library composition is frequently done by X-ray microprobe analysis in a scanning electron microscope.[32] This technique benefits from a potentially small probe size and in vacuum analysis which allows the detection of low energy X-rays from lighter elements (typical range O to U). The effects of overlapping X-ray lines from different elements is becoming less of an issue as the resolution of energy dispersive detectors improves, and is generally not an issue for high resolution wavelength dispersive detectors. Alternative techniques for determining a library's chemical composition include X-ray fluorescence (XRF)[33] and ion beam analysis (IBA).[34] Many XRF systems have a mapping capability, although the probe beam size (>100 μm) is generally much greater than that found in an SEM ($\sim$1 μm), although this need not be an issue if the composition gradient within the library is low. As for all techniques the sensitivity of a measurement will depend on the experimental conditions and, if analysis is performed in air, attenuation will reduce the sensitivity to light elements (typical range Mg to U). IBA probes range from mm size down to $\sim$1 μm for microbeam systems. Heavy elements can be mapped by proton induced X-ray emission (PIXE) and Rutherford backscattering spectrometry (RBS), which also gives depth information, and light isotopes (elements H to S) can be mapped by nuclear reaction analysis (NRA).

While the topographical structure of the surface of a library can be readily examined by SEM, its crystalline structure and the identification of specific crystallographic phases requires a technique such as X-ray diffraction (XRD).[35] This can also be mapped across the surface of a library by sequentially analysing individual points. The use of capillary optics also allows the incident X-ray beam to be focused to a small spot ($\sim$50 μm) and data collection times can be greatly reduced by using an area detector to simultaneously collect the whole X-ray powder diffractogram in a matter of minutes. Careful examination of the pattern of scattered X-rays recorded by the area detector will provide addition valuable information about grain size (relative to the beam diameter) and preferred orientation in the growth of the

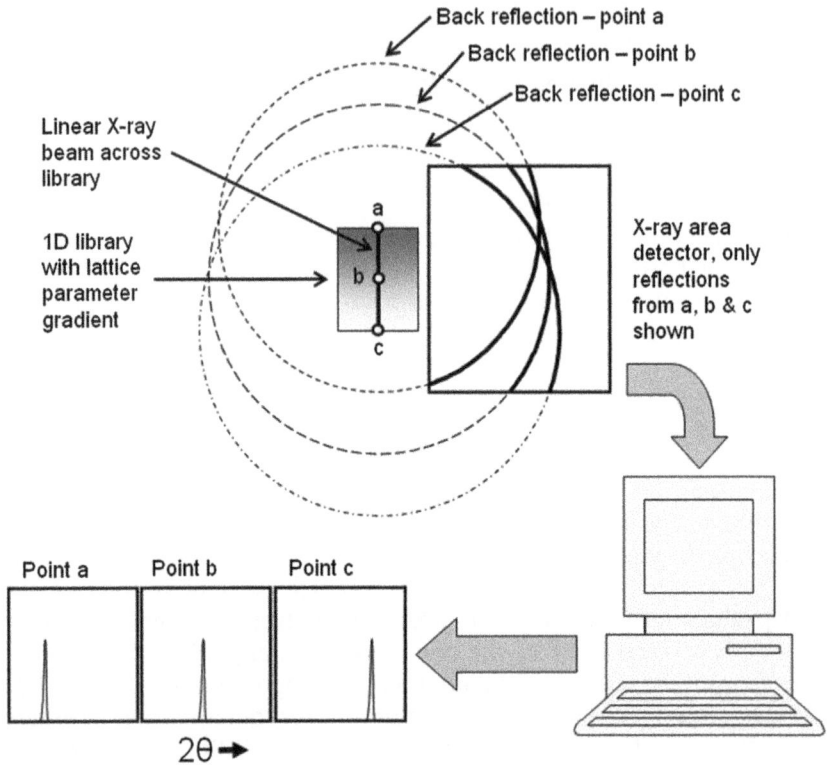

Figure 6.8   High-throughput XRD using a linear X-ray beam directed along a one-dimensional library (diffraction from three points shown). Analysis of a two-dimensional library is achieved by sweeping the linear X-ray beam across the library.

material. Recently the problem of long mapping times due to the adoption of sequential, point-by-point analysis has been addressed by sweeping a linear X-ray beam (as opposed to a discrete spot) across a library and using a mathematical algorithm to extract the X-ray powder diffractograms from separate points along the beam (see Figure 6.8).[36] The resulting two-dimensional array of diffractograms is then analysed by conventional means. For some materials, such as CZTS, it may be necessary to use a complementary technique to confirm which phases are present. The formation of this material from metallic precursors can result in a wide range of different phases that include several that have structures and lattice parameters which are practically indistinguishable from those of kesterite/stannite Cu_2ZnSnS_4 (*e.g.* cubic-ZnS, tetragonal-Cu_2SnS_3 and cubic-Cu_2SnS_3). In this case Raman microscopy has been used to identify Cu_2ZnSnS_4 through its Raman scatter peaks at 289, 339, 350 and 370 cm^{-1}.[37] Neutron diffraction has also been used to identify bulk Cu_2ZnSnS_4 as Cu^+ and Zn^{2+} have an identical number of electrons and are hence indistinguishable in atomic structure by conventional XRD.[38] However neutron scattering cross-sections are very small and this technique is impractical for thin films.

Spectrophotometry[39] and ellipsometry[40] are probably the two most common forms of optical characterisation and can provide information about a material's complex refractive index ($n* = n + ik$). This can be used to understand how the optical bandgap is affected by chemical composition and the affect a material will have when it is included as one of the layers in a multilayer stack or device. In its simplest form, transmission measurements can be used to determine the wavelength (λ) at which optical absorption begins, allowing the bandgap (E_g) to be estimated. When combined with a measurement of reflection, $n*$ can be determined. While this generally requires knowledge of the film's thickness, this too can be determined if transmission and reflection are measured over a wide spectral range and a consistent solution for $n*$ is determined. Once $n*$ is known it is a small step to determining the optical absorption coefficient ($\alpha = 4\pi k/\lambda$) along with a more accurate determination of E_g by plotting α^2 or $\alpha^{1/2}$ against photon energy. Direct transitions follow

$$\alpha^2 \propto \left(E - E_g\right), \tag{6.2}$$

and indirect transitions follow

$$\alpha^{1/2} \propto \left(E - E_g \pm E_P\right) \tag{6.3}$$

where E_P is the energy of the phonon associated with the indirect transition.

Spectroscopic ellipsometry also gives $n*$, along with film thickness, and can provide additional information about more complicated samples and anisotropic materials such as non-cubic crystals.[40] As for all techniques, the measurement area should be carefully considered, especially where reflection measurements are done over a range of angles (as in variable angle spectroscopic ellipsometry), as the footprint on the library surface will change with angle of incidence. While the above techniques are generally require sequential, point-by-point analysis, potentially faster parallel analysis is possible. This is done in imaging ellipsometry by illuminating a strip across sample surface and sweeping it in an orthogonal direction.

The electrical resistivity of a thin film is usually measured using a four point probe,[41] which consists of four linearly arranged and equispaced spring loaded electrical contacts (usually tungsten carbide) [see Figure 6.9(a)]. An electric current (I) is passed through the two outer most contacts and the voltage (V) across the innermost contacts is measured. Separating the electrical contacts for the current supply and the voltage measurement removes the effects of contact resistance and hence minimises errors for low resistivity materials. The calculation of resistivity depends on the physical size of the specimen. In the ideal case, with a film thickness (t) very much less than the probe spacing and lateral dimensions very much greater than the probe spacing, the sheet resistance (R_S, measured in $\Omega \ \square^{-1}$) and resistivity (ρ, measured in Ω m) are given by:

$$R_s = 4.5324 \frac{V}{I}$$

$$\rho = tR_s.$$

(6.4)

Step and repeat measurements can be used to map resistivity over the surface of a library. However, results should be interpreted with care when approaching the edge of the library or the boundary between regions with widely differing conductivities as the assumption of an effectively infinite sample size will breakdown. Consequently, the measurement may benefit from adopting different probe geometry where the probes are arranged in a square.[42] This minimises the overall footprint and allows measurements to be taken closer to the edge of the library before any correction is required. When taking measurements on a thin film using the square arrangement, the measured sheet resistance is twice that indicated by eqn (6.4). Automated four point probe mapping systems are commercially available, and are mainly used to assess doping uniformity on Si wafers for the semiconductor industry.

Resistivity can also be measured using a non-contact technique where a coil is held in close proximity to the library surface. An AC signal ($\sim$10 kHz) is applied to the coil and an electrical eddy current is induced in the film. This technique can again be used to create a resistivity map by moving the library beneath the measurement coil.[43,44]

Electrical conductivity (σ) is the inverse of ρ and is related to the concentrations of electrons (n_e) and holes (n_p) through their respective mobilities (μ_e and μ_p) and by:

$$\sigma = n_e e \mu_e + n_p e \mu_p$$

(6.5)

where e is the magnitude of the electronic charge. The dominant or majority carrier type can be determined several different ways depending on the degree of electrical conductivity.[45–47] Low conductivity materials can be assessed using the rectification technique that involves taking an AC measurement using a four point probe [see Figure 6.9(b)]. When the probes are brought into contact with the surface of the library, each one forms a metal-semiconductor contact or Schottky diode. During measurement an AC current is passed through contacts 1 and 2 and a DC voltmeter is applied across contacts 3 and 2. If the measured voltage is positive the majority carrier is p-type and if it is negative it is n-type. High conductivity materials are best examined using the hot probe technique where the sign of the Seebeck coefficient is used to establish carrier type. During the measurement two electrical contacts are made to the surface of the library and these are directed to a high impedance voltmeter. The introduction of a temperature gradient across the contacts causes the majority carriers to diffuse down the temperature gradient, inducing a voltage; in the case of n-type

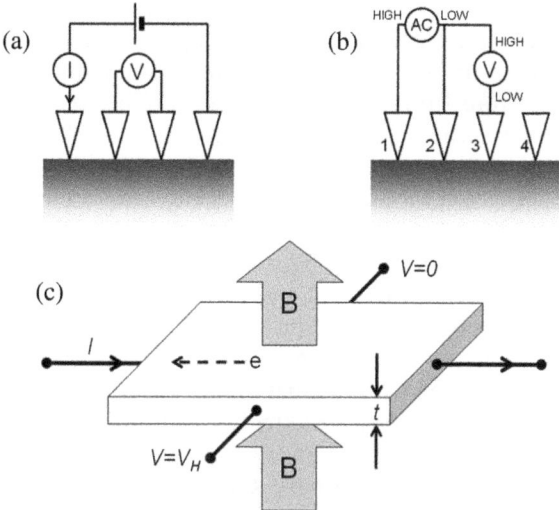

Figure 6.9 Geometry for electrical characterisation: (a) DC four-point probe for sheet resistance; (b) AC four-point probe for carrier type; and (c) Hall effect.

conductivity, for example, the contact at the highest temperature will become electrically positive. While this technique can be used to map carrier type over a library surface, care must be taken when interpreting the results as measurements can be swamped by the effects of thermal conduction from neighbouring locations and the relative change in voltage on the introduction of the thermal gradient should be used as the indicator of carrier type.

While the measured value of the Seebeck coefficient can, in theory, be used to determine carrier concentration as well as carrier type, reliable measurement is difficult. The determination of majority carrier type and both its concentration and mobility can be more readily achieved by combining electrical conductivity measurements with Hall effect measurements. During the Hall effect, the Lorenz forces acting on electrical carriers moving within a magnetic field induce a transverse voltage. The polarity of this Hall voltage (V_H) indicates the carrier type and its magnitude gives the carrier concentration (n) through:

$$V_H = \frac{I_x B_y}{net} \tag{6.6}$$

where I_x is the electrical current and B_y is the applied magnetic flux density [see Figure 6.9(c)]. The geometry of the Hall effect specimen requires the electrical contacts to be point-like and located at its edges; this is frequently done using cloverleaf shaped 'Van de Pauw' shaped specimens, which are difficult to accommodate in a combinatorial study.

The bandgap of a material can be determined by several different techniques, including the measurement of the optical absorption coefficient as discussed earlier in this section. Electrical measurements require a mechanism to excite electrons from the valence band into the conduction band; this is generally through either thermal or optical excitation. While thermal excitation is impractical for mapping combinatorial libraries, optical excitation can be applied with ease, for example, during the measurement of photoconductivity.[48] Under illumination the total electrical conductivity (σ_T) can be considered to be the sum of the photoconductivity (σ_{Ph}) plus the electrical conductivity measured in the dark (σ_D). σ_T can be determined from the current measured using a simple two point probe with a DC bias voltage applied to a pair of electrical contacts on the sample's surface, and σ_T and σ_D determined from the current measured with and without illumination. However, σ_{Ph} may be very much less than σ_T and the effect of illumination on the measured current may not be apparent. Modulating both σ_T and the measured current can be achieved by chopping the light source at a fixed frequency (usually around 10 Hz) and, by using a lock-in amplifier, the AC component of the measured current (I_{AC}) can be determined at the chopping frequency. This is equivalent to the photocurrent and is therefore proportional to σ_{Ph}. Light from a monochromater can be conveniently delivered to the area between the electrical contacts by an optical fibre and I_{AC} recorded as the wavelength of the light is swept from a long wavelength to a short wavelength (*i.e.* increasing photon energy). Selecting a region of the spectrum that straddles the bandgap energy will allow its energy to be determined as I_{AC} suddenly increases.

A range of processes is used to grow thin film semiconductors and during the evaluation of a new material system rough surfaces are common. These can confound the measurement of the optical absorption coefficient and hence the determination of the bandgap by this route. However, photoconductivity measurements are less prone to roughness effects, though they usually require the deposition of an array of electrical contacts over a library's surface.

The need to produce suitable contacts in an array across the surface of a sample provides further potential problems. As discussed in the materials characterisation sections above, the resolution needs to be considered. Geometrically, the contact size and arrangement will depend on the measurement requirements. There are measurement constraints imposed by the characterisation technique employed, such as Van der Pauw or photoconductivity and Hall techniques; the arrangement of contacts and their geometrical arrangement must be designed to yield measureable signals whilst also providing a measurement resolution that is compatible with other characterisation techniques.

As the resolution increases and measurement area decreases, the signal magnitude decreases and uncertainty increases. For resolutions approaching the micrometre scale and below a range of scanning microscopy techniques are available and include those based on scanning probe microscopy (SPM) and scanning tunnelling microscopy (STM). These can map conductivity across the surface using conductive atomic force microscopy (C-AFM), and the concentration and type of carriers with scanning capacitance microscopy (SCM), electro-luminescence mapping and electron beam induced current (EBIC) mapping techniques. These achieve high resolutions and can be used for interrogating small regions of a library to provide a greater depth of understanding; however, they are time-consuming, usually require special sample preparation and are currently not suitable tools for whole-library characterisation.

The deliberate variation of composition within a library can result in problems for the formation of the library if the complete formation of the compounds requires different process conditions that vary with the composition. This may result in a need to produce a number of libraries. However, it also has implications for the formation of electrical contacts. Formation of ohmic contacts to a material depends on the contact material and the composition, conductivity type and carrier concentration of the material to be contacted. The formation conditions can also vary with these parameters. Measurements based on the Seebeck coefficient generally assume the same material/contact for each contact. Where the material/contact varies, the local Seebeck coefficient is likely to vary. It is necessary take these variations into account if reliable data is to be extracted.

6.2.4 Device Libraries

To gain a greater understanding of a potential solar cell absorber material it may be desirable to incorporate that material into device structures and then characterise these using standard device characterisation techniques. There are a several potential problems in converting the libraries into devices. Chalcogenide thin film absorber libraries, for instance, could use $CuInSe_2$ or CdTe device designs as a starting point. It would be necessary to produce the libraries using suitable substrates such as molybdenum or TCO coated glass for superstrate or substrate configurations, respectively; such structures would need suitable buffer and window layers, and either front TCO or back metallised contacts to complete them. A functioning solar cell requires the formation of a junction and this is in turn dependent on the surface of the absorber layer and the materials used for the buffer and absorber layers as well as the interface between them. An absorber library is likely to vary in concentration of carriers and the type may also vary. In addition, the junction formation is likely to depend critically on the process conditions used.

The compositional variation inherent in an absorber material library means that the difficulty of producing practical devices from a complete library is likely to be greater than the value of the data that could be extracted. It may be possible to form functioning devices over limited regions of a library, though the device size considerations would need to take into account the issues of resolution and measurement that were discussed earlier. An alternative approach may be to consider the use of a liquid electrolyte to form the junction and contact to the material.

In general, the production of device libraries would be more appropriate to use to determine optimum device structure for an absorber layer with unvarying composition. Such a library would use gradients in thickness of the component layers (buffer, TCO) or deliberate variations in buffer or TCO layer properties to identify the combinations that yield optimum properties.

The size of device will be constrained by a balance between the resolution of the interrogation and the limitations of practical measurement. For instance, cell efficiency and current density, device capacitance and quantum efficiency measurements depend on the device area. The actual collection area may differ from the defined area if edge isolation is not complete or is measured incorrectly. As the area of a cell approaches 1 mm^2, the resolution of the parameter measured will be affected by the uncertainty in the area and this uncertainty is likely to become comparable or greater than measured parameter variations. In addition, the edges of a device can behave differently to the main area—for instance, mechanical or laser scribing is typically used to define the device area and this will cause damage resulting in different electronic and/or optical behaviour compared with the undamaged area. As the device area decreases, edge effects can increasingly contribute to the overall device behaviour.

Consider square devices with areas of 1 cm^2 and 1 mm^2. If it is assumed that scribing causes a damaged region extending up to 50 μm from the scribe edge, then the area of the device that is undamaged is 98.1 mm^2 and 0.81 mm^2, and the damaged regions represent approximately 2% and 23% of the undamaged device areas, respectively. Similar reasoning applies to the measurement of the device areas and the ability to produce contacts with equal areas reproducibly with less variation than the parameter being interrogated. As device areas decrease towards 1 mm^2, the ability to make electrical contacts to devices becomes difficult and the number of measurements increasingly large. However, larger device areas can give misleading results because they integrate response over the collection area and regions that encompass electrical shunts and thus they will either produce an average response or little or no response. In either case the value of using a library to identify optimum combinations will be reduced.

Stepped libraries of device structural variations could be used with the step size determined by the desired size of the device area. Although this approach fixes the resolution and thus reduce the value of the library approach, given the increase in uncertainty that occurs as the device size decreases, it may not be as significant a drawback as for the materials interrogation as discussed earlier. The choice of linear or stepped gradient libraries for assessing device component layer combinations will depend on many factors. The choice of library type and device size will depend on the practicality of producing and interrogating the library and the reliability of the analyses data that can be extracted.

6.3 CZTS and Cu₂ZnSnS₄

In this section, we give a brief outline to why CZTS is a strong alternative material for absorber layers in thin film solar cells. Among the many systems available are the naturally occurring adamantine minerals Cu_2ZnSnS_4 (CZTS) and $Cu_2ZnSnSe_4$ (CZTSSe), which avoid many of the sustainability problems mentioned at the start of this chapter. The use of highly abundant, low-cost and non-toxic elements is a major attraction and a dominant driving force behind the current research into CZTS. It is regarded as a highly promising absorber layer and has become the subject of much interest owing to the fact it has desirable characteristics applicable to PV, with a reported optimal direct bandgap in the range 1.45–1.6 eV and a high absorption coefficient (>104 cm^{-1}).[49-51] CZTS is analogous to the chalcopyrite material CIS with half of the In(III) being substituted iso-electronically by Zn(II) and the other half with Sn(IV); notably both are considered sustainable elements. The quaternary compound can form into one of two main crystallographic polymorph structures classified as stannite-type (space group I42m) and kesterite-type (space group I4)[52] with a tetragonal structure. The distinguishing factors are associated with how the substitution of the Zn and Cu metals are arranged on the structural sites of the unit cell, as illustrated in the two-dimensional schematic in Figure 6.10. Differentiating these two polymorphs can prove challenging. Many studies only use powder XRD, but without carrying out comprehensive single crystal structural analysis using XRD and/or neutron diffraction, the structure cannot be conclusively identified. CZTS has been found to usually occur in the thermodynamically stable kesterite phase

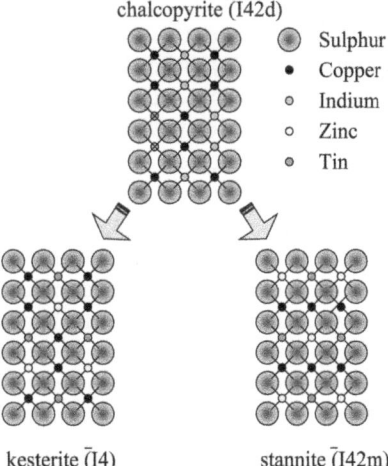

chalcopyrite ($\bar{1}$42d)

- Sulphur
- Copper
- Indium
- Zinc
- Tin

kesterite ($\bar{1}$4)   stannite ($\bar{1}$42m)

Figure 6.10 Schematic representing the evolution of the kesterite and stannite type crystal structures from the chalcopyrite structure. Derived from ref. 54.

as opposed to the stannite type.[52] Therefore, in this study the CZTS(Se) family of materials is assumed to only form as the kesterite structure.

6.3.1   Growth of CZTS

CZTS absorber layers and device structures can be grown by a wide range of techniques that can be categorised as either vacuum or non-vacuum methods. The majority of techniques involve a two-stage process where the metallic precursor elements (Cu–Zn–Sn) are initially deposited, either sequentially or simultaneously, before annealing in the presence of the chalcogen (S and/or Se) for the conversion into the semiconductor. This is known as 'sulfurisation' or 'selenisation', and 'sulfoselenisation' when both sulfur and selenium are present. A one-step process can also be used, where all materials are deposited and annealed simultaneously. Many different types of vacuum and non-vacuum technologies are available and in the following sections some of the most notable studies that represent milestones in the area of CZTS research are described.

6.3.1.1   Vacuum Deposition

Vacuum deposition mainly encompasses variants of sputtering and evaporation techniques as well as pulsed laser deposition (PLD), which are collectively known as physical vapour deposition (PVD). One of the first to work on CZTS were Ito and Nakazawa[53] in 1988, who deposited stoichiometric CZTS films and made devices using atomic beam sputtering. Tanaka *et al.*[54] deposited elemental Cu, Zn and Sn sequentially in the presence of sulfur onto a 400 °C heated quartz substrate to form CZTS thin films within 1 hour using a hybrid sputtering arrangement. However, above 450 °C significant Zn losses were experienced due to its high vapour pressure. Momose *et al.*[55] simultaneously deposited Cu, Zn and Sn metals onto soda lime glass substrates. Notably, full conversion to CZTS involved a rapid annealing process in the presence of sulfur in only 7 minutes.

Some groups have investigated the sequential deposition of each element as opposed to the co-deposition route. Fernandes *et al.*[56] fabricated CZTS precursor structures with stacking orders of Mo/Zn/Cu/Sn and Mo/Zn/Sn/Cu using DC magnetron sputtering. They found that the Mo/Zn/Sn/Cu sequence facilitated better CZTS growth and crystallinity, concluding that a Cu layer on the top prevents the elemental loss of Zn and Sn during sulfurisation by acting as a 'capping layer', although there was evidence of the secondary phase $Cu_{2-x}S$ in the form of crystallites predominantly on the surface of the film. Araki *et al.*[57] studied various stacking orders of electron beam evaporated metal layers with respect to the CZTS properties and device performance. Some very interesting results came to light: the best performing solar cell consisted of a Mo/Zn/Cu/Sn stacking order yielding device efficiency around 1.79%. This is in relatively good agreement with Fernandes *et al.*[56] as mentioned previously. They found that when Cu and Zn layers were in contact larger grain sizes were induced. Furthermore, and perhaps most importantly, when Cu was used as the bottom layer, *i.e.* in contact with Mo, formation of voids occurred between the Mo and CZTS interface and hence device performance degraded dramatically.

6.3.1.2 Non-vacuum Methods

Non-vacuum methods offer the possibility of reducing overall production and material costs with an increased opportunity for scalability and include spray pyrolysis[58] sol–gel[59] and screen printing.[60] Unsurprisingly, the work by Todorov and co-workers is mentioned first, whereby a record performing CZTSSe cell (efficiency 9.7%) was produced using a unique solution-particle process; all the constituent elements Cu, Zn, Sn and S were subjected to a 'sulfoselenide' rich atmosphere at 540 °C.[61] Guo and co-workers were the first to synthesis CZTS-based semiconductor nanocrystals using a hot-injection solution method in 2009. When subjected to Se vapour at 500 °C improved film quality, grain growth and device efficiencies were observed.[62] Electrodeposited CZTS thin films was first reported by Scragg *et al.*[64] with varying elemental metal stacked layers deposited directly onto Mo coated glass substrates. For comparative reasons, the films were fully converted within an elemental sulfur rich and also H_2S gas atmosphere at 500 °C, with the latter improving film crystallinity.

6.3.2 CZTS Device Structures and Efficiencies

Ito and Nakazawa[53] first reported photoelectric behaviour in 1988 for a CZTS/CdSnO heterostructure. Although device efficiency was not well documented, a full investigation of the p-type CZTS concluded its suitability with a direct bandgap of 1.45 eV. In 1997, Friedlmeier *et al.*[63] reported the fabrication of CZTS solar cells using a CZTS/CdS/ZnO structure, yielding a conversion efficiency of 2.3%. This record was soon overtaken by Katagiri's group,[50] with cell conversion efficiencies up to 2.63%. Further investigation and optimisation of the annealing and conversion process by this group substantially improved AM1.5 device efficiencies reaching 5.45% in 2003[65] and then their best to date of 6.8% in 2008.[66]

With increased interest and awareness of the materials potential, many research groups have since undertaken extensive studies into the fabrication and characterisation of CZTS thin film single layers and complete device structures. There has since been a remarkable improvement in device efficiencies. This is covered in depth in references.[67] Interestingly, the current record by Wang *et al.* in 2013 boasts a conversion efficiency of 12.6%.[68] This has now overtaken the previous record held by Todorov *et al.*[13] and the previous records of Barkhouse *et al.*[69] over 10% and Todorov *et al.* at 9.7% in 2010[61] (although both authors are part of the IBM group). In both cases, the device structures used a CZT(S,Se) absorber layer fabricated by a hybrid non-vacuum solution-particle approach. Also, synthesis of selenised CZTS nanocrystals *via* hot injection has shown total area efficiencies as high as 7.2% grown by Guo *et al.*[70] However, it would be desirable to grow CZTS active layers without incorporating the rare element selenium. Katagiri and co-workers maintain the current record of 6.8% by fabricating Se-free CZTS solar cells *via* an in-line sputtering and a sulfurisation process.[66] Similarly, a single-step thermal evaporation process by Wang *et al.* also yielded efficiencies up to 6.8%.[71]

The current status of kesterite-type efficiencies has shown progressive improvements but still remains substantially low compared with competing thin film technologies and certainly the theoretical limit,[22] requiring a concerted effort to fully understand and optimise material and device parameters. Interestingly, these studies have all targeted the same stoichiometric ratios and notably demonstrated that the best performing CZTS-based devices consist of Zn-rich and Cu-poor compositions, as has been well documented within the literature, irrespective of which fabrication route is pursued. The optimum compositions for the Se-free compounds were estimated to be Zn : Sn $\approx$ 1.25, Cu : (Zn+Sn) $\approx$ 0.85 and S : metal $\approx$ 1.1.[66]

The greatest challenge remains the precise control of the desired stoichiometry of the precursor elements during sulfurisation for the complete conversion to monophase kesterite CZTS. Incomplete conversion or elemental depletion during sulfurisation, as is explained later, will almost certainly induce decomposition of the pure CZTS phase and lead to the detrimental formation of binary and/or ternary metallic chalcogen phases. A greater understanding of which compositions are necessary to fabricate high-performing absorber material is required, which in turn will facilitate improved device efficiencies for ultimate commercialisation. In conjunction with this, suitably refined low-cost and scalable growth is needed. This has recently been demonstrated for electrodeposition where a CZTS solar cell with an efficiency of 7.3% was reported.[72] Furthermore, the idea of commercialising CZTS has started to come to fruition with the thin film producer Solar Frontier announcing collaboration with IBM to take CZTSSe development further.[73]

CZTS absorber layers have the added advantage of being integrated into the familiar substrate configuration, as used for CIS/CIGS devices. This multi-layer structure, as illustrated in Figure 6.11, typically starts with a glass base coated with a molybdenum electrical back contact, a p-n junction consisting of a CZTS active absorber layer (p-type) to replace materials like CIGS, and a thin CdS buffer layer (n-type). A further i-ZnO/Al:ZnO window layer and top electrical (Al) contacts complete the solar cell. Whether this configuration is the most suitable for CZTS-based devices has yet to be fully investigated. Many research groups are beginning to look into using different layers, thicknesses and substrates. Rajeshmon *et al.*[74] have, for instance, demonstrated low cell efficiencies with alternate cell structures by avoiding the usual KCN etching treatment of CZTS and further substituting the CdS buffer layer for In_2S_3.

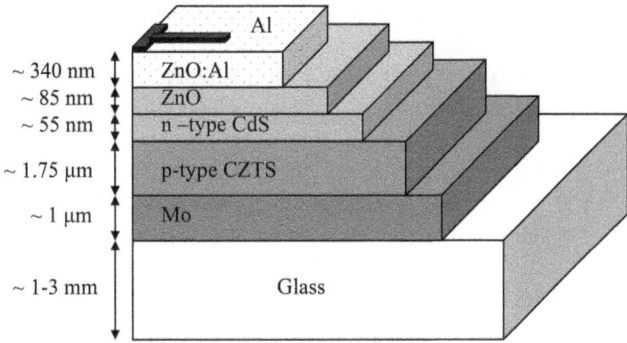

~ 340 nm — ZnO:Al
~ 85 nm — ZnO
~ 55 nm — n –type CdS
~ 1.75 μm — p-type CZTS
~ 1 μm — Mo
~ 1-3 mm — Glass
Al

Figure 6.11 Schematic cross-section of the substrate device configuration commonly used for CZTS-based solar cells. Each constituent layer is indicated with approximate thicknesses as taken from the Katagiri cell.[66]

6.3.3 Composition and Formation of CZTS

Studies of CIS and CIGS have shown that the Cu to In and Cu to (In+Ga) composition ratios play a significant role in determining the material properties suitable for solar cell fabrication and increased conversion efficiencies.[75,76] Therefore, it seems reasonable to assume similar ratios could be used to investigate kesterite films, which are represented using the universally recognised ratios Cu to (Zn+Sn), Zn to Sn, and S to metal. These have been used to understand crystal growth, phase transformation, morphology and opto-electronic properties.

Some recent reports have explored the influence of composition on CZTS formation and properties. Tanaka *et al.*[77] co-evaporated complete CZTS thin films onto stationary soda lime glass substrates at a temperature of 550 °C for 12 min, yielding films with varying Cu to (Zn+Sn) and Zn to Sn ratios ranging from 0.95 to 1.10 and 0.82 to 1.28, respectively. The overall CZTS film thickness was in the region of 310–530 nm, which is perhaps rather thin compared with most studies—notably the current world record of 12.6% uses thickness up to ~2.5 μm.[68]

In summary, the crystallinity improved with enhanced grain size as the Cu to (Zn+Sn) ratio increased and, interestingly, the structural properties were only affected by the change in Cu to (Zn+Sn) and not Zn to Sn. A further study by Babu *et al.*[78] focused on the effect of changing the Cu to (Zn+Sn) ratio in CZTS films using a four-source co-evaporation arrangement. Films were converted in a single step with a substrate temperature of 350 °C. Using a constant Zn to Sn ratio, the Cu to (Zn+Sn) ratio ranged from 0.83 to 1.15. Of particular significance in this study is the realisation that polycrystalline phase pure CZTSSe films were only evident in the 0.90–1.10 range, with the formation of secondary phases ZnSe and $Cu_{2-x}S$ occurring at 0.85 and 1.15, respectively. The optical bandgap was found to be dependent on the Cu to (Cu+Zn) ratio, with an ideal value of 1.50 eV found at 0.90–0.95.

The influence of composition in non-vacuum techniques was also explored by Tanaka *et al.*[79] using changeable sol–gel solutions. The Cu to (Zn+Sn) ratio was varied from 0.73 to 1.00 with a constant Zn to Sn ratio held at 1.15. As the ratio decreased, large grains began to develop with a shift in bandgap towards higher energies. The best performing solar cell with an efficiency of 2.03% was Cu-poor [Cu to (Zn+Sn) = 0.80]. Kumar *et al.*[80] used spray pyrolysis to investigate the concentration effect of copper salt and thiourea on CZTS formation. Conversion occurred at a substrate temperature of 350 °C and film ratios of Zn : Sn = 0.92, Cu : (Zn+Sn) = 0.79 and S : metal = 0.64 yielded a single phase kesterite structure with a bandgap of 1.43 eV. However, preventing a substantial sulfur loss remains a challenge with this particular technique. Platzer-Björkman *et al.*[81] recently studied the effects of precursor sulfur content on co-sputtered CZTS. They interestingly compared sulfurised metallic precursors with sulfur-containing precursor compounds. By sulfurising metallic precursors substantial Sn loss was observed, with the latter yielding higher quality film and improved uniformity. It was further shown that specific compositions dominated the overall device efficiencies: Zn-rich and Cu-poor/Sn-rich were highest, while near to stoichiometric material yielded low performances.

Although these studies offer an interesting and useful insight into how chemical composition effects the overall formation and characteristics of CZTS thin films, only changes within a very narrow range are targeted. In fact, most studies generally keep the Zn to Sn ratio constant while the Cu to (Zn+Sn) value incrementally deviates away from stoichiometry. Very few studies give a comprehensive overview of how CZTS formation and device performance are affected by a large elemental change over a wide spatial region. One of the main constraints is the vast number of individual samples with different stoichiometries required to undertake such a study, coupled with a large investment in time, effort and the need for advanced and automated analytical tools. As previously mentioned Tanaka *et al.*[77] showed a way of keeping the substrate stationary and deliberately changing the angle of the sputter targets to obtain a graded composition in a single run. Work by one of the authors of this chapter[82] used DC magnetron sputter deposition to grow libraries of alloy precursors that covered a much wider compositional range than used by Tanaka *et al.*,[77] with the Cu to (Zn+Sn) and Zn to Sn ratios ranging from almost zero to 4.5, and centred around Cu : Zn : Sn = 2 : 1 : 1 as required for stoichiometric Cu_2ZnSnS_4 [see Figure 6.12(b) and (c)]. Phase mapping of the library was performed using XRD complemented by Raman spectroscopy to confirm the phases identified, with composition provided by X-ray microanalysis in an SEM.

A phase composition plot for the same library as presented in Figure 6.7 is given in Figure 6.12(a). This shows that this single library covered approximately one quarter of the Cu–Zn–Sn system, with each point corresponding to a single analysis location. The figure includes the Cu–Zn–Sn phase diagram based on the theoretical 453 K (180 °C) isothermal section,[83] whereby all binary Cu–Zn and Cu_6Sn_5 phases have tie lines with solid Sn and

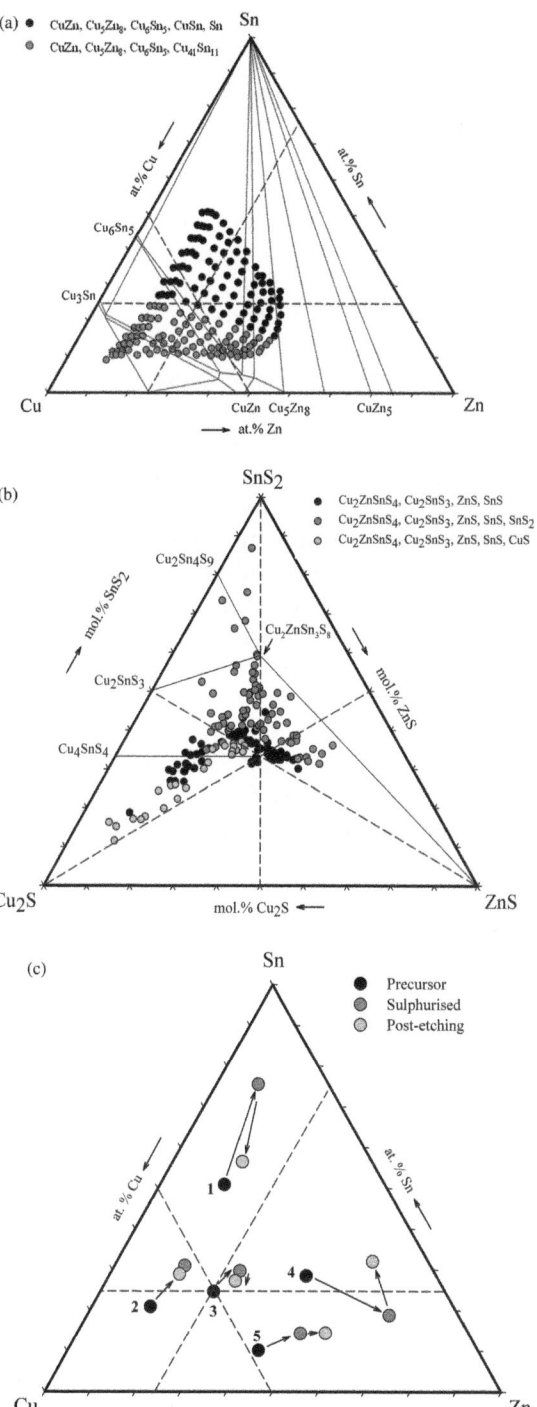

no liquid phases. The crystalline phases are strongly dependant on the composition and there is no evidence of the binary Cu_3Sn forming, but instead $Cu_{41}Sn_{11}$ occurs. However, it should be remembered that the films are unlikely to have fully reached structural equilibrium at this temperature. Figure 6.12(b) gives the results for a repeat analysis on the same library following annealing for 30 minutes in sulfur vapour at 773 K, during which the phase-pure kesterite/stannite structure forms within a small region along the Cu_2SnS_3–ZnS line system. In this study, the film crystallised as a phase mixture over the entire library, with only ternary copper–tin–sulfide phases formed as indicated in the figure. Changes in the metal content of selected points on the library are tracked in Figure 6.12(c). This shows that more metal was lost when the precursor composition moved away from the 'ideal' Cu : Zn : Sn = 2 : 1 : 1 required for stoichiometric Cu_2ZnSnS_4, suggesting that the sulfurisation process was best matched to this central region. An electrical characterisation of the library by hot probe showed that it was p-type over its entire area, confirming the results of Tanaka *et al.*,[77] and photoconductivity spectroscopy indicated that all areas on the library had a bandgap of $\sim$1.48 eV, with a film composing of Zn to Sn $\sim$2.96 and Cu to (Zn+Sn) $\sim$0.49 showing the strongest photoconductivity response. Surprisingly, this deviates substantially from the targeted composition ratios synthesised by other groups such as Katagiri *et al.*[66]

The research activity into CZTS and related materials has increased very rapidly over the past five years and new developments are being reported frequently. The current section aims to provide a brief introduction into this technology. Journals such as *Progress in Photovoltaics: Research and Applications* and *Solar Energy Materials and Solar Cells*, amongst others, provide greater detail and updates on current developments in this field than is possible in this chapter section.

6.4   Sulfosalts

Sulfosalts are a family of IV-V-VI chalcogenides that include naturally occurring silver ores[84,85] and have to date had relatively little application to photovoltaics. Having been known, in mineralogical terms, for many years, their structure has recently been defined by international committee,[86] with

---

Figure   6.12   Phase-composition plots for CZTS.[82] (a) Metallic precursors superimposed onto the ternary Cu–Zn–Sn equilibria phase diagram at 453 K[84] and (b) sulfurised precursors superimposed onto the quasi-ternary Cu_2S–ZnS–SnS_2 600 K equilibria phase diagram. (c) Ternary Cu–Zn–Sn composition plot illustrating elemental changes during processing: precursor deposition, post-sulfurisation and post-etching treatment. Note for each plot the intersection of blue dashed lines denotes stoichiometric precursor ratio (Cu : Zn : Sn = 2 : 1 : 1).

a general formula $A_xB_yC_Z$, where: A = Cu^+, Ag^+, Pb^{2+}, Sn^{2+}, Sn^{4+}, Fe^{2+}, Mn^{2+}, Hg^{2+}, Tl^+,...; and B = As^{3+}, Sb^{3+}, Bi^{3+}, Te^{4+} and C = S^{2-}, Se^{2-}, Te^{2-}.

Sulfosalts are notable for their highly complex crystallographic structure that can include a modular construction which involves a combination of interconnected sub units arranged in a 'box-work' or 'super-complex' structure.[84] This high complexity allows for high versatility through the wide range of contributing ions and variation of the dimensions of the sub-units, the connectivity between sub-units and the combinations of types of sub-units within a particular material. This gives many opportunities for engineering specific properties such as electronic bandgaps. Indeed, many sulfosalts have bandgaps that are well suited to terrestrial photovoltaics (see Table 6.1). In addition, the growth and annealing temperature of sulfosalts are generally around 200 to 350 °C, lower than the typical type conversion anneal employed for CdTe (450 °C) and the growth temperature of CIGS (500 °C). This would suggest potential for simpler and more energy efficient growth processes and therefore a shorter energy payback time. The fact that many of the sulfosalt phases are naturally occurring minerals means that they have formed and remained over geological timescales, which is promising for high stability.

As discussed in the introduction to this chapter, current successful thin film solar cells rely on scarce elements such as Te or In, and ultimately these will limit their exploitation. The economics of using sulfosalts in thin film solar cells are also presented in Table 6.1 where they are compared with more conventional materials (CdTe and CIGS) by assuming a hypothetical efficiency of 15% and a 2 μm thick absorber layer. The estimated cost (US$ m^{-2}) of each material is based on the prices of the individual elements within the absorber layer,[6] and shows that the raw materials in sulfosalt solar cells would generally account for a tiny fraction of the overall cost of a complete module, and is on the whole less than for both CdTe and CIGS cells. Some compounds are of course more costly than others, with some of the Ag compounds being significantly more expensive. However, the viability of a particular compound will of course depend on many other factors, including the both the concentrations and availabilities of the elements it is made from.

The exploitation of a particular compound can be considered, in many cases, to be dominated by the scarcity of one particular element. These 'limiting elements' are identified in Table 6.1 by normalising the weight fractions of each of the elements in a compound by its estimated global reserve.[10] The global annual production of the limiting element can then be used to estimate how well the compound would be able meet the current PV market should it be based solely on this material. For illustrative purposes this is presented in the table by assuming the 2010 global PV market (18.2 GWp[87]) and indicates that at the current rate of production the elements found in the sulfosalt compounds presented are, probably, more than capable of meeting the current market demand. A similar analysis for CdTe is more difficult because the current global production for Te is very

Table 6.1 Table showing sulfosalts from Dittrich et al.[84,85] with CdTe and CIGS shown for comparison.

Mineral	Composition	E_g/eV	Material cost/$ m^{-2}	Limiting element		
				Element	Concentration/wt%	Production demand/%
Wittichenite	Cu_3BiS_3	1.2	0.19	Bi	42	7.8
Bournonite	$PbCuSbS_3$	1.2	0.07	Sb	25	0.2
Enargite	Cu_3AsS_4	1.3	0.04	As	19	0.4
Boulangerite	$Pb_5Sb_4S_{11}$	1.3	0.07	Sb	26	0.2
Owyheeite	$Pb_7Ag_2(Sb,Bi)_8S_{20}$	1.4	0.79	Bi	34	6
Meneghinite	$Pb_{13}CuSb_7S_{24}$	1.4	0.06	Sb	19	0.2
Zinkenite	$Pb_9Sb_{22}S_{42}$	1.5	0.08	Sb	45	0.3
Andorite	$PbAgSb_3S_6$	1.5	1.29	Sb	42	0.3
Freieslebenite	$AgPbSbS_3$	1.5	3.16	Ag	20	1
Polybasite	$(Ag,Cu)_{16}Sb_2S_{11}$	1.5	7.72	Ag	52	3
Semseyite	$Pb_9Sb_8S_{21}$	1.5	0.07	Sb	28	0.2
Chalcostibite	$CuSbS_2$	1.5	0.10	Sb	49	0.3
Plagionite	$Pb_5Sb_8S_{17}$	1.5	0.07	Sb	38	0.3
Gratonite	$Pb_9As_4S_{15}$	1.6	0.02	As	11	0.3
Liveingite	$Pb_9As_{13}S_{28}$	1.6	0.02	As	26	0.6
Diantimony ditin sulfide	$Sn_2Sb_2S_5$	1.6	0.15	Sb	38	0.3
Jordanite	$Pb_{14}(As,Sb)_6S_{23}$	1.6	0.05	As	9	0.3
Stephanite	Ag_5SbS_4	1.6	10.50	Ag	68	4
Fülöppite	$Pb_3Sb_8S_{15}$	1.6	0.08	Sb	47	0.3
Fizelyite	$Pb_{14}Ag_5Sb_{21}S_{48}$	1.7	1.04	Sb	34	0.3
Miargyrite	$AgSbS_2$	1.7	4.86	Ag	37	2
Liveingite	$Pb_9As_{13}S_{28}$	1.8	0.02	As	26	0.6
Pyrargyrite	Ag_3SbS_3	1.8	8.42	Ag	60	3
Proustite	Ag_3AsS_3	1.9	8.74	Ag	65	4
Baumhauerite	$Pb_3As_4S_9$	1.9	0.02	As	25	0.6
Cadmium telluride	CdTe	1.44	2.39	Te	53	>100
CIGS	$CuIn_{0.7}Ga_{0.3}Se_2$	1.2	3.30	In	25	~50
				Ga	6	~30
				Se	49	~30

difficult to assess because of the incompleteness of the data (as discussed earlier in this chapter), with the USGS stating that US production figures are 'withheld to avoid disclosing company proprietary data'[8] and the world resources being based only on tellurium contained in copper reserves. In the calculation presented here an annual global production of 200 tonnes was assumed, based on the 2011 USGS estimate of 115 tonnes (for Canada, Japan, Peru and Russia) plus the 2009 BGS estimate for US production of 50 tonnes.[9] This was rounded up to 200 tonnes to accommodate unknown contributions from other countries such as Germany. The analysis shows that the scarcity of Te places a greater constraint on the continued use of CdTe than that for Cd, and it is quite likely that the requirements for the current PV market alone would exceed current production. For similar reasons it is not possible to say which of the elements in CIGS will have the greatest restriction on its exploitation. In, Ga and Se are all scare elements and the simple analysis presented here shows that current production of all three would be unlikely to meet all the demands of the PV market.

To date the application of sulfosalts to photovoltaic devices has been limited. Research into three candidate systems is presented in the following sections.

6.4.1 Cu–Sb–(S,Se)

The key factors for determining the choice of photovoltaic absorber materials include the energy bandgap (E_g), stability and ability to meet the needs of future multi-TW markets sustainably. To minimise absorber thickness, ideally, the bandgap should be direct, with a magnitude in the range 1.2–1.5 eV. One materials system that shows promise is that based on the sulfosalt containing Cu–Sb–S and their sulfo-selenide analogues. The binary materials of Sb_2S_3 and Sb_2Se_3 are possible candidates, but whilst Sb is significantly more abundant than indium, with typical annual global mine production above 120 000 tonnes over the last five years compared with about 500 tonnes for indium,[9] the sustainability is increased significantly by use of the Cu–Sb-chalcogenide ternary. In this section we consider the ternary compounds in the Cu–Sb–(S,Se) system.

The Cu–Sb–S materials system was the subject of a comprehensive investigation and review of the $Cu_2S–Sb_3S_2$ phases published in 1973.[88] Apart from the ternary Cu–Sb–S phases of tetrahedrite and famatinite ($Cu_{12}Sb_4S_{13}$ and Cu_3SbS_4, respectively) which are naturally occurring minerals in the copper antimony sulfur system, the chalcostibite $CuSbS_2$ phase was thought to be the only naturally occurring mineral phase in the binary $Cu_2S–Sb_3S_2$ system. This mineral is orthorhombic and was named after the Greek terms for copper and antimony; it was discovered at the Graf Jost-Christian Zeche mine, Wolfsberg, Harz, Sachsen-Anhalt, Germany in 1835.[89] There were reports of a Cu_3SbS_3 phase as early as 1890, but it was Skinner *et al.*[90] who reported synthetic Cu_3SbS_3 unambiguously in 1972. Both Skinner and co-workers and Sugaki *et al.*[88] reported the polymorph behavior of Cu_3SbS_3 that had a reversible transition temperature of between 115 °C and 125 °C; the symmetry is orthorhombic above this and below it is monoclinic. Skinner and co-workers considered it only stable above 359 °C though, by quenching, the phase was found to exist down to room temperature. In 1974, Klnup-Møller and Makovicky published the first confirmed report of this material as a naturally occurring mineral, Skinnerite.[91] The Cu–Sb–(S,Se) sulfosalt system encompasses several potential PV absorber material candidates.

6.4.1.1 $CuSb(S,Se)_2$

Early work on synthesised stoichiometric $CuSbS_2$ crystals did not imply that this material was very promising as a solar cell material. Produced using the Bridgman–Stockburger method, Wachtel and Noreika[92] reported strongly p-type behaviour for their material, within the limits of their Hall measurements. They concluded that it possessed a low mobility with a correspondingly high carrier concentration and a room temperature intrinsic bandgap of 0.28 eV,[92] confirming a prediction by Wernick and Benson.[93] However, there was uncertainty about the bandgap value for a low temperature phase due to equipment limitations and a value above 0.58 eV was predicted. The limited solid solubility of the various phases in this materials system indicated that bulk single or multicrystalline Cu–Sb–S(or Se) material would not be a strong candidate for PV applications.

More promising results were obtained from the properties of millimetre-sized single crystals of $CuSbSe_2$ and $CuSbS_2$ grown at 160 °C using a solvothermal method as reported by Zhou *et al.*[94] Optical bandgap values of 1.05 eV and 1.38 eV, respectively, were measured and both materials possessed orthorhombic symmetry with different morphologies for the different chalcogens; where some of the sulfur is replaced with selenium, the bandgap can be adjusted to the lower end of this range.

Thin film PV provides a route to lower cost solar electricity and it is in this form that Cu–Sb–S,Se materials would be likely to compete. Also, the solubility limits are known to be less rigid for polycrystalline thin films and work on thin film $CuSbS_2$ demonstrated that this material possesses many of the key features needed for PV applications. $CuSbS_2$ was produced by chemical bath deposition followed by a 400 °C heat treatment, which led to the first report of a direct optical bandgap of 1.52 eV and confirmation of p-type conductivity for this material.[95] From a weak photoresponse, the conductivity was determined to be 0.03 Ω^{-1} cm^{-1} under illumination and the resistivity in the dark was measured at 35 Ω cm. Although the material showed a relatively high conductivity, it was concluded that, by appropriate processing, it may be possible to modify the carrier concentration, mobility and bandgap of $Cu(Sb/Bi)(Se/S)_2$ thin films for solar cell applications.

Manolache *et al.*[96] selected $CuSbS_2$ as the absorber material for a three-dimensional solar cell. Thin films of Cu_3SbSe_3 were investigated for hydrogen generation by the photoelectrochemical decomposition of water. They were produced by electrodeposition of Sb_2Se_3 and Cu_2Se followed by heating at 400 °C to yield a near single phase material with a measured direct bandgap of between 1.61 and 1.68 eV. They exhibited p-type conductivity and carrier concentrations of 1.1×10^{20} cm^{-3}. Details of the processing of the material and the preparation of devices were as reported by Fernandez and Turner.[97] A Cu–Se layer was electrodeposited on a Sb–Se layer deposited on Cr plated stainless steel. The bandgap values were extracted from transmittance and specular reflectance *versus* wavelength measurements. The carrier concentration and conductivity type were extracted from capacitance–voltage and current–voltage measurements. These used layers that had electrical contacts of silver loaded epoxy and used an electrolyte to form a junction. This configuration was also used measure the bandgap *via* photocurrent *versus* wavelength measurements.[98] The feasibility of using this as an alternative absorber material to replace CuInS$_2$ is supported because CuSbS$_2$ has a similar absorption coefficient as the chalcopyrite and kesterite I-III-VI$_2$ (11-13-16$_2$) family of semiconductors, with the ionic radius of indium and antimony being almost equal. These I-V-VI (11-15-16$_2$) materials were investigated by Yu *et al.*[99] as part of an initial screening of materials for PV applications using their spectroscopic limited maximum efficiency (SLME) multidimensional approach. The group V (15) elements can exhibit both a lower (+3) and higher (+5) oxidation state. The highest absorption coefficients were found by this approach to be related to these metals exhibiting the low (+3) valency composition—similar to the group III (13) elements.

CuSbS$_2$ is a direct semiconductor, with properties that match those required for photovoltaic materials.[94] Zhou *et al.*[94] produced CuSbS$_2$ and CuSbSe$_2$ layers using a low-temperature solvothermal method for preparation of the two semiconductor from Cu, Sb and S(or Se) powders. They used a device structure similar to that of a dye cell and spray pyrolysis to deposit the CuSbS$_2$ onto a TiO$_2$ nanoporous material. Dark *I–V* characterization showed that Sb-rich material yielded rectifying behavior. Semiconductor properties were measured and the bandgap values were found to be 1.05 eV for the selenide and 1.38 eV for the sulfide. Colombara *et al.*[100] investigated the formation pathway of CuSbS and CuBiS thin films produced by sulfurisation of electrodeposited metal precursors.

In 2013, Wilman *et al.*[101] reported a 3.1% solar cell using CuSbS$_2$ layers produced by conversion of electrodeposited CuSb metallic layers *via* heating in flowing H$_2$S.

A second low valence composition was shown to be Cu$_3$SbS$_3$ and the selenide variant.[99] Maiello *et al.*[102] investigated thin films of Cu–Sb–S over a wide range of compositions using a combinatorial approach. The thin films were produced in a two-stage process in which the Cu–Sb precursors were deposited by magnetron sputtering followed by evaporation of a sulfur layer and conversion in a tube furnace. In a paper focused on conversion conditions of the metal precursors, these authors reported conversion of a predominant ternary phase of Cu$_3$SbS$_3$ from fixed composition precursors and demonstrated that the material yielded a bandgap of 1.83 eV. Using a range of S to Se ratios, the group demonstrated that the bandgap of the material could be varied from 1.83 eV for the pure sulfide to 1.38 eV for a Se-rich compound,[102] poor material adhesion preventing reliable characterization of the pure selenide. The paper reported a photoresponse for these materials and a 1.8 eV sample was processed as a solar cell, demonstrating an active device with a very low efficiency (<1%).

A number of other groups are working on Cu–Sb–S materials with most concentrating on $CuSbS_2$ compositions.[103,104] This promising material requires further work to enable it to reach a breakthrough efficiency level of ~10%.

6.4.2 Cu–Bi–S

Cu_3BiS_3 is a sulfosalt material that has been identified as a candidate material for the absorber layer in photovoltaics. It takes the mineral name, Wittichenite, after the locality of the mine in Germany where it was discovered.[105] Past studies have shown that it possesses a bandgap within the optimal range for solar radiation and has a high optical absorption coefficient—both critical characteristics for an absorbing material for use in solar cells.

If Cu_3BiS_3 is found to be suitable for use in thin film solar cells it will have two significant advantages over the established CdTe and CIGS cells. The first is relative abundance of constituent materials. In 2008 the USGS estimated world reserves of bismuth, tellurium and indium at 320 000 tonnes, 21 000 tonnes and 11 000 tonnes, respectively.[6] Mineral scarcity could limit cells using indium to a few GW[106] and cells using tellurium to 100 GW.[107] Bismuth's second advantage is that it is non-toxic; it is used as a lead replacement in solders and ammunition, and is also used in pharmaceuticals.[6,108] The toxicity level of current materials is high, so care must be taken at the manufacturing stage and again at the decommissioning stage of the cell's lifetime, further increasing costs.

In the past decade or so several groups have studied layers of Cu_3BiS_3 deposited by different methods and their studies suggest it ought to be suitable for PV applications but to date no PV cells have been made using this compound.

In the mid-1990s Nair and co-workers produced films of Cu_3BiS_3 using chemical bath deposition to deposit a multilayer of Bi_2S_3 and CuS directly onto glass. The multilayer was than annealed in air. Using XRD, the films annealed at 250–300 °C for 1 hour were shown to contain Cu_3BiS_3. The solid state reaction pathway, $6 CuS + Bi_2S_3 \rightarrow 2 Cu_3BiS_3 + 3 S$, was suggested. The films were found to have high optical absorbance in the visible region and to display p-type conductivity.[109] Later layers of Cu_3BiS_3 were grown using chemical bath deposition to deposit a single film of CuS onto a single film of Bi_2S_3; annealing at 250–300 °C again produced films of Cu_3BiS_3. These films were found to have high optical absorption in the visible region with an absorption coefficient of 4×10^4 cm^{-1} at 500 nm. They were p-type and had an electrical conductivity of 10^2 to 10^3 Ω^{-1} cm^{-1}, leading Nair *et al.* to suggest

Cu_3BiS_3 might be suitable as the p-type absorber in heterojunction solar cells.[110]

Following on from this work, in 2003 Estrella *et al.*[111] reported that films of Cu_3BiS_3 can also be formed following the solid state reaction pathway 3 CuS + Bi → Cu_3BiS_3. Bismuth was thermally evaporated onto chemically deposited CuS film and then heated to 300 °C for 1 hour. Almost complete conversion to Cu_3BiS_3 took place to give films with an absorption coefficient greater than 10^5 cm^{-1} at 650 nm and a bandgap of 1.2 ± 0.1 eV, but with a mobility lifetime product inferior to current materials used in solar cells. It was suggested the mobility lifetime product might be improved by optimising the heating process.[111]

In the mid-2000s Gerein and Haber[112] reported that Cu_3BiS_3 films could be grown using DC magnetron sputtering to create a copper and bismuth precursor on quartz substrate followed by reactive annealing in an H_2S atmosphere. Bismuth depletion was observed at high temperatures and the reaction was sluggish at low temperature. Therefore long annealing times at low temperature were required to produce Cu_3BiS_3.[112] The morphology of films using this method and another similar method using radio frequency (r.f.) magnetron sputtering to deposit metal sulfide precursors was investigated, but it was concluded that reactive annealing of sputtered precursors was not suitable for the production of Cu_3BiS_3 films smooth enough and continuous enough for use in PV.[113] However, subsequent studies showed reactive sputtering of Cu–S and Bi onto a hot substrate could produce films of phase pure Cu_3BiS_3 films that were crystalline, smooth and continuous. These films had a bandgap of 1.4 eV and an optical absorption coefficient of 10^5 cm^{-1} at 650 nm which would make them suitable for use in PV.[108,114]

More recently Mesa and co-workers prepared Cu_3BiS_3 films by reactive evaporation. In a sulfur atmosphere Bi was thermally evaporated onto a glass substrate held at 300 °C. Then Cu was thermally evaporated onto the Bi–S layer, also in a sulfur atmosphere and with heated substrate. This produced a Cu_3BiS_3 layer with no other phases identified independently of the mass ratio of Cu and Bi (within the limits of the study). Again, they had a high absorption coefficient of more than 10^4 cm^{-1} and a bandgap of 1.41 eV.[114] However, the mass ratio did have an effect on transmittance and morphology.[115]

Researchers at Bath University have used electrodeposition of copper and bismuth metallic precursors on molybdenum-coated glass. The precursors were then annealed in sulfur vapour at 450–500 °C for 30 minutes under flowing nitrogen gas. The resulting films were uniform and adherent, and had a photocurrent response.[116] Further work concerning the phase evolution of Cu_3BiS_3 films grown by this method was reported by Colombara *et al.*[100] In this study precursors were deposited either simultaneously or in layers. During heating it was found the reaction to Cu_3BiS_3 proceeds *via* the binary sulphides, but the conversion of Bi to Bi_2S_3 is much slower than Cu to CuS. In the case of the co-deposited precursor the sample must be heated slowly to ensure full conversion to Bi_2S_3 before the melting point of bismuth

(271 °C) is reached, otherwise the bismuth will melt and segregate affecting the morphology of the final film; this appeared not to happen with the layered precursors.[100,114]

One of the authors of this chapter (McCracken) is currently applying a combinatorial approach to the copper–bismuth–sulfur system. DC magnetron sputtering is used to deposit copper and bismuth metallic precursors. The sputtering targets are aligned so that the deposition is graded to give a library precursor with a lateral gradient in metal composition. Both simultaneous and layered deposition of copper and bismuth are being explored. In an attempt to covert this to the desired Cu_3BiS_3 phase, elemental sulfur is introduced by thermal evaporation directly onto the precursor. The sulfurised precursor is then sealed in an evacuated ampoule and heated. This approach has been successfully used to create films of CZTS. For Cu_3BiS_3, the heating conditions used were adapted from Gerein and Haber.[112] The sample was heated to 260 °C to avoid melting bismuth and was held at this temperature for 30 hours. Unfortunately Cu_3BiS_3 was not identified on the libraries. It was thought 260 °C could be too low to facilitate conversion to Cu_3BiS_3 using elemental sulfur. In order to heat to a higher temperature a two-step anneal was attempted. The first stage was at 250 °C to convert Bi to Bi_2S_3 without melting any bismuth metal present and, since Bi_2S_3 has a much higher melting point, a second anneal was preformed at 500 °C. A second layer of sulfur was deposited between anneals to make sure enough sulfur entered the system. XRD showed some regions of the double annealed library contained phase pure Cu_3BiS_3.

The double anneal method is being investigated in the hope of producing high quality Cu_3BiS_3 films and it may be possible to replace the double anneal with a carefully controlled single anneal. In addition a second method is being explored. Instead of thermally evaporating sulfur onto the film, a sulfur pellet is heated with the precursor in an evacuated ampoule. The pellet melts and fills the ampoule with sulfur vapour. A pellet large enough to saturate the atmosphere was used and, over a range of temperatures, preliminary results show the formation of ternary phases but not Cu_3BiS_3. There was also large degradation of films. Although discouraging, this method is simpler and removes a processing stage requiring a high vacuum. For these reasons it is being investigated in the hope that the conversion and adhesion problems can be resolved and Cu_3BiS_3 grown reliably.

6.4.3   Sn–Sb–S

The compositions of the known crystalline phases in the Sn–Sb–S system are presented in Figure 6.13(a), with the majority of the ternary phases lying along a tie line joining SnS and Sb_2S_3. These phases include $Sn_2Sb_2S_5$, which has recently been examined by Gassoumi and Kanzari,[117] who grew thin films by thermal evaporation from a $Sn_2Sb_2S_5$ source that had been prepared by annealing a stoichiometric mixture of the individual elements in an evacuated quartz tube at 600 °C. After cooling, the resulting material was identified as being phase pure $Sn_2Sb_2S_5$ by XRD; it was then powdered for use as the raw material for vacuum evaporation onto unheated glass substrates. The resulting films were found to be polycrystalline $Sn_2Sb_2S_5$ with strong (104) preferred orientation; annealing in air at temperatures up to 250 °C had little effect on the grain sizes of the films, which remained at around 20 nm. Electrically, the films were described as being highly compensated in their as-deposited state, and exhibited low resistivity and n-type conduction after annealing. Optical measurements revealed two direct bandgaps, both of which resulted in high optical absorption coefficients (10^4 to 10^5 cm^{-1}), which dropped with increased annealing temperature. The lower bandgap (1.52–1.78 eV) was attributed to electron transitions from the valence to conduction band, and the higher bandgap (1.80–1.96 eV) attributed to the valence band spilling due to the crystal field of the lattice. However, the presence of two bandgaps has also been attributed to the coexistence of two structural phases, crystalline and amorphous $Sn_2Sb_2S_5$,[118] which is consistent with the XRD results presented in their earlier study.

Gassoumi and Kanzari[119] have also examined $SnSb_2S_4$, which was found to be amorphous in the as-deposited state and crystalline following annealing in air at temperature of up to 275 °C. The thickness of the films was found to have a significant effect on electrical and optical properties, with films <600 nm in thickness undergoing a dramatic transition from semiconducting to metallic behaviour at 150 °C as demonstrated by a hysteresis in their electrical resistance. Optical measurements showed the bandgap of the as deposited material gradually reduced from 1.3 to 1.0 eV as film thickness increased from 300 to 800 nm, and dropped to below 1 eV on annealing in air.

The Sn–Sb–S system has also been examined in two combinatorial studies,[120,121] Dittrich *et al.*[120] used co-evaporation of pure elements or binary sulfides (Sb_2S_3 and SnS) to grow one-dimensional libraries on both bare and Mo-coated float glass substrates. This provided a compositional gradient of 30% over the library's length and allowed compositional changes in properties to be easily resolved. Phase identification by XRD was reported to be 'very difficult to interpret'. The Sn–Sb–S system has several phases over its full compositional range, as can be seen in Figure 6.13(b), many of which have a low symmetric structure that results in complicated diffraction patterns, with analysis confounded further by possible preferred orientation. The effect of substrate temperature was noted to have the following structural effects:

- up to 150 °C, amorphous films produced;
- from 150 to 250 °C crystalline layers with highly preferred orientation;
- above 250 °C 'statistical oriented sulfosalt structures observed'.

Optical measurements showed evidence of both direct and indirect bandgaps, with the direct bandgaps ranging from 1.3 to 2.2 eV, with an ideal 1.5 eV found at around a Sn : Sb : S ratio of 1 : 1 : 2. Results from thermo-power measurements were found to be very strongly dependant on the

S content, with S $\leq$51% appearing to give p-type conductivity whereas S >54% appeared to give n-type conductivity, which is consistent with the n-type $Sn_2Sb_2S_5$ films grown by Gassoumi and Kanzari.[122] p-type films grown on Mo-coated glass were used to make libraries of devices with a substrate structure, as usually used for CIGS devices, with a 30–50 nm thick CdS buffer layer (grown by chemical bath deposition) followed by a i-ZnO/Al:ZnO front contact. Under 1000 W m^{-2} illumination, the best device gave V_{OC} = 0.208 V, J_{SC} = 133 A m^{-2}, FF = 38% and η = 1.05%. The best results for an all sputtered Sn–Sb–S cell are V_{OC} = 0.24 V, J_{SC} = 71 A m^{-2}, FF = 56%,[85] equivalent to η = 0.95%.

The Sn–Sb–S system has recently been re-examined in a combinatorial study by Ali *et al.*,[121] during which one-dimensional libraries were grown on soda lime glass microscope slides using a similar technique to that of Dittrich *et al.*,[119] with separate Sb_2S_3 and SnS thermal evaporation sources and unheated substrates. The crystallographic structures of the libraries were assessed by XRD and were found to depend strongly on the temperature of a post-deposition anneal under Ar. As-deposited libraries had a polycrystalline structure, and libraries annealed at $\geq$150 °C were amorphous at Sb concentrations above ~25 at%. While the library was noted to be polycrystalline at lower Sb concentrations, a strong preferred orientation again prevented the identification of which specific phases were present. The transition to amorphous structure at higher annealing temperatures is somewhat unexpected; especially as high temperature annealing is usually promotes the crystallisation of amorphous films and, as previously mentioned, higher growth temperatures (>150 °C) are associated with crystalline Sn–Sb–S films.[116] In the study by Ali and co-workers,[121] spectral photoconductivity measurements gave an optical bandgap of between 1.35 and 1.4 eV across the whole compositional range, irrespective of whether the film was amorphous or crystalline, with the strongest photoresponse found at the higher Sn concentrations. Hot probe measurements generally showed n-type conductivity, with some evidence of p-type conductivity in a library annealed at 325 °C. Overall, the tendency found by Ali and co-workers for Ar annealed films to be amorphous at Sb concentrations >25 at%, and exhibit n-type conductivity and have a bandgap of around 1.3 eV, draws comparison with Gassoumi and Kanzari's results for $SnSb_2S_4$ films.[120]

Compared with other materials there is relatively little published work on potential photovoltaic materials within the Sn–Sb–S system. From what we have seen it appears that the combinatorial studies are generally consistent with those that have targeted specific phases, with the most promising composition lying around that of $Sn_2Sb_2S_5$. This gives a high optical absorption coefficient and a bandgap that is favourable for thin film terrestrial photovoltaics. However, the carrier type depends strongly on S content, with a tendency to produce n-type conductivity that would be difficult to accommodate in conventional cell structures, although p-type conductivity may be practical by producing a S deficient film. Gassoumi and Kanzari have shown that moving from $Sn_2Sb_2S_5$ to $SnSb_2S_4$ results in

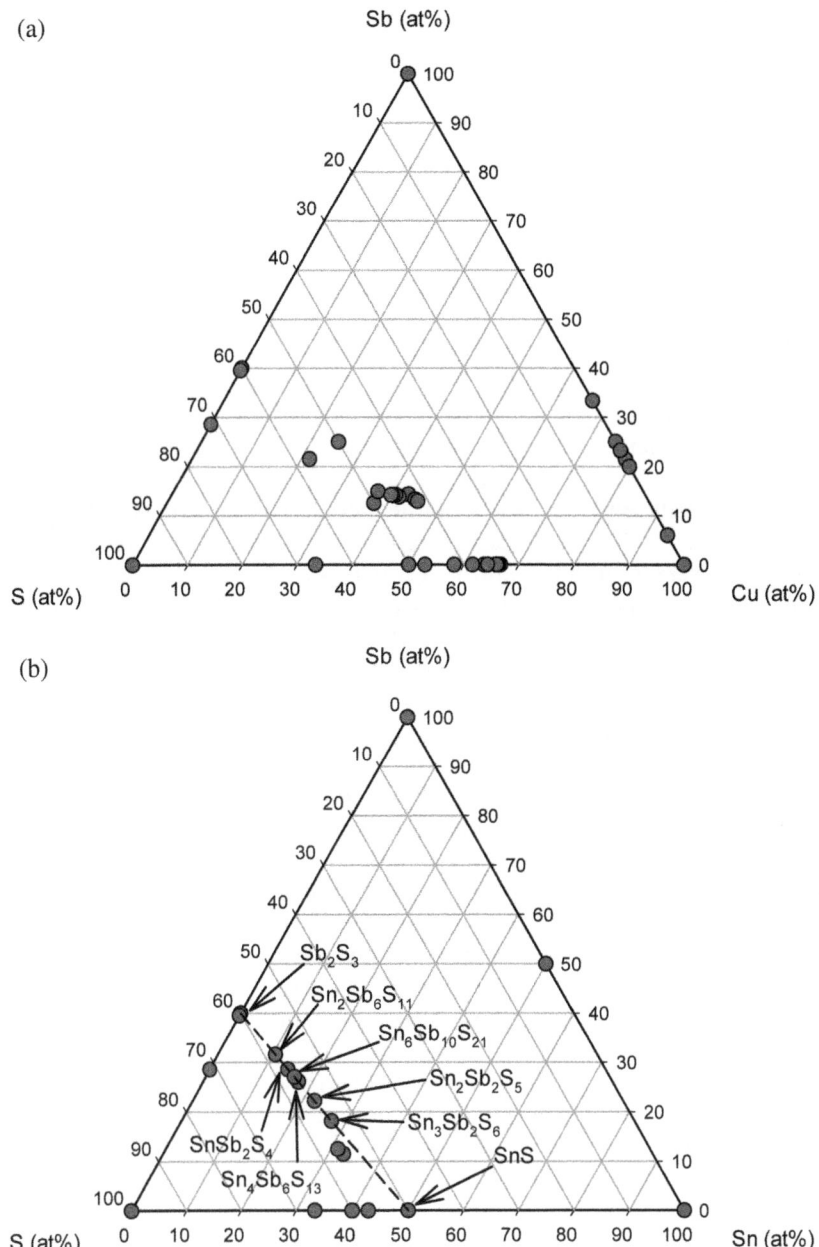

Figure 6.13 Compositions of known phases of selected ternary compounds based on data from the International Centre for Diffraction Data (ICDD). (a) Cu–Sb–S. (b) Sn–Sb–S.

a dramatic change in properties and that a higher Sb content (>25 at% according to Ali *et al.*); this may be accompanied by a tendency to an amorphous structure.[121]

6.5 Conclusions

There are many promising chalcogenide materials that have potential as absorber layers for thin film PV applications. The predicted growth in solar PV to multi terawatt scale implementation and production requires the development of materials to be focused on those that will not be constrained by production or scarcity. The rate of growth of the world PV market has decreased in the past three years largely due to the global economic down-turn following the financial crisis of 2008–2009. However, growth is begin-ning to increase again and this will need to accelerate if the goals for reducing carbon and increasing energy security through the implementation of solar PV electricity production are to be met. The timescale for change is short with many targets for major carbon reduction targeted at 2020 and 2050. To identify and develop suitable new materials within this period places additional demands on the research community to identify rapid processes to screen, identify and investigate suitable materials and devices based on these. This chapter has shown that combinatorial techniques have been used to fill this role. However, whilst these techniques are promising, the practical application of them to materials development needs to be carefully considered. The ability to select processing conditions that enable synthesis of samples with widely varying composition needs to take into account the effect across the combinatorial materials libraries. The chal-lenges include identification of process conditions which allow suitable quality of libraries to be produced that will allow their characterisation. The development of a comprehensive range of characterisation techniques is needed to enable interrogation of both materials and optoelectronic prop-erties, at suitably high resolution on thin films. The extension of the use of these techniques to device and multilayer structures will also require significant challenges to be overcome. The process has begun and the prospect of overcoming these challenges represents an exciting new prospect for the development of thin film PV materials.

The application of rapid screening has been used in the copper–zinc–tin-sulfur system. However, most of the development of this promising material for PV has relied upon conventional development approaches. With conversion efficiencies for the sulphur selenide form of the absorber already exceeding 12.5% and both vacuum and non-vacuum processing being used successfully to synthesis the material, the possibility of sustainable thin film PV is almost within our grasp. This chapter has also shown that there are other related materials in the sulfosalt family (the definition of which has been relaxed to include the selenide containing variants). From this family, the materials that have received some attention from the authors and other researchers include those based on copper–antimony and copper–bismuth, and tin–antimony sulfide. These are possible candidates for PV but these are at a very early stage of development with a relatively small research effort underway. A combinatorial approach would be a possible route to screening these materials for rapid development and the authors have used such an approach for initial studies.

The chapter has provided a brief overview of materials, sustainability issues and some new development approaches that face the field of thin film PV technologies for the challenging and exciting years ahead.

References

1. K. Zweibel, in *The Terawatt Challenge for Thin-Film PV*, National Renewable Energy Laboratory, Golden, CO, 2005, pp. 1–44.
2. A. Feltrin and A. Freundlich, *Renewable Energy,* 2008, **33**, 180–185.
3. J. E. Trancik and K. Zweibel, in *Proceedings 2006 IEEE 4th World Conference on Photovoltaic Energy Conversion*, 2006, pp. 2490–2493.
4. K. Zweibel, in *Thin Film Solar Cells: Fabrication, Characterization and Applications*, ed. J. Poortmans and V. Arkhipov, John Wiley & Sons, Chichester, UK, 2006, pp. 427–462.
5. I. Forbes and L. M. Peter, in *Materials for a Sustainable Future*, ed. T. M. Letcher and J. Scott, RSC Books, Cambridge, UK, 2012, p. 828.
6. US Geological Survey, *Mineral Commodity Summaries 2012*, US Geological Survey, Reston, VA, 2012.
7. A. Jäger-Waldau, in PV Status Report 2013, European Commission, Joint Research Centre Institute for Energy and Transport, EUR 26118 EN, Publications Office of the European Union, Luxembourg, 2013.
8. M. W. George, Tellurium, in *Mineral Commodity Summaries 2013*, US Geological Survey, Reston, VA, 2005, pp. 164–165, available at http://minerals.usgs.gov/minerals/pubs/commodity/selenium/mcs-2013-tellu.pdf.
9. T. J. Brown, R. A. Shaw, T. Bide, E. Petavratzi, E. R. Raycraft, A. S. Walters, Tech. and A. C. MacKenzie, *World Mineral Production 2007–11*, British Geological Survey, Nottingham, UK, 2013.
10. C. Mikolajczak, *Availability of Indium and Gallium*, The Indium Corporation, 2009.
11. C. S. Tao, J. Jiang and M. Tao, *Sol. Energy Mater. Sol. Cells,* 2011, **95**, 3176–3180.
12. Ad-hoc Working Group of Defining Critical Raw Materials, *Critical Raw Materials for the EU*, European Commission, Brussels, 2010.
13. T. K. Todorov, J. Tang, S. Bag, O. Gunawan, T. Gokmen, Y. Zhu and D. B. Mitzi, *Adv. Energy Mater.,* 2013, **3**, 34–38.
14. R. A. Potyrailo and I. Takeuchi, *Meas. Sci. Technol.,* 2005, **16**, 1–4.
15. L. Asinovski, D. Beaglehole and M. T. Clarkson, *Phys. Status Solidi A,* 2008, **205**, 764–771.
16. R. H. Bube, *Photovoltaic Materials*, Imperial College Press, London, 1998.
17. A. E. Rakhshani, *Solid-State Electron.,* 1986, **29**, 7–17.

18. B. P. Rai, *Sol. Cells,* 1988, **25**, 265–272.
19. T. L. Chu and S. S. Chu, *Solid-State Electron.,* 1995, **38**, 533–549.
20. A. Ennaoui, S. Fiechter, C. Pettenkofer, N. Alonso-Vante, K. Bilker, M. Bronold, C. Höpfner and H. Tributsch, *Sol. Energy Mater. Sol. Cells,* 1993, **29**, 289–370.
21. Regulation (EC) No 1907/2006 of the European Parliament and of the Council of 18 December 2006 concerning the Registration, Evaluation, Authorisation and Restriction of Chemicals (REACH), establishing a European Chemicals Agency, *Off. J. Eur. Union,* 2006, **L396**, 3–280.
22. W. Shockley and H. J. Queisser, *J. Appl. Phys.,* 1961, **32**, 510–519.
23. C. Wadia, A. P. Alivisatos and D. M. Kammen, *Environ. Sci. Technol.,* 2009, **43**, 2072–2077.
24. A. Ennaoui, S. Fiechter, G. Smestad and H. Tributsch, in *Energy and the Environment into the 1990s: Proceedings 1st World Renewable Energy Congress*, Pergamon Press, Oxford, 1990, vol. 1, pp. 458–464.
25. J.-C. Zhao, *Prog. Mater. Sci.,* 2006, **51**, 557–631.
26. M. Ohring, *Materials Science of Thin Films*, Academic Press, London, 2nd edn, 2001.
27. J. A. Haber, N. J. Gerein, T. D. Hatchard and M. Y. Versavel, in *Proceedings of 31st IEEE Photovoltaic Specialists Conference*, 2005, pp. 155–158.
28. S. S. Mao, *Appl. Phys. A,* 2011, **105**, 283–288.
29. D. Lincot, *Thin Solid Films,* 2005, **487**, 40–48.
30. O. Goldstein, *The Production and Characterisation of Inorganic Combinatorial Libraries*, PhD thesis, Cranfield University, Shrivenham, UK, 2010.
31. Supplied by J. Kurt Lesker Ltd. 15/16 Burgess Road Hastings, East Sussex TN35 4NR, England, www.lesker.com.
32. J. Goldstein, D. E. Newbury, D. C. Joy, C. E. Lyman, P. Echlin, E. Lifshin, L. Sawyer and J. R. Michael, *Scanning Electron Microscopy and X-ray Microanalysis*, Springer, New York, 2003.
33. B. Beckhoff, B. Kanngießer, N. Langhoff, R. Wedell and H. E. Wolff, *Handbook of Practical X-ray Fluorescence Analysis*, Springer, Berlin and London, 2006.
34. Y. Wang and M. A. Nastasi, *Handbook of Modern Ion Beam Materials Analysis*, Cambridge University Press, Cambridge, 2nd edn, 2010.
35. B. D. Cullity and S. R. Stock, *Elements of X-ray Diffraction*, Prentice Hall, Upper Saddle River, NJ, 2001.
36. S. Roncallo, O. Karimi, K. D. Rogers, D. W. Lane and S. A. Ansari, *J. Appl. Crystallogr.,* 2009, **42**, 174–178.
37. P. A. Fernandes, P. M. P. Salomé and A. F. da Cunha, *J. Alloys Compd.,* 2011, **509**, 7600–7606.
38. S. Schorr, *Sol. Energy Mater. Sol. Cells,* 2011, **95**, 1482–1488.
39. O. S. Heavens, *Optical Properties of Thin Solid Films*, Dover Publications, New York, 1991.

40. H. Fujiwara, *Spectroscopic Ellipsometry: Principles and Applications*, John Wiley & Sons, Chichester, UK, 2007.

41. J. Chan, *EECS 143 Four-Point Probe*, University of California, Berkeley, EECS, 1994; modified by P. Friedberg, 2002.

42. S. B. Catalano, *IEEE Trans. Electron Devices*, 1963, **10**, 185–188.

43. G. L. Miller, D. A. H. Robinson and J. D. Wiley, *Rev. Sci. Instrum.*, 1976, **47**, 799–805.

44. www.lehighton.com.

45. D. K. Schroder, *Semiconductor Material and Device Characterization*, John Wiley & Sons, Hoboken, NJ, 3rd edn, 2006.

46. Keithley Instruments, *Measuring the Resistivity and Determining the Conductivity Type of Semiconductor Mater. Using a Four-Point Collinear Probe and the Model 6221 DC and AC Current Source*, Application Note Series Number 2615, Keithley Instruments, Inc., Cleveland, OH, 2005.

47. G. Golan, A. Axelevitch, B. Gorenstein and V. Manevych, *Microelectron. J.*, 2006, **37**, 910–915.

48. N. V. Joshi, *Photoconductivity: Art, Science, and Technology*, CRC Press, Boca Raton, FL, 1990.

49. S. Chen, X. G. Gong, A. Walsh and S.-H. Wei, *Appl. Phys. Lett.*, 2009, **94**, 041903.

50. H. Katagiri, K. Saitoh, T. Washio, H. Shinohara, T. Kurumadani and S. Miyajima, *Sol. Energy Mater. Sol. Cells*, 2001, **65**, 141–148.

51. K. Tanaka, N. Moritake and H. Uchiki, *Sol. Energy Mater. Sol. Cells*, 2007, **91**, 1199–1201.

52. S. R. Hall, J. T. Szymanski and J. M. Stewart, *Can. Mineral.*, 1978, **16**, 131–137.

53. K. Ito and T. Nakazawa, *Jpn. J. Appl. Phys.*, 1988, **27**, 4.

54. T. Tanaka, T. Nagatomo, D. Kawasaki, M. Nishio, Q. Guo, A. Wakahara, A. Yoshida and H. Ogawa, *J. Phys. Chem. Solids*, 2005, **66**, 1978–1981.

55. N. Momose, M. T. Htay, T. Yudasaka, S. Igarashi, T. Seki, S. Iwano, Y. Hashimoto and K. Ito, *Jpn. J. Appl. Phys.*, 2011, **50**, 01BG09.

56. P. A. Fernandes, P. M. P. Salomé and A. F. da Cunha, *Semiconduct. Sci. Technol.*, 2009, **24**, 105013.

57. H. Araki, A. Mikaduki, Y. Kubo, T. Sato, K. Jimbo, W. S. Maw, H. Katagiri, M. Yamazaki, K. Oishi and A. Takeuchi, *Thin Solid Films*, 2008, **517**, 1457–1460.

58. T. Prabhakar and J. Nagaraju, in *Proceedings 35th IEEE Photovoltaic Specialists Conference (PVSC)*, 2010, pp. 001964–001969.

59. N. Moritake, Y. Fukui, M. Oonuki, K. Tanaka and H. Uchiki, *Phys. Status Solidi C*, 2009, **6**, 1233–1236.

60. Z. Zhou, Y. Wang, D. Xu and Y. Zhang, *Sol. Energy Mater. Sol. Cells*, 2010, **94**, 2042–2045.

61. T. K. Todorov, K. B. Reuter and D. B. Mitzi, *Adv. Mater.*, 2010, **22**, E156–E159.

62. Q. Guo, H. W. Hillhouse and R. Agrawal, *J. Am. Chem. Soc.*, 2009, **131**, 11672–11673.

63. T. M. Friedlmeier, N. W. Wieser, T. H. Dittrich and H. W. Schock, in *Proceedings 14th European Photovoltaic Specialists Conference*, 1997, 1242.

64. J. Scragg, P. Dale and L. Peter, *Electrochemistry Commun.*, 2008, **10**, 639–642.

65. H. Katagiri, K. Jimbo, K. Moriya and K. Tsuchida, in *Proceedings of 3rd World Conference on Photovoltaic Energy Conversion*, 2003, vol. 2873, pp. 2874–2879.

66. H. Katagiri, K. Jimbo, S. Yamada, T. Kamimura, W. S. Maw, T. Fukano, T. Ito and T. Motohiro, *Appl. Phys. Express.*, 2008, **1**, 2.

67. H. Wang, *Int. J. Photoenergy*, 2011, **2011**, 10.

68. W. Wang, M. T. Winkler, O. Gunawan, T. Gokmen, T. K. Todorov, Y. Zhu and D. B. Mitzi, *Adv. Energy Mater.*, 2014, **4**, 1301465, doi: 10.1002/aenm.201301465.

69. D. A. R. Barkhouse, O. Gunawan, T. Gokmen, T. K. Todorov and D. B. Mitzi, *Prog. Photovoltaics: Res. Appl.*, 2012, **20**, 6–11.

70. Q. Guo, G. M. Ford, W.-C. Yang, B. C. Walker, E. A. Stach, H. W. Hillhouse and R. Agrawal, *J. Am. Chem. Soc.*, 2010, **132**, 17384–17386.

71. K. Wang, O. Gunawan, T. Todorov, B. Shin, S. J. Chey, N. A. Bojarczuk, D. Mitzi and S. Guha, *Appl. Phys. Lett.*, 2010, **97**, 143508.

72. S. Ahmed, K. B. Reuter, O. Gunawan, L. Guo, L. T. Romankiw and H. Deligianni, *Adv. Energy Mater.*, 2012, **2**, 253–259.

73. Solar Frontier, Solar Frontier and IBM Sign Agreement to Develop CZTS Solar Cell Technology [online], press release, Solar Frontier, Tokyo, available at http://www.solar-frontier.com/eng/news/2010/C002161.html [accessed 19 October, 2010].

74. V. G. Rajeshmon, C. S. Kartha and K. P. Vijayakumar, in *AIP Conference Proceedings*, 2011, vol. 1349, pp. 683–684.

75. W. Li, S. Cohen, K. Gartsman, R. Caballero, P. van Huth, R. Popovitz-Biro and D. Cahen, *Sol. Energy Mater. Sol. Cells*, 2012, **98**, 78–82.

76. A. Chirilă, S. Buecheler, F. Pianezzi, P. Bloesch, C. Gretener, A. R. Uhl, C. Fella, L. Kranz, J. Perrenoud, S. Seyrling, R. Verma, S. Nishiwaki, Y. E. Romanyuk, G. Bilger and A. N. Tiwari, *Nat. Mater.*, 2011, **10**, 857.

77. T. Tanaka, A. Yoshida, D. Saiki, K. Saito, Q. Guo, M. Nishio and T. Yamaguchi, *Thin Solid Films*, 2010, **518**, S29–S33.

78. G. Suresh Babu, Y. B. Kishore Kumar, P. Uday Bhaskar and S. Raja Vanjari, *Sol. Energy Mater. Sol. Cells*, 2010, **94**, 221–226.

79. K. Tanaka, Y. Fukui, N. Moritake and H. Uchiki, *Sol. Energy Mater. Sol. Cells*, 2011, **95**, 838–842.

80. Y. B. K. Kumar, P. U. Bhaskar, G. S. Babu and V. S. Raja, *Phys. Status Solidi A*, 2010, **207**, 149–156.

81. C. Platzer-Björkman, J. Scragg, H. Flammersberger, T. Kubart and M. Edoff, *Sol. Energy Mater. Sol. Cells*, 2012, **98**, 110–117.

82. K. J. Hutchings, *High Throughput Combinatorial Screening of Cu-Zn-Sn-S Thin Film Libraries for the Application of Cu₂ZnSnS₄ Photovoltaic Cells*, PhD thesis, Cranfield University, Shrivenham, UK, 2012.

83. C.-Y. Chou and S.-W. Chen, *Acta Mater.*, 2006, **54**, 2393–2400.

84. H. Dittrich, A. Bieniok, U. Brendel, M. Grodzicki and D. Topa, *Thin Solid Films*, 2007, **515**, 5745–5750.

85. H. Dittrich, A. Stadler, D. Topa, H.-J. Schimper and A. Basch, *Phys. Status Solidi A*, 2009, **206**, 1034–1041.

86. Y. Moëlo, E. Makovicky, N. N. Mozgova, J. L. Jambor, N. Cook, A. Pring, W. Paar, E. H. Nickel, S. Graeser, S. Karup-Møller, T. Balic-Žunic, W. G. Mumme, F. Vurro, D. Topa, L. Bindi, K. Bente and M. Shimizu, *Eur. J. Mineral.*, 2008, **20**, 7–46.

87. http://www.solarbuzz.com.

88. A. Sugaki, H. Shima and A. Kitakaze, *Phase Relations of the Cu₂S-Sb₂S₃ System*, Yamaguchi University, Japan, 1973.

89. mindat.org, Chalcostibite [online], http://www.mindat.org/min-983.html.

90. B. J. Skinner, F. Luce, D. and E. Makovicky, *Economic Geology*, 1972, **67**, 924–938.

91. S. Klnup-Møller and E. Makovicky, *Am. Mineral.*, 1974, **59**, 889–895.

92. A. Wachtel and A. Noreika, *J. Electron. Mater.*, 1980, **9**, 281–297.

93. J. H. B. Wernick and K. E. Benson, *Phys. Chem. Solids*, 1957, **3**, 157–159.

94. J. Zhou, G.-Q. Bian, Q.-Y. Zhu, Y. Zhang, C.-Y. Li and J. Dai, *J. Solid State Chem.*, 2009, **182**, 259–264.

95. Y. Rodríguez-Lazcano, M. T. S. Nair and P. K. Nair, *J. Cryst. Growth*, 2001, **223**, 399–406.

96. S. Manolache, A. Duta, L. Isac, M. Nanu, A. Goossens and J. Schoonman, *Thin Solid Films*, 2007, **515**, 5957–5960.

97. A. M. Fernandez and J. A. Turner, *Sol. Energy Mater. Sol. Cells*, 2003, **79**, 391–399.

98. A. Bansal, T. Deutsch, J. E. Leisch, S. Warren, J. A. Turner and A. M. Fernández, in *Proceedings of the 2001 DOE Hydrogen Program Review*, NREL/CP-570-30535, National Renewable Energy Laboratory, Golden, CO, 2001, pp. 347–358.

99. L. Yu, R. S. Kokenyesi, D. A. Keszler and A. Zunger, *Adv. Energy Mater.*, 2013, **3**, 43–48.

100. D. Colombara, L. Peter, K. Rogers and K. Hutchings, *J. Solid State Chem.*, 2012, **186**, 36–46.

101. S. Wilman, I. Shigeru, I. Yuta, H. Takashi and M. Michio, *Thin Solid Films*, 2014, **550**, 700–704.

102. P. Maiello, G. Zoppi, R. W. Miles, N. Pearsall and I. Forbes, *Sol. Energy Mater. Sol. Cells*, 2013, **113**, 186–194.

103. F. Al-Saab, B. Gholipour, C. C. Huang, D. W. Hewak, A. Anastasopoulos and B. Hayden, in *Proceedings of the 9th Photovoltaic Science, Applications and Technology Conference (PVSAT-9)*, Solar Energy Society, Abingdon, UK, 2013, p. 99.

104. E. Peccerillo and K. Durose, in *Proceedings of the 9th Photovoltaic Science, Applications and Technology, Conference (PVSAT-9)*, Solar Energy Society, Abingdon, UK, 2013, p. 187.

105. Wittichenite Mineral Data [online], available at http://webmineral.com/data/Wittichenite.shtml#.U6rNpbH-tbp.

106. A. G. Aberle, *Thin Solid Films*, 2009, **517**, 4706–4710.

107. S. Price and R. Margolis, 2008 Solar Technologies Market Report, National Renewable Energy Laboratory, Golden, CO, 2010.

108. N. J. Gerein and J. A. Haber, *Chem. Mater.*, 2006, **18**, 6297–6302.

109. P. K. Nair, S. M. Nair, H. Hu, H. Ling, R. A. Zingaro and E. A. Meyers, in *Proceedings SPIE 2531, Optical Materials Technology for Energy Efficiency and Solar Energy Conversion XIV*, 1995, pp. 208–219.

110. P. Nair, L. Huang, M. Nair, H. Hu, E. Meyers and R. Zingaro, *J. Mater. Res.*, 1997, **12**, 651–656.

111. V. Estrella, M. T. S. Nair and P. K. Nair, *Semiconduct. Sci. Technol.*, 2003, **18**, 190.

112. N. Gerein and J. Haber, in *Proceedings 31st IEEE Photovoltaic Specialists Conference 2005*, 2005, pp. 159–162.

113. N. Gerein and J. Haber, *Chem. Mater.*, 2006, **18**, 6289–6296.

114. F. Mesa and G. Gordillo, *J. Phys.: Conf. Ser.*, 2009, 012019.

115. F. Mesa, A. Dussan and G. Gordillo, *Phys. Status Solidi C*, 2010, **7**, 917–920.

116. P. Dale, A. Loken, L. Peter and J. Scragg, *ECS Trans.*, 2009, **19**, 179–187.

117. A. Gassoumi and M. Kanzari, *J. Optoelectron. Adv. Mater.*, 2009, **11**, 414–420.

118. I. Gaied, A. Gassoumi, M. Kanzari and N. Yacoubi, *J. Phys.: Conf. Ser.*, 2010, **214**, 012127.

119. A. Gassoumi and M. Kanzari, *J. Optoelectron. Adv. Mater.*, 2010, **12**, 1052–1057.

120. H. Dittrich, K. Herz, J. Eberhardt and G. Schumm, in *Proceedings 14th European Photovoltaic Solar Energy Conference*, 1997, pp. 2054–2057.

121. N. Ali, A. Hussain, S. T. Hussain, M. A. Iqbal, M. Shah, I. Rahim, N. Ahmad, Z. Ali, K. Hutching, D. Lane and W. A. A. Syed, *Curr. NanoSci.*, 2013, **9**(1), 149–152.

122. A. Gassoumi and M. Kanzari, *Thin Films Chalcogenide Lett.*, 2009, **6**, 163–170.

III–V Solar Cells

JAMES P. CONNOLLY[*a] AND DENIS MENCARAGLIA[b]

[a]Nanophotonic Technology Centre, Universidad Politécnica de Valencia, Camino de Vera s/n, 46022 Valencia, Spain; [b]Laboratoire de Génie Électrique de Paris, LGEP, UMR 8507 CNRS-Supélec, Université Pierre et Marie Curie, Université Paris-Sud, 11 rue Joliot-Curie, Plateau de Moulon, 91192 Gif-sur-Yvette Cedex, France
*E-mail: connolly@ntc.upv.es

7.1 Introduction

Upward trends in energy costs are a powerful driver in the development of new energy sources and efforts to reduce the relative costs of a range of technologies. This is the case for the III–V semiconductor compounds, which are traditionally[1] an expensive photovoltaic (PV) technology whilst also being the most efficient, with corresponding advantages and disadvantages.

Principal among the disadvantages are the relatively complex synthesis and device fabrication, and corollary issues such as availability of relatively rare elements (In, Ga).[2] These two points are largely responsible for the higher cost. Among the advantages, however, are a number of materials characteristics which help make III–V solar cells the most efficient photovoltaic materials available at present. The main reason for this is the flexible combination of a range of materials from binary to quaternary compounds, with a corresponding flexibility of bandgap engineering. More significantly, a number of these compounds interact strongly with light, since they largely retain direct bandgaps and correspondingly high absorption coefficients, and therefore also tend to radiate light efficiently. This is a class of materials

RSC Energy and Environment Series No. 12
Materials Challenges: Inorganic Photovoltaic Solar Energy
Edited by Stuart J C Irvine
Published by the Royal Society of Chemistry, www.rsc.org

which therefore features most of the opto-electronically efficient semiconductors.

With these advantages, the III–V semiconductors are a flexible group of materials well suited for opto-electronic applications. They are therefore good materials for high efficiency solar cells using the basic single junction concepts developed since the early days of photovoltaics. The bandgap engineering aspect allows this class of cells to be tailored to different spectra (*e.g.* global, direct concentrated or space spectra) and their corresponding applications. Moreover, they are ideally suited for the fundamental development of new concepts because the flexible bandgap engineering properties allow new designs to be investigated.

The overall result of these considerations is that niche applications requiring high efficiency or fundamental research have largely driven III–V photovoltaic development to date. Historically, the first and most important niche application has involved space applications where low weight and hence high efficiency combined with reliability is the prime concern. This is currently being supplemented by terrestrial applications using the cost-reducing solar concentrator technologies.

The following sections address some materials aspects of III–V solar cell development with a focus on design for maximum efficiency resting on the flexibility afforded by this family of materials.

7.2   Materials and Growth

7.2.1   The III–V Semiconductors

The III–V semiconductors are based on group III (boron group) and group V (nitrogen group) elements as illustrated in Figure 7.1. This common diagram[3] locates the most interesting compounds for photovoltaic applications in terms of their bandgaps and lattice constants as reported in a range of sources.[4–7] Also shown is the terrestrial solar spectrum AM1.5 allowing us to see the III–V materials in the context of available power, and demonstrating good coverage of almost the entire terrestrial spectrum.

The most commonly used substrates are first GaAs and then InP, which incidentally possess bandgaps near the ideal for solar conversion. Furthermore, materials compatible with these substrates are the most technologically important. These are first of all lattice matched compounds, which can be grown without strain relaxation and associated defects reducing device performance. The dashed arrows in Figure 7.1 indicate substrate lattice parameters and we include here the two main group IV semiconductors Ge and Si, which provide possible routes to lower cost fabrication *via* heterogeneous growth. We note in passing that Ge is close to a lattice match with GaAs, and that a small amount of In to GaAs can allow exact lattice matching to low-cost Ge substrates. The substrate is therefore frequently used in triple designs as the lowest bandgap component. In addition to the binary compounds, a rich family of ternary and quaternary compounds are available

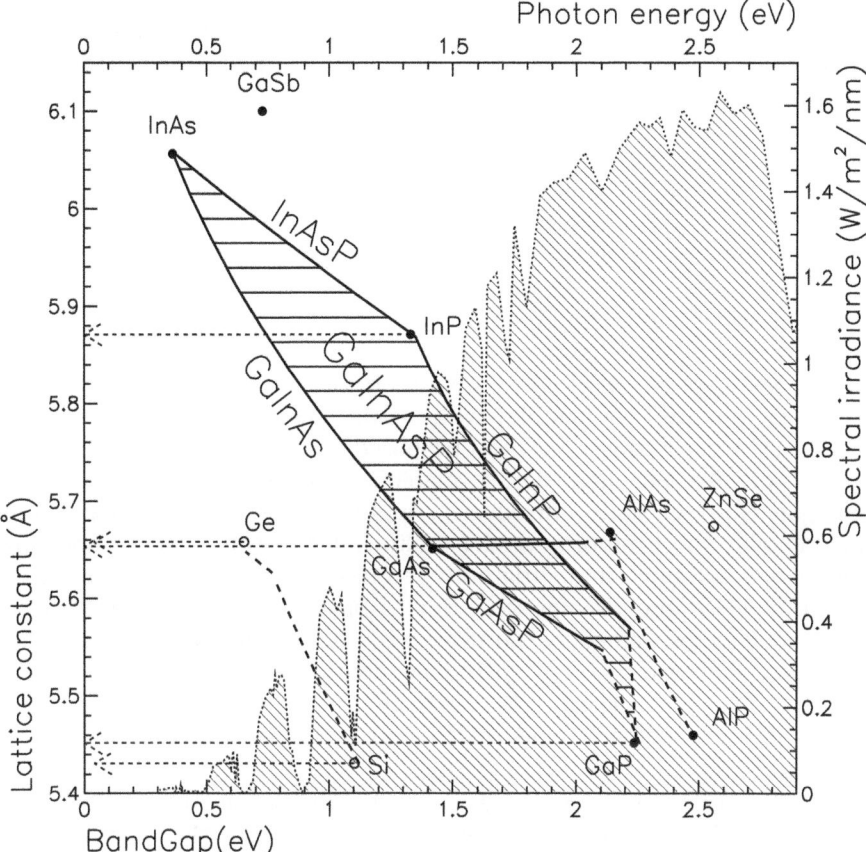

Figure 7.1 Significant III–V semiconductors in terms of 300 K lattice parameters and bandgaps, with the horizontal dashed area indicating the range of GaInAsP compositions, and positioning common III–V and group IV substrates in the context of the solar spectrum (shaded area upper, right axis).

by substitution of a range of fractions of different group III and group V atoms while maintaining a stoichiometric III–V ratio.

Considering first the GaAs materials family, the ternary compound Al_x-$Ga_{1-x}As$ is historically the first extensively studied material.[8–10] It retains a direct bandgap greater than GaAs over the greater part of the compositional range and remains nearly lattice matched to GaAs across the entire range of Al fractions (Figure 7.1). Residual strain can be essentially eliminated with the addition of a small amount of phosphorus allowing lattice matched $AlAs_{0.96}P_{0.04}$ growth on GaAs. However, it suffers from materials issues related primarily to Al and associated recombination DX centres,[9–11] which limit device efficiencies unacceptably. Furthermore, these compounds are increasingly unstable and highly reactive with increasing Al composition.

As a result, despite the seemingly ideal bandgap engineering potential of these compounds, their use in photovoltaic cells is largely limited to window layers, tunnel junctions and some work on heterojunction AlGaAs/GaAs concepts.[12] Nevertheless, there exist other less reactive Al ternary and quaternary compounds with gaps greater than GaAs. The chief amongst these for photovoltaic applications is AlInP lattice matched to GaAs. It is an important window layer and usually preferably to AlGaAs due to its lower reactivity.

Considering phosphides, the ternary compound $Ga_{0.515}In_{0.485}P$ is also lattice matched to GaAs. The gap of this material varies as a function of sub-lattice ordering: an ordered group III lattice yields a direct gap of 1.96 eV, which is reduced, depending on the degree of disorder, by up to 0.5 eV.[13]

For bandgaps lower than GaAs, there is a shortage of attractive III–V compounds compatible with GaAs. The quaternary solution, $In_{1-x}Ga_xN_yAs_{1-y}$ nitrides, has been proposed[14] as the addition of just a few percent of nitrogen allows lattice matching and an ideal third junction bandgap. However, it also introduces crippling material defects that lead to unacceptably short minority carrier lifetimes for reasons that are not fully understood, although interstitial nitrogen has been shown to play a role.[15] But despite slow progress for some time, a breakthrough has recently been achieved with the pentenary GaInNAsSb by Sabnis and colleagues at Solar Junction, who have reported an independently verified world record efficiency of 43.5% at 400 suns.[16] Despite this impressive result, details are unavailable and the performance of this novel pentenary dilute nitride is ill defined.

Following this impressive result, a new 44.7% four-junction record[17] was achieved in 2013. This avoided nitride materials and instead used wafer bonding to combine two independently grown dual-junction structures. The structures consist of a GaInP/GaAs tandem wafer bonded to a GaInAsP/GaInAs tandem, achieving the 44.7% record in a direct spectrum for a concentration of 297 suns. This is currently the absolute record efficiency albeit at some cost in materials and in fabrication complexity.

Concerning substrates of InP, only the ternary $In_{0.53}Ga_{0.47}As$ of bandgap approximately 0.72 eV is lattice matched to it. As we will see this is a non-ideal bandgap combination for multi-junction designs. More fundamentally, InP, despite its near ideal band structure and corresponding limiting efficiency of 31%, achieved just 22% two decades ago with no certified progress since,[18] although related work[19] continues on cells lattice matched to InP substrates. This is due partly to inherent performance issues and to the fact that InP is a relatively dense and rather brittle material,[20,21] therefore posing handling difficulties and making industrial low-cost development challenging.

The overall conclusion is that the quaternary compound GaInAsP is currently the most important materials family, including as it does compounds lattice matched to all the major substrates in use. It comprises as subsets the three important ternary phosphides—GaAsP, GaInP and

InAsP—as well as the all-important GaInAs materials family, essential in a wide range of applications.

This overview of materials leads us to the conclusion that compounds of the GaInAsP family on GaAs and Ge substrates are the most promising. The following sections give an overview of some progress in the development of designs based on these materials.

7.2.2 Growth Methods

A brief mention of III–V growth[22] is key to understanding the cost of these materials. To start with, wafer growth is by standard single crystal boule fabrication usually by one of two methods.

The first is the Czochralski method, where a single crystal seed of the material in a known crystal orientation is placed in contact with a melt comprising a liquid solution of the same material, which may be encapsulated to prevent the evaporation of some species, and in particular As. The crystal is pulled slowly from the melt producing a single crystal ingot or boule.

The similar float zone and Bridgman alternatives consist of moving the melt away from the seed, rather than pulling the seed and crystal away from the melt. The recrystallisation occurs behind the moving heater and associated melt zone, resulting again in a single crystal boule. This method has one advantage in that impurities are expelled from the melt at the interface with the crystal. By this means, very high purity crystals can be achieved by repeated passes with the impurities segregated in the section of the boule furthest from the seed.

The boule is subsequently mechanically cut into wafers. Following this, surface treatment such as polishing produces the final single crystal wafers ready for further processing. The processing may consist of direct conversion into devices. For solar cells, the process consists primarily of diffusion or ion implantation of doping profiles defining the junction and enabling photovoltaic action, followed by additional essential features such as metal contacting which is an art we will not explore further here.

The steps described so far allow fabrication of single junction solar cells. More sophisticated structures are made using such wafers as growth substrates by further epitaxial growth techniques.

The first class is the relatively low-cost chemical vapour deposition methods. The dominant variant is atmospheric pressure metal organic vapour phase epitaxy (MOVPE) or metal organic chemical vapour deposition (MOCVD). This is a technique whereby metal organic precursor gases, optionally including dopant species, are flowed through a growth chamber. The precursors impinge on the wafer, placed on a temperature controlled stage, leading to epitaxial growth at a rate of the order of microns per hour. A simple example is trimethylgallium (TMGa) and arsine (AsH_3) in a H_2 carrier react $[Ga(CH_3)_3 + AsH_3 \rightarrow GaAs + 3\ CH_4]$ forming epitaxial GaAs monolayers.

In principle any number of sources can be attached to a growth reactor. Switching between these enables the layer-by-layer growth of heterogeneous semiconductors within limits set by material properties of strain and reactivity, and material-specific residual background levels that may accumulate in the reactor.

Further techniques such as low pressure chemical vapour deposition (LPCVD) are variations on the same theme, each with its strengths and weaknesses. Overall, however, MOCVD is much used due to its relatively low cost and the monolayer control achievable in the best conditions.

The second higher cost growth method is molecular beam epitaxy (MBE). In this ultra-high vacuum technique, ultrapure precursor solids are placed in radiatively heated graphite Knudsen cells attached to the growth chamber containing the substrate. Opening shutters on the cells allows a molecular beam to be emitted from the cell at a temperature controlled rate. This beam impinges on the temperature controlled substrate stage, which may be angled to adjust growth modes and conditions, and rotated to optimise growth uniformity. A range of Knudsen cells are usually attached to a MBE reactor in order to deposit heterogeneous structures on a single substrate, with the same limitations due to geometry and materials properties. Here variants again exist, for example, gas-source molecular beam epitaxy (GSMBE) or metal organic molecular beam epitaxy (MOMBE).

7.2.3 Heterogeneous Growth

The layer deposition methods outlined above allow excellent two-dimensional control of different materials, but there are fairly tight limitations on the heterostructures that can be grown. The first, which we will not mention in detail, is that certain species with high sticking coefficients, for example, have an unfortunate tendency to haunt growth chambers (*e.g.* by dynamically adsorbing and desorbing from their surfaces). This can seriously contaminate subsequent layers and must be avoided by growth chamber purges which significantly increase machine downtime and deposition cost. Furthermore, ideal growth conditions differ for different materials. In particular, different growth temperatures are routinely needed for different materials, but must be carefully optimised to take account of different thermal coefficients of expansion, of solubility, and therefore of elemental species migration. The greatest difficulty in this class is generally dopant diffusion, as in the well-known case of highly mobile Zn diffusion in AlGaAs/GaAs.

Finally, an all-important heterogeneous growth consideration is the lattice constant. Sequentially growing layers with different lattice constants in the same stack gives rise to strain. The total strain energy increases with the thickness of the layers deposited. Above a limit known as the Matthews–Blakeslee[23] critical thickness, the cumulative strain energy density at the heteroface becomes greater than the bond energy and the total system energy releases strain potential energy by breaking bonds in the interface region. This is strain relaxation, which generates dislocations that seriously compromise cell performance and even structural integrity. The Matthews–Blakeslee limit is a function of materials parameters, the lattice constant and elasticity tensors. For example, no more than approximately 351 Å or so of $In_{0.01}Ga_{0.99}As$, that is about 60 monolayers, can be grown on GaAs before the limit is exceeded and misfit dislocations are generated.

However, solutions to the lattice misfit problem have been implemented. One is strain compensation in multilayer structures with alternating compressive and tensile strain layers. Another is to allow the layer to relax. It is seen that after a sufficient further layer thickness, the dislocation density can return to reasonably low levels, sufficient for some device applications. This is known as the virtual substrate or relaxed buffer technique by metamorphic growth. Finally, a variation on this is the graded buffer growth technique,[24] which has been used with some success in multi-junction solar cells. With this method the composition is varied in incremental steps, restricting the dislocations in each case and again producing an effective virtual substrate.

7.3 Design Concepts

Improving design starts with understanding losses. In solar cells under illumination these include extrinsic losses such as reflection or external resistance, and intrinsic, such as optical and electrical transport losses. The optical losses result from poor light–matter interaction and, first of all, inefficient light absorption. The transport losses can be described under the umbrella of finite carrier lifetimes and corresponding recombination loss *via* a range of channels. Understanding and addressing these loss issues can be achieved most reliably *via* numerical modelling,[25-27] which can deliver exact solutions to analytically intractable problems. These are obtained at the expense, however, of some physical understanding, though this can be recovered by sweeping large parameter spaces. These numerical methods are usually required for complex materials and structures.

Understanding can also be gained *via* analytical methods by applying approximations resulting from exploring physical processes in limiting cases. In this case precisely the reverse is true in that greater understanding is achieved at the expense of physical accuracy though this may, however, be recovered by refining the theoretical picture. Furthermore, this approach is well suited to crystalline materials and structures symmetrical enough to lend themselves to analytical methods, as is the case with many III–V designs.

The following sections follow the second route and apply analytical models to a range of scenarios. We first develop a picture of the dominant sources of efficiency loss and investigate some examples of solutions that may address these losses.

7.3.1 Light and Heat

We now examine the fundamental losses of light absorption and heat dissipation in order to define the basic concepts of solar cell design and the answers that III–V materials may bring to the issue. Of the first category of optical losses mentioned earlier, we start with the transparency of the cell to photons with energies below its bandgap. This obvious and important fact in cell design is illustrated in Figure 7.2, which shows the fraction of the incident AM1.5G spectrum with energy below gap that is transmitted through a cell. The resulting transparency loss as a function of bandgap is illustrated in Figure 7.3. This loss is small for low gap materials as expected, and rises with increasing bandgap. InP, for example, is subject to a 26.6% transparency loss.

The next fundamental loss is thermalisation (Figure 7.2), whereby carriers photo-excited with energies greater than the bandgap E_g rapidly thermalise, mainly *via* collisions with the lattice, establishing a steady state minority carrier population with a quasi-Fermi level (QFL) near the band-edge. The photogenerated carriers are harvested with a fixed energy close to that of the lowest energy photons absorbed, wasting the remainder largely as heat. The resulting loss is shown in Figure 7.3 and this time shows an unsurprising high thermalisation loss for low bandgaps. For InP, again, the loss is a 26.7%, nearly identical to the transparency loss. This symmetry is consistent with the fact that InP is close to the optimum bandgap, as we will see with more exact methods. Together, the transparency and thermalisation loss mechanisms lead to a total maximum efficiency of 46.7%, as shown on Figure 7.3 by the solid line combining both loss mechanisms.

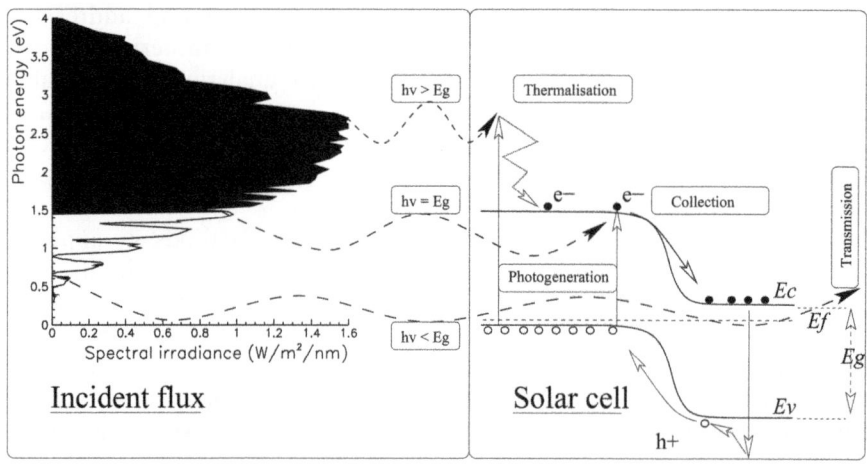

Figure 7.2   Illustration of losses with respect to the AM1.5G spectrum in a GaAs cell, showing transparency loss for photons with less energy than E_g, and thermalisation loss for electrons and holes absorbing photons with energy greater than E_g.

For a an ideal GaAs cell with no further losses, integrating the potential power as described shows that the total maximum efficiency with only thermal and transmission losses is 45.1%. This limit is set by incident energy losses of 24.8% by thermalisation and 30.1% through transparency, which is consistent with the slightly higher bandgap of GaAs compared with InP.

Finally, the best case cell efficiency from this analysis is 48.8% for a bandgap of 1.13 eV—surprisingly close to silicon.

Having set out the basic mechanisms illustrating the trade-off between greater absorption and greater thermal loss, we now develop a more accurate picture of efficiency limiting mechanisms in solar cells, in order to address the resulting issues.

7.3.2 Charge Neutral Layers

The first transport loss is the well-known Shockley[28] injection current in the dark, whereby majority carrier electrons and holes diffuse from an n or p region across the built-in potential or junction bias under a concentration gradient and against the junction potential. They diffuse into a charge neutral region where they are minority carriers and therefore recombine, giving rise to a net current.

The diffusion or injection rate is a function of how long the diffusing carriers remain as minority carriers before recombining—the faster they recombine, the faster they are replaced, thereby increasing the injection current. This is characterised by hole minority carrier lifetimes τ_n in the n-doped charge neutral region and likewise τ_p for electrons in the p-doped charge neutral region or, *via* the Einstein relationships $L_n = [\tau_n D_n]^{1/2}$ and $L_p = [\tau_p D_p]^{1/2}$ in terms of electron and hole diffusion constants D_n and D_p, respectively, and diffusion lengths L_n, L_p across charge neutral widths x_p

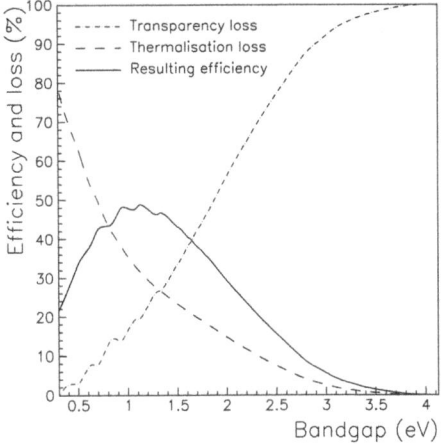

Figure 7.3 Effects of thermal and transmission losses on single junction solar cell efficiency as a function of cell bandgap considering no other losses.

and x_n. The complete expression for the dark current density at bias V can be expressed as:[29]

$$J_S(V) = q\left(e^{\frac{qV}{K_B T}} - 1\right) \begin{bmatrix} \dfrac{n_{ip}^2}{N_A} \dfrac{D_n}{L_n} \left(\dfrac{\dfrac{S_n L_n}{D_n} \cosh\dfrac{x_p}{L_n} + \sinh\dfrac{x_p}{L_n}}{\dfrac{S_n L_n}{D_n} \sinh\dfrac{x_p}{L_n} + \cosh\dfrac{x_p}{L_n}} \right) \\ \\ + \dfrac{n_{in}^2}{N_D} \dfrac{D_p}{L_p} \left(\dfrac{\dfrac{S_p L_p}{D_p} \cosh\dfrac{x_n}{L_p} + \sinh\dfrac{x_n}{L_p}}{\dfrac{S_p L_p}{D_p} \sinh\dfrac{x_n}{L_p} + \cosh\dfrac{x_n}{L_p}} \right) \end{bmatrix} \qquad (7.1)$$

where n_{ip} is the intrinsic carrier concentration in the p layer doped at a level N_A, of surface recombination velocity S_n, and corresponding parameters n_{in} and N_D in the n doped layer with its recombination velocity S_p.

The Shockley injection formalism, operating under the impetus of a concentration gradient, does not differentiate between bulk transport recombination mechanisms (whether radiative or non-radiative). The lifetimes follow an inverse sum law of contributions from a range of recombination mechanisms, the most important of which are radiative transitions across the gap and non-radiative recombination with phonon emission. The latter usually dominates in charge neutral layers as a consequence of doping as we will see later in this discussion. We will also see that, although the Shockley injection does not discriminate between radiative and non-radiative processes, explicitly modelling the upper limit of the radiative recombination can enable us to define an explicitly non-radiative Shockley injection level.

The photocurrent from these layers can be evaluated using standard one-dimensional analytical methods in the depletion approximation.[30] We repeat them briefly here to show the complementarity of light and dark solutions and increased model reliability that results, and the understanding that this yields.

Neglecting possible optical reflections at the back contact, the generation rate G at position x is determined by the Beer–Lambert law relating incident flux F, reflectivity R and absorption coefficient α as:

$$G(x, \lambda) = F(1 - R)\alpha e^{-\alpha x} \qquad (7.2)$$

The resulting photocurrent collected from the charge neutral fraction of the p layer can be evaluated from the excess minority carrier concentration Δn_p, relative to equilibrium in the dark. This is determined, in the absence of an electric field term in the charge neutral layers, from the balance of generation and recombination, expressed in terms of Δn_p in the p layer as:

$$\frac{d^2 \Delta n_p}{dx^2} - \frac{\Delta n_p}{L_n^2} = -\frac{G(x)}{D_n} \qquad (7.3)$$

which can be solved with appropriate boundary conditions given by surface recombination and the depletion approximation.[3] A similar expression for Δp_n yields excess minority carrier concentration in the charge neutral n layer. The photocurrent from both charge neutral layers is given by the gradient, with the appropriate sign, taken at the p and n layer depletion edges, of the minority carrier concentrations. The sum of these two photocurrents defines the charge neutral layer contribution to the total solar cell photocurrent, J_{PH}.

7.3.3 Space Charge Region

The space charge region (SCR) non-radiative recombination dark current can be expressed in terms of hole and electron diffusion profiles extending across it. This is the Shockley–Read–Hall (SRH) formalism,[31] which can be expressed analytically in terms of carrier densities n and p as a function of depletion layer extending from x_1 to x_2 giving the following SRH recombination current density:

$$J_{SRH}(V) = q \int_{x_1}^{x_2} \left(\frac{p(x)n(x) - n_i^2}{\tau_n(p(x) + p_t) + \tau_p(n(x) + n_t)} \right) dx \qquad (7.4)$$

This current describes the non-radiative recombination considering only mid-gap trap levels (the most efficient for recombination) and a space charge layer with trapped electron and hole densities n_t and p_t, and electron and hole non-radiative lifetimes τ_n and τ_p, respectively.

In the space charge region under illumination, the injected majority profiles are perturbed by a small population of free carriers collected at the depletion edges, and of free carriers photogenerated in the SCR. The transport of these excess free carriers is dominated by drift and lifetimes much greater than the short transit time across the depletion layer. For this reason, the photocurrent contribution from the space charge region is assumed to be equal to the integral of the generation rate over that region. The sum of this SCR photocurrent and the charge neutral contributions defines the total photocurrent J_{PH}.

7.3.4 Radiative Losses

The last loss mechanism we consider is the radiative loss which applies in some measure to both charge neutral and space charge regions. This is the loss that is a direct consequence of absorption, and sets the fundamental limit on the efficiency of solar cells.[32]

The generalised Planck equation expresses light emitted by a grey body[33] as a function of absorption, geometry and chemical potential or QFL separation of recombining species. It defines the total current density J_{RAD} corresponding to the emitted luminescent flux at bias V from a radiative emitter as an integral over the photon energy E and surface S as:[34]

$$J_{RAD}(V) = q \int_0^{\infty} \left(\frac{2n^2}{h^3 c^2} \left(\frac{E^2}{e^{(E - q\Delta\varphi)/kT} - 1} \right) \int_S \alpha(E, \theta, s) \, dS \right) dE \qquad (7.5)$$

where n is the refractive index of the grey body, $\Delta\Phi$ is the QFL separation (the difference between hole and electron QFLs) and the other symbols have their usual meanings. The absorptivity $\alpha(E,\theta,s)$ is the line integral over position through the different layers of the cell along the optical path of radiation at angle θ with the normal exiting or entering surface S, the total emitting surface in three dimensions.[33] Therefore, J_{RAD} is minimised by reducing S, for example, by coating the cell with reflective materials except on the absorbing fraction of the cell's surface facing the Sun. This incidentally increases light trapping and is closely related to photon recycling concepts.

The spatial variation of the QFL separation is a function of material quality: In the high mobility SCR, it is essentially equal to the applied bias, and constant. In defective (*e.g.* heavily doped) charge neutral layers, injected carriers have short lifetimes. Their density therefore decreases exponentially away from the SCR edge, which is equivalent to a QFL separation tending to zero. In thin, high purity charge neutral material, however, lifetimes are long and the diffusion length may be significantly greater than the charge neutral layer thickness. In this case, it is reasonable to assume, as did Araujo and Martí,[32] the upper limiting case of a constant quasi-Fermi level separation equal to the applied bias across the entire device and a consequently higher radiative recombination current.

These points are important in determining how close to the radiative limit cells are operating. They do so by placing a maximum possible upper limit on the radiative character of charge neutral layers in the case that these layers have lossless transport. This limit can be expressed by evaluating eqn (7.5) with an absorptivity path integral $\alpha(E,\theta,s)$ across the charge neutral layers. In this way, the QFL separation and absorption of different layers and their position determines their contribution to the total luminescence. In this way we can express the upper limit J_{RAD}^{cn} on charge neutral luminescence and corresponding recombination current.

The minority carrier transport is fixed by the carrier continuity eqn (7.6), the solution of which yields the charge neutral contribution to the photocurrent J_{PH}. Knowing the total Shockley injection, J_S, and the *upper* limit on its radiative fraction J_{RAD}^{cn} we can define the *lower* limit of the non-radiative fraction of the Shockley injection J_S^{cn} as the remainder. That is, the lower limit of the non-radiative fraction of the Shockley injection is $J_S^{cn} = J_S - J_{RAD}^{cn}$.

To summarise these points, the photocurrent and Shockley formalisms complement the radiative limit. The combination of these three formalisms enables us to formulate an explicitly non-radiative modification of the generic Shockley injection level and to define a radiative recombination fraction as a function of bias. This is defined as the radiative fraction of the total recombination as follows:

$$\eta_{RAD}(V) = \frac{J_{RAD}}{J_{SRH} + J_{RAD} + J_S^{cn}} \qquad (7.6)$$

where the definition relies on the explicitly non-radiative J_S^{cn} in the subscript to avoid double-accounting for the charge neutral radiative recombination current which is already included in J_{RAD}.

This important point allows us to clarify an issue whereby dark current measurements and their diode ideality factors cannot differentiate between non-radiative and radiative limit recombination regimes in the dark. These regimes, and hence the radiative recombination fraction defined above, remain relevant in the light, *via* the superposition principle that we will come to below.

The significance of these remarks on explicit radiative recombination fraction is that, as we mentioned earlier, the radiative limit is the most efficient operating regime of solar cells, in which the only loss is the re-emission of light at an intensity partly determined by the absorption coefficient as shown in the Planck non-black-body equation. It is instructive to look at the resulting maximum conversion efficiency shown in Figure 7.4. The ideal bandgap of 1.35 eV for the AM1.5G spectrum is remarkably close to InP, which is wholly consistent with the approximate analysis of thermal and transparency losses seen earlier. A similar analysis can be applied to multiple junction designs as we will see subsequently.

7.3.5 Resulting Analytical Model

The sum of contributions from the charge neutral p and n zones and the space charge regions gives the total photocurrent density J_{PH}. This defines the external quantum efficiency including reflection loss (QE) as the ratio of collected carriers to number of incident photons at a given wavelength, that is, the probability that a photon incident on the solar cell gives rise to a charge carrier collected at the cell terminal. Finally, the light current density (J_L) under applied bias, assuming superposition of light and dark currents is given by:

$$J_L(V) = J_{PH} - (J_S + J_{SRH} + J_{RAD}) \tag{7.7}$$

where we use the photovoltaic sign convention of positive photocurrent.

This light current density enables us to evaluate solar cell figures of merit such as the short circuit current density $J_{SC} = J_L(0)$, the maximum power point V_{MP} and fill factor (FF) in the standard manner.[30] Effects of parasitic resistance are included when modelling real data in the usual manner, that is, a series resistance defining a junction bias, and a parallel resistance and associated shunt current reducing the photocurrent.

To put this model in context of other work it is, firstly, equivalent to the classic Henry model for single to multi-junction cells[35] giving 31% efficiency for a single junction in the radiative limit, as seen above (Figure 7.4). It also agrees with further development concluding with Araujo and Martí[32] and references therein. These authors consider an optimum radiatively efficient design with unit QE, no non-radiative losses, and emission losses restricted to the solid angle subtended by the Sun; they calculate a limiting efficiency

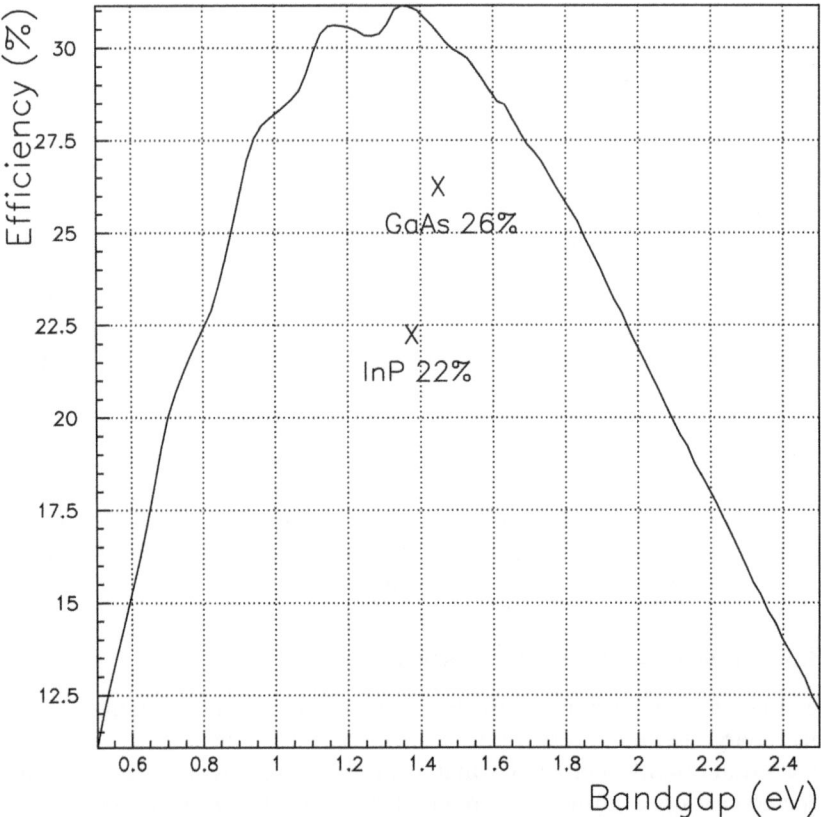

Figure 7.4 Single junction solar cell ideal conversion efficiency in the radiative limit
as a function of bandgap showing best single junction results to date for
two key III–V semiconductors. Both have bandgaps close to the optimum
of 1.35 eV with a potential conversion efficiency of 31.1% for an ideal cell
with only radiative recombination losses.

of 40.7% for a single unit gap cell. The method used here is in good agreement, giving 40.1% in the same conditions.

A further contextual issue is that of real data and the capacity of the model
to fit them. A result of this modelling approach is the additional constraints
placed on variable parameters by the light and dark mechanisms. For
example, the minority carrier transport properties are constrained by their
specification of photocurrent collection, and with a symmetry that reflects
the minority carrier origin of both effects, are constrained by the Shockley
injection current. This results in fewer free parameters and a more exact
understanding of efficiency limiting processes.

The only remaining free parameters are parallel and series resistive
losses, and non-radiative lifetimes for electrons and holes in the space
charge region. Both of these, however, are adequately constrained by the

dark current data and in their respective bias ranges constitute single parameter fits.

The examples given so far have considered only AM1.5G solar spectra. The same principles hold for other spectra which we introduce here, such as the AM1.5 direct AOD and spatial AM0 spectra, together with the light concentration frequently used in the field.

7.3.6   Single Junction Analyses

To illustrate the concepts developed above, we look at two examples of single junction solar cells. The first is a well-characterised non-ideal 20% efficient pin structure (extrapolated to 5% shading) comprising a nominally undoped intrinsic i layer between emitter and base, while the second is a record 25% GaAs np solar cell[36] with less available data but showing a slightly different and superior operating regime, and also with 5% shading reported. Figure 7.5 shows the spectral response data and model for both example cells. The first notable difference is the significant intrinsic region contribution in the 20% pin cell (as would be expected) and the good fit resulting from the use of the measured reflectivity (not shown) in the modelling.

The 25% np cell shows a slightly inferior fit [Figure 7.5(b)], resulting from the need to calculate the front reflectivity for a dual layer MgF_2/ZnS anti-reflection coating as described by Kurtz *et al.*[36] (using 120 nm MgF_2 and 65 nm ZnS thickness rather than the inconsistent 6.5 nm quoted in the reference). It shows a negligible SCR contribution and a significantly higher short wavelength response than the 20% cell [Figure 7.5(a)]. This is despite the less promising np geometry, for which the short wavelength QE is dominated by less efficient hole minority carrier collection in the n-type emitter layer. This is related to the main novelty of this cell, which is the use of a 30 nm GaInP window on a thin 0.1 μm n-type emitter. The novel window is responsible for very low emitter–window recombination velocity allowing a thin emitter without excessive spreading resistance and high collection efficiency in this thin n-type layer.

Figure 7.6 shows the complementary modelling in the dark for both example cells, using the transport parameters consistent with the QE modelling. The left and right axes show the dark current contributions and resulting radiative efficiency, respectively. The 20% cell never reaches radiative dominance, the radiative share of recombination reaching about 18% as the cell approaches flat band and the effects of series resistance start to appear [Figure 7.6(a)]. More importantly, the radiative fraction at the maximum power point V_{MP} under one sun illumination is just 0.1%, showing overwhelmingly non-radiative dominance in this cell. Although dark IV data for the 25% efficient cell are not available to similarly high bias, the modelling of the lower efficiency cell strengthens the analysis of what data are available [Figure 7.6(b)]. Additionally, the fit is not as exact, which is attributable in part to less precise knowledge of reported cell geometry and in particular cell grid coverage reported as approximately 5%.[36]

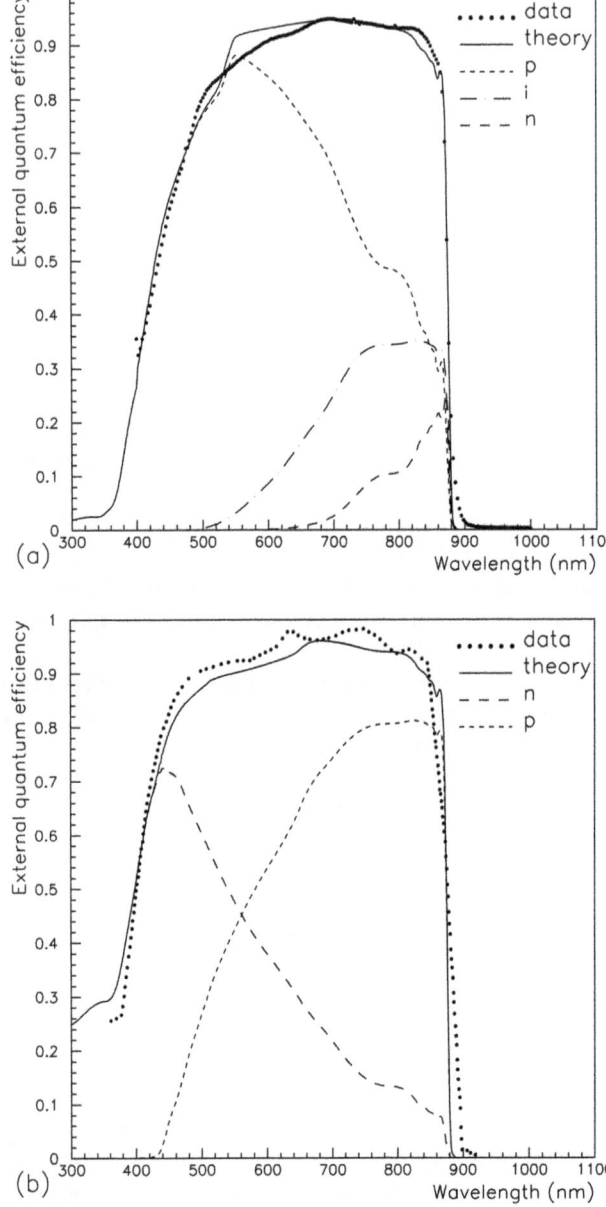

Figure 7.5 Spectral response of GaAs cells showing (a) a 20% efficient pin structure and (b) a record 25% efficient pn cell.

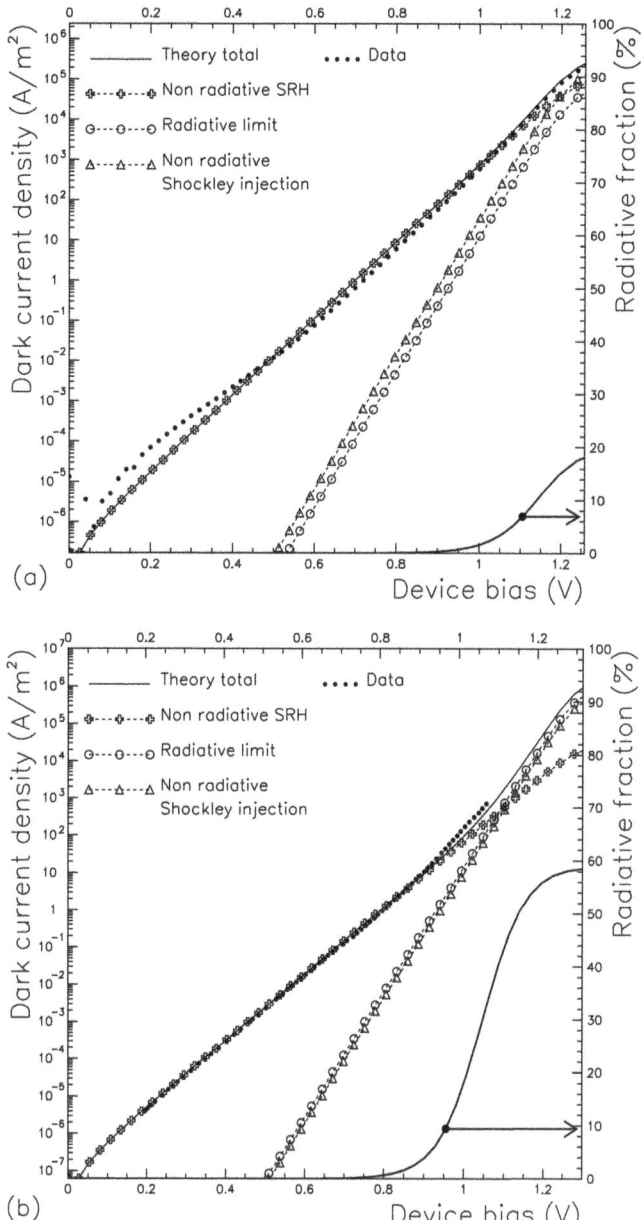

Figure 7.6 Dark current of (a) a 20% efficient pin GaAs cell and (b) the record 25% efficient pn GaAs cell. The higher conversion efficiency for the pn cell is consistent with its higher radiative recombination fraction: at high bias approaching flat band, these recombination fractions reach 58% and 18%, respectively, as indicated on the right axis of both figures.

Bearing these issues in mind, the overall agreement, summarised in Table 7.1, is nevertheless close. Comparing the two cells, we note that the radiative current density is comparable, if slightly higher in the pn as might be expected in the light of the differences in geometry and shading which are such as to have little bearing on the net rate. However, the non-radiative SRH rate is much greater in the pin structure, despite the electron and hole SCR lifetime of 10 ns as opposed to just 2 ns in the 25% pn cell. This apparent contradiction between the longer lifetime in the less radiatively efficient pin cell and the lower non-radiative recombination rate in the pn cell has two causes: The most important is the obvious longer lifetime in the undoped i layer of one structure which means less dopants, or equivalently a lower defect density, and a longer lifetime as shown by the modelling. The lower non-radiative injection current in the more efficient pn, on the other hand, is explained in part by the superior performance of the n-type charge neutral layer in the pn case as a result of the novel window layer at the time of publication.

The overall conclusion is that the modelling is consistent with available light and dark data, and suggests that the 25% record cell is just about radiatively dominated but only at high bias. That is, the explicitly radiative recombination from the SCR and charge neutral layers accounts for 58% of the total as the cell approaches flat band. In addition, series resistance is negligible in this case, reflecting the high quality GaInP of window layer design and consequent high conductivity of the solar cell surface layers with little loss of photogeneration. This represents the highest radiative efficiency this cell can conceivably attain at the high current levels obtained at high illumination levels under concentration. More practically, and more importantly, is the situation at the maximum power bias $V_{MP} = 0.91$ V (Table 7.1). At this bias under one sun illumination, the radiative recombination fraction is 4%. This is far higher than the less efficient 20% cell, and yet still overwhelmingly non-radiatively dominated.

There emerges a consistent picture of the physical phenomena developed in describing these high purity crystalline solar cells. This is that the dark current and light current modelling consistency leads to constrained modelling which reveals detailed information concerning the operational regime of solar cells.

One conclusion of looking at the radiative fraction in the high bias regime where ideality 1 starts to dominate is that a solar cell with an ideality of 1 may be far from the radiative limit. It may in fact only ever asymptotically

Table 7.1 Record GaAs cell parameters published by Kurtz[36] for AM1.5G compared with analytical model results

	$J_{sc}/A\ m^{-2}$	V_{mp}/V	V_{oc}/V	$FF/\%$	$Efficiency/\%$
Kurtz[32]	285	NA	1.05	85.6	25.0 ± 0.8
Model	278	0.91	1.05	82.7	24

approach the radiative limit as doping levels in the charge neutral layers are decreased, hence reducing the doping related defect density and non-radiative recombination rate. In this low doped case, however, the overall cell efficiency drops due to a significant reduction of the built-in potential relative to the cell bandgap. An optimum can be estimated with the analytical methods described. A detailed analysis is beyond the scope of this chapter, but we can say that the optimum is a trade-off between high doping levels and efficient transport. High doping ensures a high junction potential and lower injection. But shorter neutral layer lifetimes imply both higher injection levels and lower collection efficiency. A proper optimisation in terms of these competing processes ensures high collection efficiency together with a low Shockley injection, consistent with tending asymptotically towards the radiative limit.

7.3.7 Conclusions

This analytical overview of solar cell performance has examined the trade-off between thermal and transparency losses, and suggests that reducing these important losses is a promising strategy. The more detailed analysis of radiative and non-radiative losses has shown a more realistic and significantly lower achievable efficiency with a single bandgap. Analysis of an efficient published cell shows an interesting point, which is that solar cells with ideality factors tending towards 1 at high bias are not necessarily tending towards a regime dominated by the highest potential efficiency radiative recombination limit.

These two points analysing single junction performance and loss set the stage for designs going beyond the single junction design in the following sections.

7.4 Multi-junction Solutions

7.4.1 Theoretical Limits

In order to reduce the fundamental losses illustrated in Figures 7.2 and 7.3, we must first absorb all incident photons and yet arrange things such that all these photons are absorbed close to the band-edge. These conflicting requirements could be resolved by reshaping the spectrum either by means of an intermediate filter absorbing all incident photons and re-emitting light in a narrow spectrum or ideally as a monochromatic beam. This method, and its variants up and down conversion,[37,38] needs only a single junction accepting the reshaped spectrum but predictably suffers from efficiency losses in the spectral conversion.

Another option is spectral splitting,[39] whereby the spectrum is separated into different, ideally monochromatic beams, which are absorbed by solar cells with appropriate bandgaps tuned to the part of the spectrum they are designed to convert to electrical power. This is a multiple cell solution where, for most applications, the sub-cells would be connected in parallel, or in series with an equal series current constraint.

This concept of spectral splitting finally leads us to a simpler solution, that is developing the notion of sub-cells to achieve a similar result by arranging the sub-cells optically in series, each acting as an optical filter to those underneath it. As shown in Figure 7.7, the first cell to see the spectrum converts and filters the high energy photons, and so on through ideally an infinite number of junctions. This is the multi-junction solar cell and is ideally suited to III–V solar cell materials since, as we saw at the beginning of this chapter, these materials cover the greater part of the solar spectrum.

The multi-junction solution raises the problem of how to connect the sub-cells. The mechanically stacked solution is to place them in series optically, and contact them individually in parallel or even completely independently. This has the advantage of allowing the combination of arbitrary materials which may be lattice and current mismatched. However, the complexity resulting from the multiple connections and optically efficient stacking means that this technique is limited to concentrator arrays.[1]

Another solution is the monolithic series connected design, illustrated in Figure 7.7 for a tandem cell. This scheme requires a constant series current constraint through all sub-cells and is therefore limited by the lowest photocurrent contribution. The other cells are forward biased away from their maximum power point until the series current constraint is met, resulting in a loss in output power. It also implies compatible growth in principle, regarding lattice constants and growth methods.

A final design issue is the reverse diode presented by the series connection of subsequent sub-cells, illustrated in Figure 7.7, whereby the pn junction sub-cells are inevitably connected by a reverse biased np junction acting as a blocking diode. This must be short-circuited by a highly doped tunnel

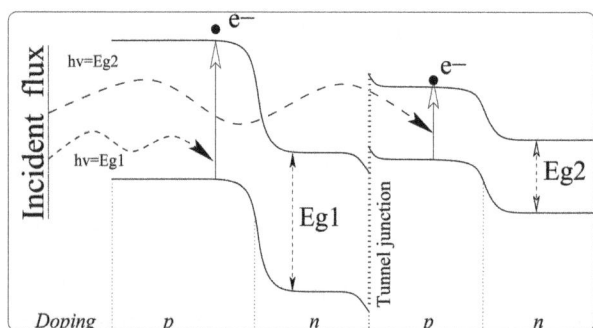

Figure 7.7   Monolithic tandem multi-junction solar cell band diagram solution to thermalisation losses. The series connection of pn junctions requires constant series current and a tunnel junction in order to cancel the current blocking np junction formed between pn sub-cells. The design raises materials compatibility issues for monolithic cell growth, in particular lattice constants.

junction[40] allowing majority carriers to flow unimpeded between sub-cells, thus completing the circuit.

A simplified model sufficient for our purposes is given by Demassa and Knott[41] and allows calculation of tunnel junction characteristics in terms of bulk materials parameters of the layers defining the tunnel junction, in particular effective masses, permittivity, and doping levels. A brief calculation, which we will not describe in detail, shows that good tunnel junction materials must possess a high density of states and must be degenerately doped. Solar cells impose further constraints, which are that the tunnel diode must be as thin as possible and possess a high bandgap in order to remain optically thin, since any light absorbed in these degenerately doped layers does not contribute to photovoltaic action. A properly designed tunnel junction is ohmic up to a limiting current and may be treated as a series resistance, together with an associated optical loss, which is the approach used in the modelling presented here.

To conclude this brief mention of ohmic tunnel junctions, we note, referring to Figure 7.1, that AlGaAs and GaInP, lattice matched to GaAs, are good candidates for tunnel junctions because of their high bandgaps. Another material mentioned earlier is AlInP lattice matched to GaAs, which constitutes another indirect high bandgap material for tunnel junctions and window layers as we shall see subsequently.

7.4.2 Material Limitations

The radiative efficiency limit for an infinite number of sub-cells[32] is about 86%. More practically, we find that a tandem cell with two junctions may reach 42.2% without concentration. This is illustrated by the efficiency contour (Figure 7.8) in terms of upper and lower gaps assuming only radiative losses and therefore a perfect, lossless tunnel junction. Non-ideal bandgap combinations may, however, reach efficiencies close to this. Examining at the contour shows a tandem efficiency that is relatively insensitive to change in bandgap as long as both gaps are varied simultaneously. For example, a tandem with gaps (0.8, 1.6 eV) will perform roughly the same as one with gaps (1.19, 1.78 eV), with an efficiency of about 42%— both only slightly lower than the absolute maximum. Nevertheless, considering the ideal limit first, the upper and lower bandgaps for the absolute maximum tandem efficiency of 42.2% are 1.63 eV and 0.957 eV (Figure 7.8).

For the most promising substrate, GaAs, Figure 7.1 shows two materials with the higher gap (1.63 eV) that are lattice matched to GaAs. The first, $Al_xGa_{1-x}As$, remains direct from $x = 0$ up to $Al_{0.49}Ga_{0.51}As$, a bandgap ranging from 1.424 to about 2 eV and including the 1.63 eV cell for a composition of approximately $x = 0.17$. The second possible material is in the $Ga_xIn_{1-x}As_yP_{1-y}$ family, which ranges from 1.424 eV (GaAs) to 1.9 eV (that is, $x = 0.51$, $y = 0.49$) with a continuous range of group III and group V compositions lattice matched to GaAs. This same flexibility raises another important advantage, which is the possibility of lattice matching this quaternary to Ge, and even to Si substrates.

Ironically, this phosphide material has been much studied on InP substrates for telecommunications applications,[7] but has received little attention on GaAs substrates because of the availability of AlGaAs which is historically well-established, and easier to grow[6] despite its non-ideal minority carrier characteristics. The consequence is that materials knowledge is largely restricted to the lattice matched ternary endpoint $Ga_{0.51}In_{0.49}P$, and that lattice matched quaternary materials are simply expressed by a linear interpolation as $(GaAs)_{1-z}(Ga_{0.51}In_{0.49}P)_z$.

These considerations suggest that this quaternary materials family is worthy of greater attention. However, in the current state of knowledge, the quaternary composition $(GaAs)_{0.8}(Ga_{0.51}In_{0.49}P)_{0.2}$ has the correct direct gap of 1.63 eV for our ideal tandem structure, for which we need to identify a lower gap 0.957 eV material.

For this lower gap, however, there is no lattice matched candidate. Using GaAs, the lowest available gap, as the lower gap sub-cell of a tandem cell yields an ideal efficiency limit of 38%. This dictates an ideal upper sub-cell bandgap of 1.95 eV, obtainable with AlGaAs but approaching the indirect transition for this material where the recombination associated with the DX centre corresponding to the L indirect valence band minimum becomes

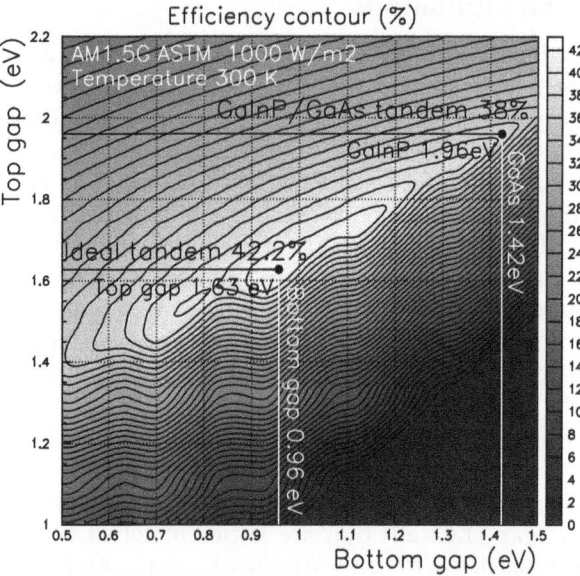

Figure 7.8   Ideal dual junction solar cell maximum conversion efficiency in the radiative recombination limit as a function of higher and lower junction bandgaps, showing the absolute maximum of 42.2% and the highest conversion efficiency of 38% achievable with a GaInP on GaAs tandem.

increasingly important.[11] For these reasons, AlGaAs is generally not considered as a candidate for tandem cells and we will not consider it further. The GaAs based tandem, however, remains a viable design with a compatible phosphide material which is nearly ideally matched by GaInP and has the potential to reach 38% efficiency.

Coming back to the ideal tandem efficiency limit of 42%, the closest material with the correct lower sub-cell bandgap of 0.957 eV is $In_{0.43}Ga_{0.57}As$. This compound is lattice mismatched to GaAs substrates, with a critical thickness of just 8 nm after which misfit dislocations result in a serious penalty in cell efficiency.

Figure 7.9 shows another step on the road to the 86% limit consisting of three junctions, demonstrating the result of a numerical optimisation of the three gap system efficiency in the radiative limit. In this case, the efficiency contour is much sharper than for the tandem case. Any deviation from the ideal brings a rapid decrease in efficiency that cannot be corrected to the same degree by adjusting the other two bandgaps. The ideal sub-cell bandgaps found for the triple multi-junction cell are 1.91, 1.37 and 0.94 eV, which together give an efficiency of 48%. Referring to Figure 7.1, we find

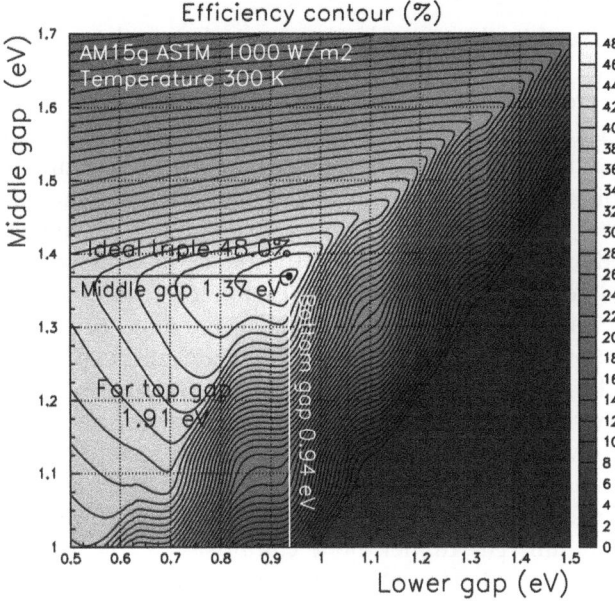

Figure 7.9   Ideal triple junction solar cell maximum conversion efficiency in the radiative recombination limit showing a section through the triple gap bandgap/efficiency volume by fixing the highest gap component at its ideal value of 1.91 eV. The world record triple at gaps 1.87, 1.40 and 0.67 eV of conversion efficiency 40.7% under a different, concentrated spectrum cannot be shown on this contour given its non-ideal top sub-cell bandgap.

again that the most promising materials belong to the GaInAsP family and indeed are significantly closer to simple ternary compounds. The upper gap is well approximated by $Ga_{0.515}In_{0.485}P$ with of gap 1.9 eV which is lattice matched to GaAs. The middle gap cell at 1.37 eV corresponds to $In_{0.05}Ga_{0.95}As$ which has a critical thickness of 70 nm on a GaAs substrate. Finally the lowest gap varies little compared with the tandem case and therefore presents the same problem as in that case, that is, the lack of lattice matched lower gap materials.

This survey of the materials available in III–V semiconductors for ideal tandem and triple junction cells shows that even this wide range of materials requires some additional tricks to circumvent materials issues resulting from mismatched materials. Our brief exploration of available materials leads us to conclude, somewhat counterintuitively given the range of materials available, that the most immediately promising materials for both tandem and triple junction designs are GaAs and GaInP, and an as yet undetermined lower gap material for the triple. The following sections examine some solutions to these issues and the record-breaking multi-junction cells that have been fabricated as a result.

7.4.3   A Tandem Junction Example

The previous sections have provided some understanding of the sources of efficiency loss, how to moderate them, and candidates amongst the III–V materials where these ideas may be put into practise.

Among the large body of work on multi-junction cells (see for example Andreev[1]), we now present an analysis of a tandem consisting of the GaAs/GaInP combination we mentioned earlier with a theoretical maximum efficiency of 38%. This has been attempted by a number of groups. One of the first achieved over 30% under a concentration of 100–200 suns in 1994.[42] We focus on a later result from 1997 by Takamoto *et al.*[43] to whom we are indebted for quantum efficiency and dark current data. The Takamoto paper reports over 30% efficiency under a global unconcentrated spectrum.

The full devices structures are available in Takamoto's paper and we mention only the main points here. The device structure chosen is n on p with an AlInP window layer. Like the Kurtz single junction cell, the n-type emitter is heavily doped and only 50 nm thick, whereas the lightly doped base (hence with good minority carrier transport) is over an order of magnitude thicker at 0.55 μm. The tunnel diode comprises of n and p doped InGaP layers of 30 nm each doped 10^{25} m^{-3} (Zn) and 0.8×10^{25} m^{-3} (Si), respectively, and sandwiched between higher gap AlInP cladding layers. These are intended, in part, to reduce the common problem of dopant diffusion from the highly doped tunnel junction. The device is completed by a GaAs bottom junction with an InGaP back surface minority carrier reflector.

Figure 7.10(a) shows the modelled QE again assuming a double layer MgF_2/ZnS anti-reflection coat and showing a good fit overall. The breakdown of different regions clearly demonstrates one strength of the design, that is,

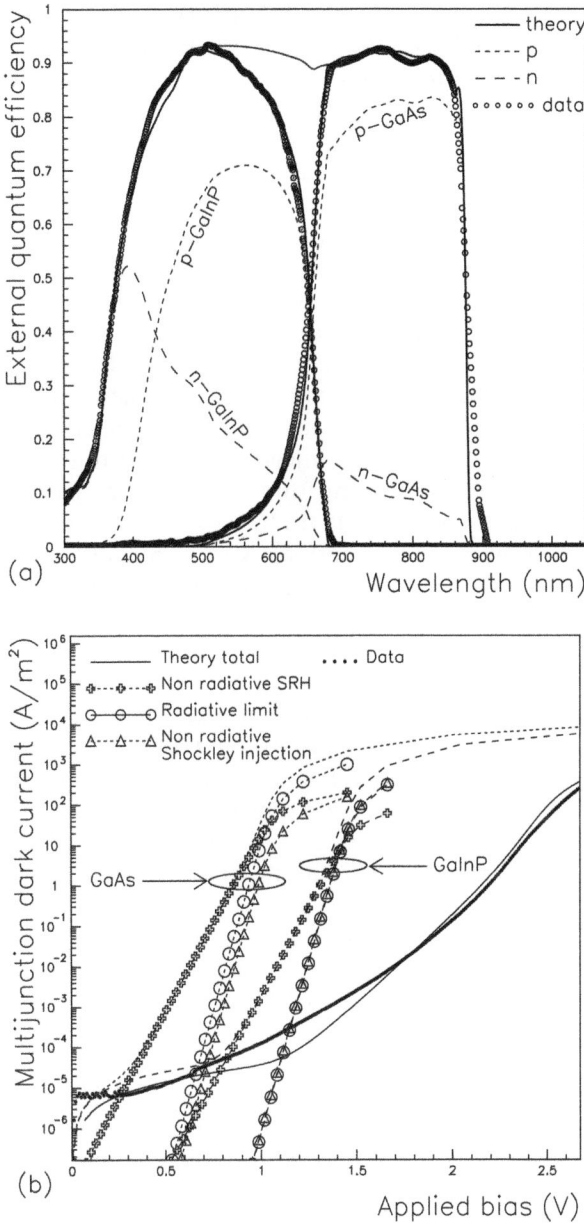

Figure 7.10 QE data and modelling showing the detail of layer contributions (a) and dark current data and modelling (b) for a high efficiency GaInP/ GaAs tandem (Japan Energy Corp.[39]).

the dominance of photocurrent produced by p-type layers which benefit from better electron minority carrier transport than the n-type layers. In this n on p geometry, the more efficient p layer dominance is achieved by the use of a thin n layer, combined with good interfaces between the GaAs n layer and the tunnel junction cladding, and the topmost n-type GaInP layer and the AlInP window. A further important point to note is the significant QE of the bottom GaAs cell above the GaInP bandgap. This means the top cell is some way from being opaque near its bandgap. This is another design feature of this cell. In order to ensure current continuity in the light of the imperfect bandgap combination imposed by the GaAs as discussed above, the top cell is made thinner so as to ensure current continuity at the expense of a slightly greater thermalisation loss in the GaAs cell.

Figure 7.10(b) shows the corresponding dark current fit together with individual sub-cell and overall tandem dark currents. The tandem current is determined by the sub-cell current–voltage characteristics by adding sub-cell biases at constant current assuming an ohmic tunnel junction. As mentioned earlier, this estimates the Shockley injection from the transport parameters and calculates the radiative current from the cell geometry and absorption coefficient, both validated by the QE fit. In the case of the dark current there is, in this case, an imperfect fit and signs of systematic error at low bias, which cannot be elucidated further without separate dark IV curves for both sub-cells. At higher bias, however, agreement is sufficient to indicate that the higher gap GaInP sub-cell is approximately 40% radiative and the bottom GaAs sub-cell approximately 65% radiative, comparable if slightly better than the Kurtz single junction cell reviewed previously.

The model results are compared with the published data in Table 7.2. In this case, the model slightly underestimates experimental data, due in part to underestimating the short circuit current density by about 2%, but more importantly because of overestimating the dark current at high bias by a factor of up to 2 in the region of the V_{oc} at 2.5 V. This can be seen in the underestimate in V_{oc} in particular.

Concerning the dark current overestimate, the radiative current can only be overestimated if there is a net reduction of the luminescence. The luminescence from the front surface is small due to the critical angle for total internal reflection at the front surface of 18% for this design. Most of the luminescence (of the order of 90%) is therefore lost to the substrate where it is effectively absorbed. The only scenario for a significant overestimation of the radiative current is therefore the presence of a reflective rear surface. This

Table 7.2   Tandem cell published parameters for AM1.5G compared with analytical model results, showing reasonable agreement but with an underestimated V_{oc} due to an overestimated dark current

	$J_{sc}/A\ m^{-2}$	V_{mp}/V	V_{oc}/V	$FF/\%$
JEC[39]	142.5	2.49	85.6	30.3
Model	139.5	2.32	87.0	29.4

is ruled out by the good quality of the QE fit near the band-edge with the nominal structure [Figure 7.10(a)], which has no rear reflector.

The only remaining possibility is that the minority carrier transport properties in the charge neutral layers are better than standard values tabulated as a function of composition in the literature, as used by the model.

However, despite these issues we can conclude that the tandem solar cell operates in a regime which is consistent, and slightly better than, the single junction GaAs cells. That is, the GaAs sub-cell 12% radiative fraction at the maximum power point reported in Table 7.3 is greater than the corresponding 4% reported above in Section 7.3.6 for earlier single junction values, which is consistent with marginally superior V_{MP}, V_{oc} and FF (*cf.* Table 7.1). The GaInP sub-cell is comparable if slightly better at 13% radiative fraction at its operating bias V_{MP}. To put this performance in context, this 30% record tandem operates at an impressive 78% of the 38% ideal radiative limit for this bandgap combination and at 71% of the 42% ideal radiative limit with no materials restrictions.

7.4.4 Record Efficiency Triple Junction

The examples discussed in previous sections are in terms of an AM1.5G global spectrum without concentration. The more complex multi-junction cells, and increasingly single junction cells, are usually reported in terms of the AM0 spectrum just outside the Earth's atmosphere, or the terrestrial direct beam AOD spectrum[44] at AM1.5. Table 7.4 gives a summary of ideal triple junction characteristics calculated in the radiative limit for these three spectra without concentration. The AM0 spectrum gives the lowest conversion efficiency despite the highest power since this is the broadest spectrum, and as such an ideal cell loses more power below its bandgap in the infrared.

Table 7.3 Modelled sub-cell radiative recombination fraction at respective maximum power points showing non negligible radiative recombination levels

Tandem sub-cell	V_{mp}/V	Radiative fraction $\eta_{RAD}(V_{mp})/\%$
GaInP	1.34	13
GaAs	0.926	12

Table 7.4 Ideal triple cells gaps and efficiencies for standard global, direct, and near-earth space spectra

Spectrum	Top gap/eV	Mid gap/eV	Low gap/eV	$J_{sc}/A\ m^{-2}$	V_{oc}/V	Efficiency/%
AM1.5G (1000 W m^{-2})	1.91	1.37	0.94	167.0	3.20	47.9
AOD (913 W m^{-2})	1.86	1.34	0.93	157.3	3.09	47.6
AM0 (1354 W m^{-2})	1.84	1.21	0.77	235.2	2.84	44.0

The materials limitations are more stringent than in the tandem case. This is first because the triple junction efficiency is more sensitive to variations in bandgap and because the ideal bandgaps are further from those of available lattice matched materials.

In this context, these designs and their higher efficiencies have led to the development of lattice matched and heterogeneous growth III–V cells on Ge substrates. An example consisting of GaInP, GaInAs and Ge substrate sub-cells is provided by King et al.,[45] whose paper includes the detailed cell structure together with lattice matched and mismatched cells. The lattice matched material looks at the optimum material quality option, whereas the lattice mismatched, metamorphic option is intended to approach the ideal sub-cell bandgaps more closely. In addition, this paper covers a further interesting degree of freedom that we have mentioned above, which is the use of group III sub-lattice disorder to control the bandgap and thereby the current matching in the triple junction. The band structure is not explicitly stated by King et al. but is approximately (1.87, 1.40, 0.67) eV for lattice matched GaInP, $Ga_{0.99}In_{0.01}As$ and Ge sub-cells. In a little more detail, the optimal bandgaps in the ideal limit for a Ge substrate sub-cell are 1.88 eV for the top GaInP sub-cell and 1.33 eV GaInAs middle gap sub-cell for a 0.67 eV Ge sub-cell and substrate. The ideal one sun AM1.5G efficiency for this structure is 45.5%.

The GaInP may be engineered to match this with judicious use of composition and ordering mentioned above. The ideal $Ga_{0.955}In_{0.055}As$ cannot, but the tolerable critical thickness or nearly 3 μm for this layer is the reason the metamorphic route is investigated by King et al.[45]

The net difference between the two cases is relatively minor, however, with similar performance within margin of error. In this discussion, therefore, we limit the analysis to the lattice matched case as a direct progression from the previous single junction and tandem cases examined. The main interest from the point of view of this discussion is the investigation of ideality 1 and 2 mechanisms reported in the paper for both lattice matched and mismatched triple structures. The method used by King et al.[45] is the probing of J_{SC} and V_{oc} as a function of cell illumination intensity. Subject to the assumption of the superposition principle mentioned earlier, this yields the dark current, and is provided for the upper GaInP and middle GaInAs sub-cells but not the low gap Ge sub-cell. Concerning earlier discussion of the meaning of ideality 1 regimes, the paper explicitly defines the ideality $n = 1$ regime as the regime where the dark current is dominated by the Shockley injection current and concludes that the sub-cells increasingly approach the ideality 1 regime at high bias, consistent with the modelling reported here.

Figure 7.11(a) and (b) show the modelled QE for each sub-cell, assuming as before a calculated reflectivity consistent with the published data. The modelling uses transport data from the literature,[4-7] validated by the good QE fit. As illustrated by the sub-cell light IV in Figure 7.12, the model shows that, as reported by King et al.,[45] the Ge substrate bandgap is significantly below the optimum and therefore this sub-cell over-produces current. In

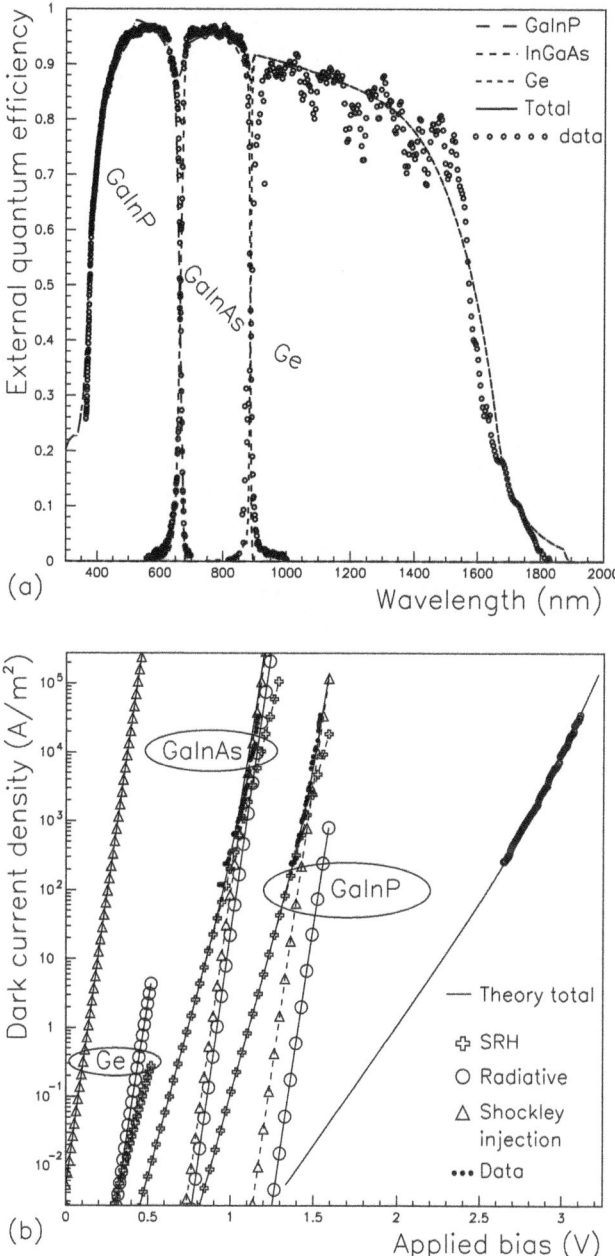

Figure 7.11 QE data and modelling (a) for the three sub-cells of a GaInP/GaAs/Ge record triple junction cell[41] and (b) dark current data and modelling for two of the sub-cells and the combined triple junction response showing the recombination regimes described in the text.

consequence, it is forced into forwards bias in order to decrease its net current and to achieve current parity and continuity with the other sub-cells. It therefore operates at lower efficiency at a bias beyond its maximum power point. The other consequence of a non-ideal lower bandgap is excessive thermalisation in this Ge sub-cell and consequent efficiency loss.

The dark current fitting is shown in Figure 7.11(b) for available sub-cell data and the overall experimental tandem response. The lack of Ge data requires the use of transport data from the literature for this material which is nevertheless validated by the good QE fit. The relatively short range of available dark current data for the GaInAs junction, GaInP junction and overall triple junction device nevertheless covers the transition between the $n \approx 2$ and $n \approx 1$ idealities, together with the experimental maximum power point V_{MP} and V_{oc} reported.[45] For these three sets of data, the model provides a good fit subject to the proviso of short datasets. With regards to efficiency, the light current theory and experimental data are shown in Figure 7.12, and corresponding conversion efficiency parameters given in Table 7.5, for aperture area efficiency as reported,[45] which disregards shading losses.

The first conclusion from these results is that the dark current data available are sufficient to analyse the operating regime of this cell at the maximum power point. The combined modelling of the GaInP and GaInAs

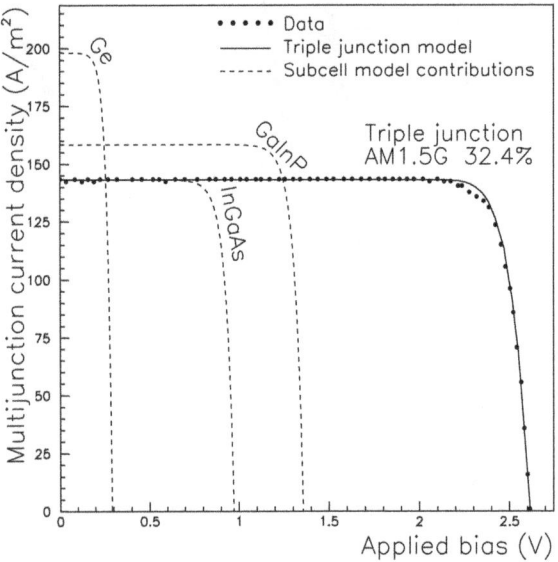

Figure 7.12 Triple junction light IV data and modelling showing good agreement and a structure current limited by the middle GaInAs cell which is thereby reverse biased away from its maximum power point, while the overproducing Ge cell is forward biased away from its V_{mp}, as is the GaInP sub-cell to a lesser extent.

Table 7.5 Comparison of modelling and published parameters for the triple junction[41] lattice matched GaInP/GaInAs/Ge design under the AM1.5G spectrum compared with modelling. Good agreement overall is observed despite the assumptions made regarding the Ge subcell

	$J_{sc}/A\ m^{-2}$	V_{mp}/V	V_{oc}/V	$FF/\%$	$Efficiency/\%$
King[41]	143.7	2.62	2.30	85	32.0
Model	143.2	2.62	2.31	86.0	32.4

sub-cells in Figure 7.12 and Table 7.5 shows that the range of dark current data and modelling covers the illuminated maximum power point bias V_{MP}. It also covers the transition from non-radiative dominated ideality at low bias to the $n = 1$ ideality marking the Shockley injection regime which may be radiatively dominated. Taken as a whole, the modelling and experiment for this triple show a close match in the critical transition from non-radiative dominated high ideality behaviour at low bias to $n = 1$ behaviour at high bias. In order to explore radiative recombination contribution, Table 7.6 shows the radiative fraction of the total recombination current in each sub-cell at its maximum power point. This shows that the radiative fraction of the sub-cells is significantly lower than the single junction (4% radiative at V_{MP}) and tandem junction records (GaInP and GaInAs operate at 12% and 13% radiative fraction respectively at their V_{MP} points).

The analysis shows quantitatively that this cell, even when favoured by reporting active area efficiency, is significantly further from the ideal limit than the single and tandem junction cells seen earlier. This is to some extent reflected in its proportionally greater difference with the 45.5% limiting efficiency for the sub-cell bandgaps (1.88, 1.33, 0.67 eV) in the same conditions: That is, the cell is operating at 70% of the ideal limit within constraints set by the Ge substrate, and at 66% of the unconstrained radiative limit for an ideal triple junction cell which is the 48% figure we discussed earlier illustrated in Figure 7.9.

7.4.5 Conclusions

The fundamental loss mechanisms described above can be addressed by bandgap engineering and *via* the flexible range of materials available in the III–V family. However, materials limitations still impose compromise. These include flexible design at the expense of material purity (the metamorphic route), or greater material quality at the expense of design (homomorphic lattice matched design).

A study of single, tandem and triple junctions shows impressively increased efficiencies, which are tantalisingly close to the radiative limit. However, increasing the number of junctions brings a diminishing rate of efficiency improvement along the lines highlighted in the earlier discussion on thermalisation losses. In addition, practical issues move the design

Table 7.6 Sub-cell radiative fraction of the total recombination current at their respective maximum power points for the triple junction device showing overall strong non-radiative dominance

Sub-cell	V_{oc}/V	V_{mp}/V	Radiative fraction $\eta_{RAD}(V_{mp})/\%$
GaInP	1.35	1.12	0.5
GaInAs	0.95	0.84	0.5
Ge	0.29	0.22	2

further from the radiative limit, due partly to increasingly complex manufacturing and lower materials quality.

These considerations are best summarised by the results under one sun AM1.5G illumination showing a manufactured tandem operating at 71% of the radiative limit *versus* 66% for a triple junction cell: figures that show potential for improvement. On that note, we find a comparable triple junction cell has achieved 40.7% under concentrated sunlight.[45] Further triple junction results from Stan *et al.*[46] are a sign that, as mentioned above, there is scope for significant improvement with these high efficiency strategies, among which is the trend towards radiative dominance.

A final point to note is that the simple use of increased light concentration allows operation in a regime closer to radiative dominance—as long as the cell material is sufficiently pure to deliver a ideality $n = 1$ that is explicitly dominated by radiative recombination, as opposed to one dominated by non-radiative Shockley injection recombination pathways in the charge neutral layers.

7.5 Remarks on Nanostructures

The bulk semiconductor structures described so far show significant scope for improvement as we have demonstrated on a technological front. Other approaches using the flexibility of the III–V materials have more fundamental potential going beyond the limitations of pn multi-junctions devices.

The first key concept as we saw earlier remains the equilibrium population of majority and minority carriers leading to thermalisation loss, and the delivery of all carriers at a single potential. A second key concept is the symmetry of bulk materials resulting in spatially homogeneous properties, such as homogeneous emission over all solid angles. Lifting this homogeneity as mentioned earlier allows reduction of radiative losses, for example, and opens the possibility of further gains by reducing structural symmetry. This question of symmetry overlaps with the concept of meta-materials in the general sense: geometric arrangement of available materials such as to modify their combined properties.

To conclude this chapter we mention briefly some concepts in III–V solar cell research touching on these issues and with a common theme. This is the modification of bulk materials properties by manipulating materials and geometries on the nano scale, and thereby creating spatially inhomogeneous materials properties.

A design addressing these issues is the hot carrier cell[47] which uses two concepts of slowed carrier thermalisation and energy-selective carrier extraction. The thermalisation rate is decreased by phonon emission rate reduction as a result of modifying the phonon density of states, for example, with the use of nanostructures on the quantum scale in two[48] or three dimensions: quantum dots (QDs) and quantum wells (QWs).[49] The second is achieved by modifying the carrier density of states at or near the contacts and allowing only a narrow energy spectrum of carriers being transported to the contacts. Again, this is achievable by structures that provide well defined energy bands which are, again, QWs and QDs.

This mention of hot carrier cells, and manipulating carrier energy distributions emphasises the promise of quantum confined structures and the relevance of III–V materials to this field. There is a rich and fascinating body of research on related issues ranging over quantum wires, quantum dots and quantum wells.

We conclude with some comments on quantum well solar cells (QWSCs), a concept that has been developed nearly exclusively in III–V materials for a number of years, and reviewed recently by Barnham *et al.*[50] The QWSC is a pin structure with lower gap quantum wells sandwiched between higher gap barriers in the undoped intrinsic i region and higher gap doped p and n layers. It was initially proposed and studied in the AlGaAs materials system as a means of extending the absorption of a solar cell whilst keeping a junction potential and hence a V_{oc} determined largely by the higher gap bulk regions enclosing the quantum wells. It transpires, however, that the loss mechanisms mentioned above, and ultimately the fundamental Planck radiative efficiency limit, are still determined largely by the well material, that is, the lower bandgap.

The design has led to a lively debate summarised in the recent review[50] and a number of phenomena going beyond the bulk semiconductor operating regime. The first is signs of reduced QFL separation in the wells, implying decreased carrier populations and decreased recombination relative to bulk samples. Secondly, luminescence studies have shown signs of hot carrier populations in the wells.[51] Finally, however, these effects have not, to date, lead to verifiable operation in efficiency regimes going beyond the bulk.

On the materials front, practical materials advantages of QWSCs have been identified. The first and most practical is the development of the strain balancing technique, whereby alternating quantum well and barrier layers, typically made of GaInP and GaInAs, are grown lattice mismatched on a GaAs substrate but with thicknesses below the critical thickness. The alternating tensile and compressive strain in wells and barriers leads to strain balancing with no net generation of defects. An arbitrary number of quantum well/barrier periods may be grown in principle. The resulting design is the strain balanced quantum well solar cell (SBQWSC).[52]

A second and more fundamental advance is the operation of these cells in the radiative recombination limit.[53] The recombination in this design may be up to 90% radiatively dominated as a consequence of the lower quantum well gap being located in the high purity, high mobility i region as discussed earlier.

As a consequence of this radiative dominance, the design symmetry may be further manipulated to restrict emission by fabricating mirrors on all surfaces oriented away from the incoming solar spectrum (essentially the back surface of the cell). The fundamental difference between this approach applied to a bulk cell and to the SBQWSC is the absorption range of the bulk charge neutral layers encasing the space charge region. For the bulk cell, luminescence emitted and reflected from the back cell is partly re-absorbed by the charge neutral layers, and some of this is lost *via* non-radiative recombination according to fractions as calculated earlier in this chapter. For an SBQWSC, on the other hand, the doped and therefore lossy charge neutral layers are transparent to the dominant radiative recombination loss because of their greater bandgap. As a result, the non-radiative loss pathway for these photons *via* the charge neutral layers is cut off. The only remaining pathways for the luminescence are reabsorption in the radiatively dominated quantum wells or re-emission towards the source of incident radiation, that is, the Sun. This scenario represents the closest achievable design to the fundamental efficiency limit whereby the only recombination loss is radiative emission, restricted to the solid angle of acceptance of the incoming radiation.

On a practical level, a detailed analysis[53] demonstrates intrinsic radiative dominance in QWSCs as opposed to non-radiative dominance in bulk cells. The same work further reveals the interesting switch in behaviour in QWSCs coated with back-surface mirrors. The dark current (and V_{oc}) of the QWSC is dominated by the radiative low-gap quantum well layers. The dark current of the mirror-backed cell, however, is dominated by non-radiative recombination in the higher gap charge neutral bulk regions of the cell.

To place this in context, results by Quantasol achieved a world record efficiency of 28.3% under concentration,[54] and other unpublished results have since achieved efficiencies over 40% for multi-junction QWSCs since Quantasol moved into management by JDSU.

To conclude this overview of future concepts, these record nanostructured efficiencies illustrate some of the more fundamental routes forward in solar cell research. These bring together known concepts of multi-junction solar cells together with novel physical concepts of heat and light management. In both these concepts, III–V materials remain key for the flexibility of design this family of materials allows.

7.6 Conclusions

Ongoing changes in global energy supply have moved III–V materials from niche applications towards the mainstream. The principal reasons for this are fossil fuel supply, volume fabrication reduction in costs, and increasing cell efficiency due to the greater flexibility of cell design possible. These benefits, together with concentrating photovoltaics, justify the complex fabrication processes.

In order to appreciate the relevance of flexible materials, an investigation of high efficiency strategies is necessary. We find that fundamental light and thermal management is key; analytical analysis allows us to explicitly quantify this, together with other material-related loss mechanisms. An analysis of bulk III–V pn and pin solar cells shows interesting and contrasting contributions from radiative and non-radiative losses, and how closely the cells approach the radiative limit. The analysis emphasises that bulk single gap cells are inherently non-radiatively dominated. An analysis of record multi-junction cells gives an understanding into how thermal management has increased efficiency to date. The relatively low radiative efficiency in these designs, however, emphasises that there is significant potential for improvement.

A brief note on high efficiency nanostructured concepts, some on the verge of commercial success, underlines the relevance of the III–V materials family for future concepts. As time goes by and energy costs inevitably increase, the importance of these materials seems set to remain central to development of sustainable energy supply.

References

1. V. M. Andreev, in *Practical Handbook of Photovoltaics: Fundamentals and Applications*, ed. T. Markvart and L. Castañer, Elsevier Science, Oxford, 2003, pp. 417–434.
2. G. Phipps, C. Mikolajczak and T. Guckes, *Renewable Energy Focus*, 2008, **9**(4), 56, 58–59.
3. S. M. Sze and K. K. Ng, *Physics of Semiconductor Devices*, 7th edn, John Wiley & Sons, Chichester, UK, 2007.
4. T. P. Pearsall, *GaInAsP Alloy Semiconductors*, John Wiley & Sons, Chichester, UK, 1982.
5. S. Adachi, *J. Appl. Phys.*, 1989, **66**(12), 6030–6040.
6. I. Vurgaftman, J. R. Meyer and L. R. Ram-Mohan, *J. Appl. Phys.*, 2001, **89**, 5815.
7. O. Madelung, *Semiconductors: Data Handbook*, Springer, Berlin and London, 3rd edn, 2004.
8. D. E. Aspnes, *Phys. Rev. B,* 1976, **14**(12), 5331–5343.
9. P. Bhattacharya, *Semicond. Sci. Technol.*, 1988, **3**, 1145–1156.
10. S. Adachi, *Properties of Aluminium Gallium Arsenide*, EMIS Datareviews Series no. 7, INSPEC, 1993.
11. J. C. Bourgoin, S. L. Feng and H. J. von Bardeleben, *Phys. Rev. B*, 1989, **40**(11), 7663–7670.
12. V. M. Andreev, V. P. Khvostikov, V. R. Larionov, V. D. Rumyantsev, E. V. Paleeva and M. Z. Shvarts, *Semiconductors,* 1999, **33**(9), 976.
13. W. Zhang, M. Zhang, M. Chen, D. Jiang and L. Wang, in *Proceedings ISES Solar World Congress 2007*, 2009, vol. 4, pp. 996–999.

14. D. J. Friedman, J. F. Geisz, S. R. Kurtz and J. M. Olson, *J. Cryst. Growth,* 1998, **195**, 409–415.
15. J. F. Geisz and D. J. Friedman, *Semiconduct. Sci. Technol,* 2002, **17**, 769–777.
16. V. Sabnis, H. Yuen and M. Wiemer, in *Proceedings 8th International Conference on Concentrating Photovoltaic Systems CPV-8*, American Institute of Physics, College Park, MD, 2012, vol. 1477, pp. 14–19.
17. A. W. Bett, S. P. Philipps, S. Essig, S. Heckelmann, R. Kellenbenz, V. Klinger, M. Niemeyer, D. Lackner, F. Dimroth, in *Proceedings 28th European PV Solar Energy Conference and Exhibition*, 2013, session 1AP.1.1.
18. M. A. Green, K. Emery, Y. Hishikawa and W. Warta, *Prog. Photovoltaics: Res. Appl.,* 2009, **17**, 320–326.
19. H.-J. Schimper, Z. Kollonitsch, K. Möller, U. Seidel, U. Bloeck, K. Schwarzburg, F. Willig and T. Hannappel, in *Proceedings 20th European PV Solar Energy Conference and Exhibition*, 2005, pp. 492–494.
20. F. Himrane, N. M. Pearsall and R. Hill, in *Proceedings 18th IEEE Photovoltaic Specialists Conference*, Institute of Electrical and Electronics Engineers, New York, 1985, pp. 338–343.
21. T. P. Pearsall. *Properties, Processing and Applications of Indium Phosphide*, Institution of Electrical Engineers, London, 2000.
22. R. A. Stradling and P. C. Klipstein, *Growth and Characterisation of Semiconductors*, Adam Hilger, Bristol and New York, 1990.
23. C. Kittel, *Introduction to Solid State Physics*, Wiley, New York, 7th edn, 1996.
24. R. R. King, D. C. Law, K. M. Edmondson, C. M. Fetzer, G. S. Kinsey, H. Yoon, R. A. Sherif and N. H. Karam, *Appl. Phys. Lett.,* 2007, **90**, 183516.
25. S. M. Hubbard, C. D. Cress, C. G. Bailey, R. P. Raffaelle, S. G. Bailey and D. M. Wilt, *Appl. Phys. Lett.,* 2008, **92**, 123512.
26. M. Hermle, G. Létay, S. P. Philipps and A. W. Bett, *Prog. Photovoltaics: Res. Appl.,* 2008, **16**, 409–418.
27. A. Trellakis, T. Zibold, T. Andlauer, S. Birner, R. K. Smith, R. Morschl and P. Vogl, *J. Comput. Electron.,* 2006, **5**, 285–289.
28. W. Shockley, *Electrons and Holes in Semiconductors*, Van Nostrand, Princeton, NJ, 1950.
29. M. E. Nell and A. M. Barnett, *IEEE Trans. Electron Devices,* 1987, **ED-34**(2).
30. J. Nelson, *The Physics of Solar Cells*, Imperial College Press, London, 2003.
31. (a) W. Shockley and W. T. Read, *Phys. Rev.,* 1952, **87**, 835; (b) R. N. Hall, *Phys. Rev.,* 1952, **87**, 387.
32. G. L. Araujo and A. Martí, *Sol. Energy Mater. Sol. Cells,* 1994, **33**, 213.
33. J. Nelson, J. Barnes, N. Ekins-Daukes, B. Kluftinger, E. Tsui and K. Barnham, *J. Appl. Phys.,* 1997, **82**, 6240.
34. J. P. Connolly, I. M. Ballard, K. W. J. Barnham, D. B. Bushnell, T. N. D. Tibbits and J. S. Roberts, in *Proceedings 19th European PV Solar Energy Conference and Exhibition*, 2004, pp. 355–359.
35. C. H. Henry, *J. Appl. Phys.,* 1980, **51**(8), 4494.

36. S. R. Kurtz, J. M. Olson and A. Kibbler, in *Proceedings 21st IEEE Photovoltaic Specialists Conference*, Institute of Electrical and Electronic Engineers, New York, 1990, pp. 138–140.

37. T. Trupke, M. A. Green and P. Wuerfel, *J. Appl. Phys.*, 2002, **92**(3), 1668–1674.

38. T. Trupke, M. A. Green and P. Wuerfel, *J. Appl. Phys.*, 2002, **92**(7), 4117–4122.

39. A. G. Imenes and D. R. Mills, *Sol. Energy Mater. Sol. Cells*, 2004, **84**, 19–69.

40. L. Esaki, *IEEE Trans. Electron Devices*, 1976, **ED-23**(7), 644–647.

41. T. A. Demassa and D. P. Knott, *Solid-State Electron.*, 1970, **13**, 131.

42. D. J. Friedman, S. R. Kurtz, K. A. Bertness, A. E. Kibbler, C. Kramer, J. M. Olson, D. L. King, B. r. Hansen and J. K. Snyder, in *Proceedings 1st World Conference on Photovoltaic Energy Conversion*, Institute of Electrical and Electronic Engineers, New York, 1994, vol. 2, pp. 1829–1832.

43. T. Takamoto, E. Ikeda, H. Kurita and M. Ohmori, *Appl. Phys. Lett.*, 1997, **70**(3).

44. K. A. Emery, D. Myers and S. Kurtz, in *Proceedings 29th IEEE Photovoltaic Specialists Conference*, Institute of Electrical and Electronic Engineers, New York, 2002, pp. 840–843.

45. R. R. King, D. C. Law, K. M. Edmondson, C. M. Fetzer, R. A. Sherif, G. S. Kinsey, D. D. Krut, H. L. Cotal and N. H. Karam, in *Proceedings 4th World Conference on Photovoltaic Energy Conversion*, Institute of Electrical and Electronic Engineers, New York, 2006, pp. 760–763.

46. M. Stan, D. Aiken, B. Cho, A. Cornfeld, J. Diaz, V. Ley, A. Korostyshevsky, P. Patel, P. Sharps and T. Varghese, *J. Cryst. Growth*, 2008, **310**, 5204–5208.

47. R. T. Ross and A. J. Nozik, *J. Appl. Phys.*, 1982, **53**, 3813–3818.

48. B. K. Ridley, *J. Phys C: Solid State Phys.*, 1982, **15**, 5899–5917.

49. S. K. Shrestha, P. Aliberti and G. J. Conibeer, *Sol. Energy Mater. Sol. Cells*, 2010, **94**, 1546–1550.

50. K. W. J. Barnham, I. M. Ballard, B. C. Browne, D. B. Bushnell, J. P. Connolly, N. J. Ekins-Daukes, M. Führer, R. Ginige, G. Hill, A. Ioannides, D. C. Johnson, M. C. Lynch, M. Mazzer, J. S. Roberts, C. Rohr and T. N. D. Tibbits, in *Nanotechnology for Photovoltaics*, ed. L. Tsakalakos, CRC Press, Boca Raton, FL, 2010, ch. 5.

51. J. P. Connolly, D. C. Johnson, I. M. Ballard, K. W. J. Barnham, M. Mazzer, T. N. D. Tibbits, J. S. Roberts, G. Hill and C. Calder, in *Proceedings 4th International Solar Concentrators for the Generation of Electricity or Hydrogen (ICSC-4)*, 2007, pp. 21–24.

52. K. W. J. Barnham, B. C. Browne, J. P. Connolly, J. G. J. Adams, R. J. Airey, N. J. Ekins-Daukes, M. Führer, V. Grant, K. H. Lee, M. Lumb, M. Mazzer, J. S. Roberts and T. N. D. Tibbits, in *Proceedings 25th European PV Solar Energy Conference and Exhibition/5th World Conference on Photovoltaic Energy Conversion*, 2010, pp. 234–240.

53. J. P. Connolly, in *Advanced Solar Cell Materials, Technology, Modeling, and Simulation*, ed. L. Fara and M. Yamaguchi, IGI Global, Hershey, PA, 2012, ch. 5.

54. Record verified at Frauenhofer ISE testing centre and reported in the press (*e.g.* http://www.nanotech-now.com/news.cgi?story_id=33957 [accessed 26 June, 2014]).

CHAPTER 8

Light Capture

STUART A. BODEN* AND TRISTAN L. TEMPLE

Nano Research Group, Electronics and Computer Science, University of Southampton, Highfield, Southampton SO17 1BJ, UK
*E-mail: sb1@ecs.soton.ac.uk

8.1 Introduction

The first step in achieving efficient conversion of light to electricity in a photovoltaic (PV) device is to ensure that as much of the incident light as possible is transmitted through to the active layer and absorbed there. This chapter focuses on reducing the optical losses that are detrimental to this efficient 'capturing' of light. These losses fall into two main categories:

1. Reflectance from the top surface.
2. Incomplete absorption of light.

These two optical loses are being tackled by the development of antireflection (AR) and light trapping (LT) schemes, respectively, and this chapter explores a variety of implementations of AR and LT ranging from thin film coatings, through texturing on various length scales, to the exploitation of localized surface plasmons in metal particle arrays. The mechanisms through which these techniques operate are explained and the fabrication methods being used to experimentally and commercially realize these approaches are described. The effectiveness and feasibility of each technique for wide scale adoption in PV applications is also discussed.

RSC Energy and Environment Series No. 12
Materials Challenges: Inorganic Photovoltaic Solar Energy
Edited by Stuart J C Irvine
© The Royal Society of Chemistry 2015
Published by the Royal Society of Chemistry, www.rsc.org

8.2   The Need for Antireflection

For normal incidence, the reflectance, R, from an interface between two materials with refractive indices of n_1 and n_2 respectively is given by:[1]

$$R = \left| \frac{n_2 - n_1}{n_2 + n_1} \right|^2 \tag{8.1}$$

For absorbing materials, the reflectance is given by replacing the n terms in eqn (8.1) by a complex refractive index, ň, where:

$$ň = n - ik \tag{8.2}$$

The imaginary component, k, describes absorption in the material and is often referred to as the extinction coefficient. The reflectance is then given by:

$$R = \left| \frac{ň_2 - ň_1}{ň_2 + ň_1} \right|^2 = \frac{(n_2 - n_1)^2 + (k_2 + k_1)^2}{(n_2 + n_1)^2 + (k_2 + k_1)^2} \tag{8.3}$$

Eqn (8.3) shows that the larger the difference between the refractive indices of the two materials, the greater the reflectance at the interface between them. The real part of the refractive index for silicon, n, ranges from 3.5 to 6.9 over wavelengths in the range 300 to 1200 nm, which is high compared with materials such as glass and ethylene vinylacetate (EVA) that typically surround a solar cell in a module, for which n is approximately 1.5 (Figure 8.1). This leads to a normal incidence reflectance for an air–silicon interface, which is important for laboratory cells, of between 31 and 61% and for an EVA–silicon interface, found in encapsulated cells, of between 16 and 46% (Figure 8.1). In one analysis, it has been shown that the reduction in

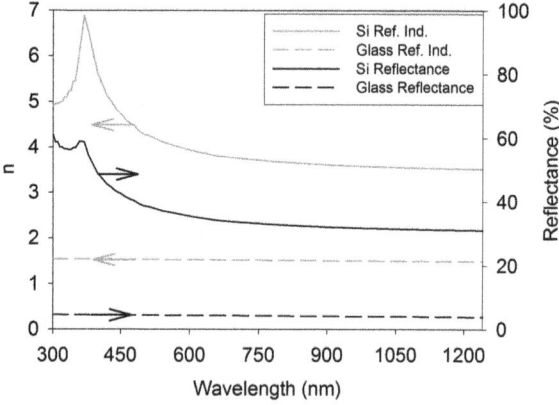

Figure 8.1   Variation of real part of refractive index (n) for silicon and glass with wavelength and reflectance spectra for silicon–air and glass–air interfaces. Silicon data taken from ref. 3, B270 crown glass data taken from ref. 4.

short-circuit current due to reflectance losses if no AR scheme is employed is approximately 36.2% for a laboratory crystalline silicon (c-Si) cell and 19.5% for an encapsulated cell.[2] There is therefore a need for effective methods of reducing reflectance losses in solar cells.

8.3   The Need for Light Trapping

Light transmitted into the semiconductor must be absorbed to generate an electron–hole pair, which can then be split and extracted to produce current. Incomplete absorption of the solar spectrum reduces the maximum achievable efficiency of a solar cell. The fraction of light absorbed as it travels through a semiconductor layer of thickness l is given by the Beer–Lambert law:

$$A = e^{-\frac{4\pi k}{\lambda_0} l} \qquad (8.4)$$

Here, λ_0 is the free space wavelength and k is the imaginary component of the refractive index known as the extinction coefficient, as discussed above. The likelihood of a photon being absorbed is a function of the extinction coefficient, the wavelength and the thickness of material it passes through. Silicon is an indirect bandgap semiconductor and so has a low extinction coefficient in the visible and near infrared (NIR) (Figure 8.2). Fewer than 50% of photons with wavelength larger than 625 nm are absorbed after a single pass through 2 µm of silicon. Increasing the thickness to 200 µm considerably improves absorption, but it is still low near the bandgap.

Solar cells cannot be made arbitrarily thick because excited carriers need to be extracted from the semiconductor. Increasing the thickness of the device results in higher photon absorption, but also increases the chance of photoexcited carriers recombining at defects or impurities within the semiconductor. Light trapping increases the effective optical path length without

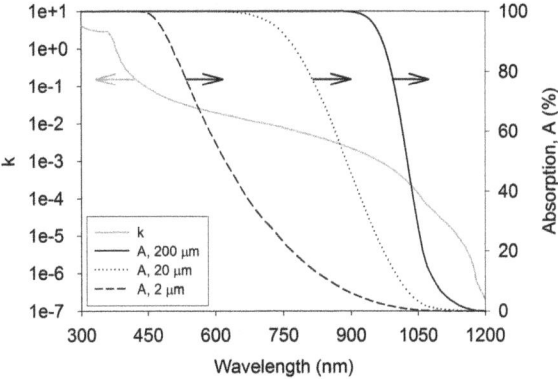

Figure 8.2   Variation of the imaginary part of refractive index (k) for silicon and the absorption of light after passing through 2, 20 and 200 µm of silicon.

affecting the physical thickness of the semiconductor, which enables the use of thinner and lower quality semiconductor layers. A useful figure of merit for light trapping schemes is the path length enhancement factor, Z, given by:

$$Z = \frac{\text{effective optical thickness}}{\text{actual thickness}} \tag{8.5}$$

The maximum path length enhancement that can be achieved in a conventional solar cell is $4n^2$, where n is the refractive index of the semiconductor.[5] This is known as the ergodic limit or the Yablonovitch limit, and is around 50 for silicon in the NIR. Enhancements larger than the Yablonovitch limit are only possible when the active layer is substantially thinner than the incident wavelength, *i.e.* within the wave-optics regime.[6]

8.4 Mechanisms

Various mechanisms are exploited by the different techniques for AR and LT implemented in photovoltaic devices. These are introduced before specific examples of their applications are discussed.

8.4.1 Antireflection

8.4.1.1 Destructive Interference

In electromagnetic wave theory when two or more light waves overlap, according to the principle of superposition, the resultant electric field intensity is equal to the vector sum of the intensities from each wave. Consequently, if two waves are half a cycle out of phase with respect to one another, superposition results in the cancelling out of their electric field intensities and the resultant intensity is zero. This is called destructive interference and is the fundamental mechanism behind the common usage of simple thin film coatings to reduce reflection at an interface. Thin film antireflection coatings (ARCs), consisting of one (single-layer ARC or SLAR), two (double-layer ARC or DLAR) or more than two (multi-layer ARC) layer(s) of materials with a refractive index in between those of the two media either side of the interface, cause destructive interference between beams reflected from each interface. This results in a reduction in the intensity of the reflected beams of a particular wavelength for which the coating is designed.

8.4.1.2 Multiple Reflections

Texturing with either regular, geometric features such as pyramids or more irregular, random features can lead to a reduction of reflection through a multiple reflection mechanism. A manifestation of this occurs when light reflected from one part of a textured surface is directed onto a different part of the surface and so is incident more than once on the surface of the solar cell, resulting in more light being coupled into the cell (Figure 8.3).

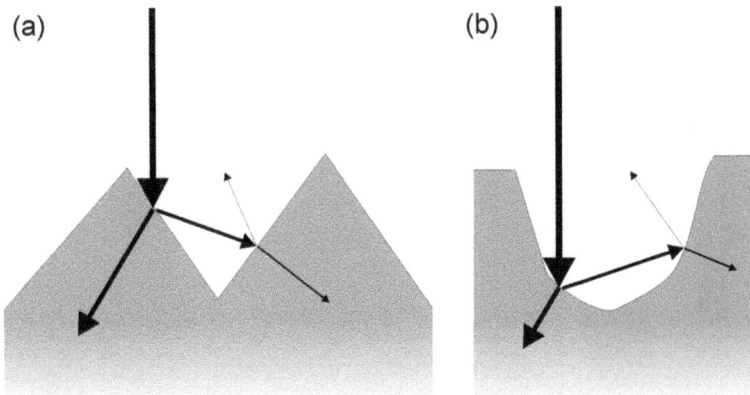

Figure 8.3 Diagrams of multiple reflection AR mechanism with two different texture types: (a) pyramids; and (b) bowl-like features. From ref. 7.

8.4.1.3 Graded Index

Removing the step change in refractive index at an interface by texturing it with features on the subwavelength scale can be used to minimize reflection. Most of the solar spectrum lies within the visible range (0.4–0.7 μm). When light from the Sun interacts with structures of dimensions much below this range, it behaves as if it were encountering a homogeneous medium whose optical properties are a weighted spatial average of the profile's optical properties. If the subwavelength features are regular in size and arrangement, we have what is known as a zero order diffraction grating because all the higher orders are evanescent and only the zero order propagates.[8] Consequently, a surface textured with ridges smaller than the wavelength of light will interact with the light as if it had a single-layer ARC of refractive index governed by the ratio between the ridges and channels [Figure 8.4(a)]. Likewise, a stepped profile will act as a multilayer ARC [Figure 8.4(b)] and a tapered profile will behave as an infinite stack of infinitesimally thin layers, introducing a gradual change in refractive index from one medium to the other [Figure 8.4(c)]. This effectively smooths the transition between one medium and another, ensuring that incident light does not encounter a sudden change in refractive index which would cause a proportion to be reflected.

8.4.2 Light Trapping

8.4.2.1 Reflection

The most basic light trapping scheme simply consists of a highly reflective back contact. This increases the optical path length by 2 ($Z = 2$), because light that would ordinarily pass through the cell only once is reflected from the back contact and so passes through the cell a second time. A reflectance of greater than 98% is achievable using an aluminium back contact with

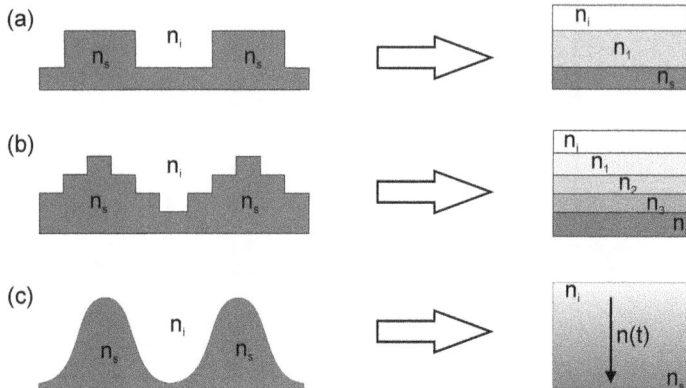

Figure 8.4 Subwavelength feature profiles and their analogous refractive index profiles: (a) ridges; (b) staircase; and (c) nipple-like protuberances.

a dielectric layer of appropriate thickness between the back contact and silicon.[3] Texturing the back surface with micron-scale regular geometric features (pyramids, inverted pyramids, grooves) can enhance light trapping by changing the direction in which light propagates through the cell. The effective optical path length is increased *via* two mechanisms: Firstly, light deflected away from the normal to the cell may undergo total internal reflection and so will be forced to pass through the cell multiple times. Secondly, the distance across a cell for light travelling at a more oblique angles will be greater than that of light travelling perpendicular to the cell surface.

8.4.2.2   Refraction

Changing the direction along which light propagates through a cell can also be achieved through a refraction mechanism, by texturing the top surface. The texturing again consists of regular geometric features, which incline the surface at angles away from the plane of the cell. Light refracted through the textured surface will thus be forced to travel at oblique angles to the cell surface, increasing the amount trapped and absorbed.

8.4.2.3   Diffraction

By reducing the size and periodicity of regular geometric features on the front or back of a cell, diffraction can be invoked to trap light within the cell. A surface covered in a periodic array of features with a period on the scale of the wavelength of light will introduce periodic variations in phase in an incident wavefront. Interference leads to a series of intensity maxima and minima distributed in various directions. This is a form of diffraction and the periodic array of features is known as a diffraction grating.[9] By creating a reflection/transmission grating on the back/front surface of a solar cell, light can be forced to travel at oblique angles through the cell. For thin film solar cells, light is only allowed to propagate in a discrete number of modes parallel to the film and diffraction gratings can be designed to generate diffraction orders which couple to the allowable modes, increasing the optical path length and therefore the amount of light absorbed.

8.4.2.4 Scattering

An external scattering layer can be used to transmit light into a solar cell across a range of angles. Light passing through a transparent medium can be scattered by variations in refractive index, for example voids or embedded dielectric particles, that have dimensions of the order of the wavelength or larger. In a thick non-absorbing scattering layer, light will undergo multiple scattering events before eventually being emitted at a random angle from the top surface. This structure can be used as a diffuse reflector and any photons reflected beyond the critical angle will be trapped within the semiconductor layer by total internal reflection. An ideal diffuse reflector is non-absorbing and has a Lambertian angular distribution of reflection, such that the reflection intensity is proportional to the cosine of the angle.[10,11]

Metal nanoparticles embedded in a dielectric medium can also be used as an external scattering layer. The optical cross-section of metal nanoparticles typically exceeds the geometric cross-section, and so a two-dimensional (2D) planar array is sufficient to interact with all incident photons. This extremely strong optical interaction is due to the excitation of localized surface plasmons (LSPs), which are resonant oscillations of conduction electrons. Light couples to an LSP, which then decays either radiatively, resulting in scattering, or non-radiatively, resulting in absorption. Absorption can be minimized by correct choice of nanoparticle parameters, but cannot be entirely eliminated. Metal nanoparticles situated near a semiconductor layer scatter preferentially into waveguide propagating modes.[12,13]

Some examples of the implementation of AR and LT mechanisms are now explored in more detail.

8.5 Thin Film Antireflection Coatings

Practically all solar cell designs incorporate some form of thin film to improve the transmission of light across the various interfaces within a cell/module by exploiting destructive interference. When deposited at an interface involving the active component of the cell, the thin film can also provide electrical passivation, reducing surface recombination and so improving the quantum efficiency of a cell.

8.5.1 Optical Considerations

A single-layer ARC (SLAR) can be designed by tuning its thickness so that light reflected from the coating–substrate interface travels a distance equal to half a wavelength more than light reflected from the air–coating interface. When the two waves meet again, one will be 180° out of phase with the other and so they will destructively interfere. For an SLAR of refractive index n_1 and

(a) $n_0 < n_1 < n_s$ **(b)** $n_0 < n_1 < n_2 < n_s$

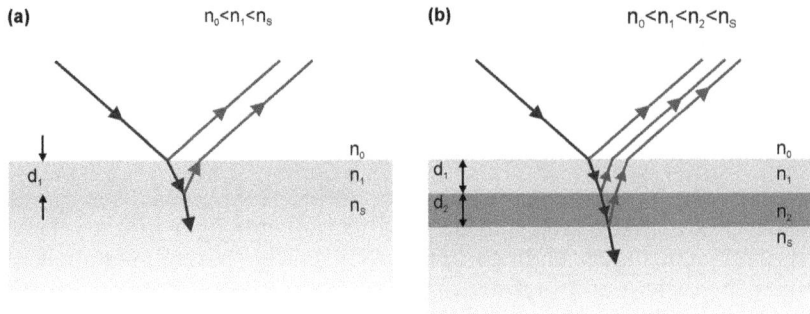

Figure 8.5 Thin film AR coatings: (a) single layer (SLAR); and (b) double layer (DLAR). From ref. 7.

thickness d_1 on a substrate of refractive index n_s [Figure 8.5(a)], optimum antireflection for light of wavelength λ, at normal incidence from air, is achieved when:

$$n_1 d_1 = \frac{\lambda}{4} \tag{8.6}$$

For complete destructive interference, the amplitudes of the waves must be equal, *i.e.* from eqn (8.6) (neglecting absorption):

$$\frac{n_1 - n_0}{n_1 + n_0} = \frac{n_s - n_1}{n_2 + n_1} \tag{8.7}$$

Solving eqn (8.7) for n_1 leads to the second condition for optimum SLAR performance:

$$n_1 = \sqrt{n_0 n_s} \tag{8.8}$$

DLARs operate on the same principles as SLARs, but with two minima in the reflectance spectra, they can be designed to exhibit a low reflectance over a broader wavelength range. This has the added benefit that the AR effect is less sensitive to variations in layer thickness compared with SLARs. For a double-layer ARC (DLAR), it can be shown that the optimum refractive indices, n_1 and n_2, where layer 1 is the top layer [Figure 8.5(b)], are calculated using:

$$n_1^3 = n_0^2 n_s \quad \text{and} \quad n_2^3 = n_0 n_s^2 \tag{8.9}$$

The corresponding optimum thicknesses, d_1 and d_2, are given by:

$$d_1 = \frac{\lambda}{4n_1} \quad \text{and} \quad d_2 = \frac{\lambda}{4n_2} \tag{8.10}$$

For solar cell applications, the design wavelength is usually chosen to be at the peak in the solar spectrum, *i.e.* $\lambda \approx 600$ nm. Taking a typical c-Si cell as an example, n_s at 600 nm is 3.941, which leads to the optimum refractive index

Table 8.1 Optimum refractive index and thickness values calculated using eqn (8.6)–(8.10) for SLARs and DLARs at air–Si EVA–Si and air–glass interfaces for light with a wavelength of 600 nm

	Air–silicon	EVA–silicon	Air–glass
SLAR n_1	1.985	2.431	1.225
SLAR d_1 (nm)	75.6	61.7	122.5
DLAR n_1	1.580	2.070	1.145
DLAR n_2	2.495	2.856	1.310
DLAR d_1 (nm)	95.0	72.5	131.0
DLAR d_2 (nm)	60.1	52.5	114.5

Note: Refractive indices of 3.941 for silicon and 1.5 for glass and EVA were used in the calculations.

and thickness values for SLARs and DLARs at the air–silicon, EVA–silicon and air–glass interfaces given in Table 8.1.

Research in this area has focused on growing materials with refractive index values as close as possible to these optima. This is more difficult for the air–glass interface because few suitable materials exist with the required low refractive index. Magnesium fluoride (MgF_2), with a refractive index of 1.38 is commonly used for AR on camera lenses and has been explored as an AR material for solar glass. However, as deposition involves expensive vacuum techniques for which large area uniformity and stability is difficult to achieve, such an approach has not been widely adopted by PV manufacturers. Experimental PV devices including CdS/CdTe[14,15] and $CuInS_2$/$CuGaS_2$[16] solar cells have employed an evaporated thin film of MgF_2 for antireflection on the glass substrate, reducing the average top surface reflectance from 5.6% to 4.3% in one case.[16] The coating material has been investigated for PV devices for operation in space but questions remain as to the stability of the film.[17] Low refractive index fluorinated polymers have also been studied as possible SLAR coatings for glass.[18,19] This approach is more compatible with large-scale PV module manufacturing because cheaper spray coating techniques can be used to deposit the material which, when dry, forms a chemically resistant, hard-wearing AR surface layer.[20,21]

For an air–silicon interface, the optimum refractive index for a single-layer ARC is approximately 1.99. Cerium oxide (CeO) is identified by Zhao and Green[22] as having a near optimum refractive index of 1.953. Cerium dioxide (CeO_2) has a higher refractive index with values of between 2.467 and 2.780 obtained for radio frequency (r.f.) sputtered films at various deposition temperatures.[23] It is therefore more suitable for the bottom layer in DLARs. Similarly, zinc sulfide (ZnS), with a refractive index of ≈ 2.33 is also used in DLARs,[22] most notably in combination with MgF_2 for the 24.7% efficient passivated emitter and rear locally diffused (PERL) cell.[24] Absorption in the ultraviolet (UV) is relatively high for ZnS, but it is still an effective ARC material because of the low solar irradiance in this part of the spectrum. Stoichiometric silicon nitride (Si_3N_4), grown by low-pressure chemical vapour deposition (LPCVD), has a refractive index between 1.91 and 2.17 over the 300–1240 nm wavelength range.[25] It also has negligible absorption over this range, an obvious additional requirement for any ARC, and has been used in buried contact solar cell technology.[26] Depositing silicon nitride using plasma-enhanced chemical vapour deposition (PECVD) offers more flexibility in refractive index. By changing the NH_3/SiH_4 flow ratio, Nagel et al.[4] grew SiN_x layers with refractive indices measured at 632.8 nm of between 2.62 and 1.85. Encouraged by the wide range of refractive index values obtainable with this material, researchers investigated the possibility of SiN_x DLARs.[27,28] Unfortunately, the extinction coefficient of the SiN_x films increases with increasing silicon content and so the performance of such DLARs is limited by absorption in the bottom layer.

Another commonly used ARC material with variable optical properties is titanium dioxide (TiO_2). In this case, instead of varying the stoichiometry, the refractive index can be tuned by altering the density and phase of the material through different deposition conditions and by post-deposition sintering.[29] TiO_2 can exist as three different phases. Low deposition temperatures lead to an amorphous film; increasing the deposition temperature or annealing above 350 °C results in the metastable crystalline phase of anatase, with a maximum (single crystal) refractive index at $\lambda = 600$ nm of 2.532.[30] Sintering above 800 °C leads to the formation of the stable crystalline rutile phase, with a maximum refractive index at $\lambda = 600$ nm of 2.70.[31] The film density also increases with sintering temperature and time, leading to an increase in refractive index. In the presence of high concentrations of water vapour, porous films with lower refractive index values can be formed. Richards et al.[32] used ultrasonic spray hydrolysis to create an anatase TiO_2 film with a refractive index of 2.44, which is near optimum for a SLAR at an EVA–silicon interface. Using atmospheric pressure chemical vapour deposition (APCVD), Richards[29] also showed that thin films of TiO_2 with refractive indices (at $\lambda = 600$ nm) varying from 1.52 for a porous film to 2.63 for a dense, rutile film can be grown. This remarkable range of refractive indices for one material inspired the design of a TiO_2 DLAR, with a dense, high n bottom layer and a porous, low n top layer. Near optimum values for an air–silicon interface of $n_1 = 1.52$ and $n_2 = 2.489$ were achieved. For an encapsulated cell, the optimum n_2 is too high at 2.856 to be achieved with even the most dense TiO_2 layer and the best obtained was a DLAR with $n_1 = 1.95$ and $n_2 = 2.63$. The modelled weighted reflectance (which is the reflectance weighted by the photon flux density of the solar spectrum) of an encapsulated surface with this DLAR is 7%, which is impressively low considering that this includes 4.3% absolute reflectance from the glass cover.

Silicon dioxide is non-absorbing over the wavelength range 300–1240 nm, but it has a refractive index of 1.46 which is too low for good performance in encapsulated cells. However, it has been widely used as an SLAR for unencapsulated laboratory cells, for example, the original buried-contact designs featured a SiO_2 SLAR.[33] It also has a refractive index close to the optimum for the top layer of an unencapsulated DLAR.

8.5.2 Surface Passivation

The main benefit of using thermally grown SiO_2 for an ARC is that it confers excellent surface passivation to silicon. Defect states within the bandgap cause recombination losses in silicon and the highest concentrations of defect states are found at the surface due to the presence of non-terminated bonds. Growing a thermal oxide, or other suitable passivating layer, results in termination of these defect states and so a reduction in recombination losses. An additional mechanism also operates whereby positive charges within the oxide repel holes from the surface and so hinder carrier recombination.[34]

Many of the other materials considered above, namely ZnS, TiO_2 and Si_3N_4, provide negligible surface passivation[4,27,35,36] and so are often combined with a thin layer of SiO_2. Surface recombination velocities of less than 10 cm s^{-1} have been reported for a silicon surface passivated with a thermally grown oxide.[37]

PECVD silicon nitride films can also be very effective for surface passivation, with a record low value of surface recombination velocity of 4 cm s^{-1} being reported.[38] Alas, the best passivating nitride films are those with a high silicon content and so a correspondingly high extinction coefficient which leads to unacceptable absorption in the film.[4] There are some reports of good passivation with silicon-poor SiN_x films,[39] but these films suffer from a poor thermal stability, with passivation properties degrading during high temperature processing. A very thin layer of SiO_2 can be used in conjunction with SiN_x to overcome this problem.[39,40]

8.5.3 Other Thin Film Considerations

A high chemical resistance is a requirement of thin film coatings for solar cells because cleaning, etching and metal plating processes involving a variety of acids and bases are often performed after the coating has been deposited. Most of the films mentioned above exhibit excellent chemical resistance, with notable exceptions being the susceptibility of SiO_2 to HF and ZnS to any water-based solutions. TiO_2 is highly resistant to chemical attack in its crystalline phases of anatase or rutile, but its amorphous phase exhibits poor chemical resistance.[41] Phase changes and general thermal expansion coefficient mismatches between layers can also lead to stress-induced cracking and peeling of the films during high temperature processing. For industrial cells, cost also needs to be considered. Growing high quality thermal oxide requires heating to around 1000 °C for extended periods of time. In the BP Solar buried contact process, for example, Si_3N_4 is deposited at temperatures of 700–800 °C for 0.5–1.5 hours per batch.[54] These massive thermal budgets substantially increase the cost of the cell. Films that can be deposited at lower temperatures, for example, sputtered or PECVD deposited SiN_x and TiO_2 films deposited by spray pyrolysis or APCVD are therefore advantageous.

Thin film coatings are used throughout PV as a means of antireflection but they are not a perfect solution. Optical properties of films are heavily dependent on deposition conditions and fabricating thin films with the optimum refractive indices and mechanical and thermal properties is challenging. The performance of single-layer ARCs is very wavelength dependent and only optimum for the particular wavelength to which the thickness of the coating is tuned. This does not pose a problem for laser applications but for solar cells, which operate with light from a broad frequency range, it is not ideal. The use of multilayer ARCs addresses this issue but these coatings suffer from other problems. Thermal expansion coefficient mismatches and general poor interface properties can cause thin films to detach from the substrate, a problem that increases with increasing number of layers. Another issue with layered ARCs is that diffusion of material can cause mixing between the layer(s) and the substrate, and so degradation of the ARC properties and electronic properties of the cell.

8.6   Micron-scale Texturing

Reflectance from a surface can be reduced by texturing with features on the micron scale and so enabling the mechanism of multiple reflections. Texturing on this scale also results in refraction or scattering of the transmitted light and so confers a light trapping effect. For monocrystalline silicon, anisotropic, alkaline etching is used to create arrays of geometric features (pyramids and grooves). For polycrystalline silicon, isotropic etching and direct ablation texturing methods are employed, resulting in more randomly textured surfaces.

8.6.1   Alkali Etching: Pyramids and Grooves

The most common form of surface texturing for monocrystalline Si solar cells involves forming arrays of micron-scale pyramids by taking advantage of the different etch rates of the silicon crystal planes in alkali etchants.[10,42] A weak alkaline solution will anisotropically etch silicon, with the $\langle 100 \rangle$ planes etching at a higher rate than $\langle 111 \rangle$ planes. Such an etch treatment on a wafer with a $\langle 100 \rangle$ surface will result in the formation of facets on $\langle 111 \rangle$ planes, creating pyramids, inverted pyramids or grooves (Figure 8.6) which, through the multiple reflection mechanism, will afford a reduction in the overall reflectance.

8.6.1.1   Maskless Texturing

The use of anisotropic etching for solar cells was first reported by Haynos and colleagues, who textured the top surface of Comsat cells.[43] Their texturing scheme used hydrazine hydrate as a selective etchant and relied on random nucleation, forming arrays of randomly arranged micron-scale pyramids on the surface. These cells achieved energy conversion efficiencies under

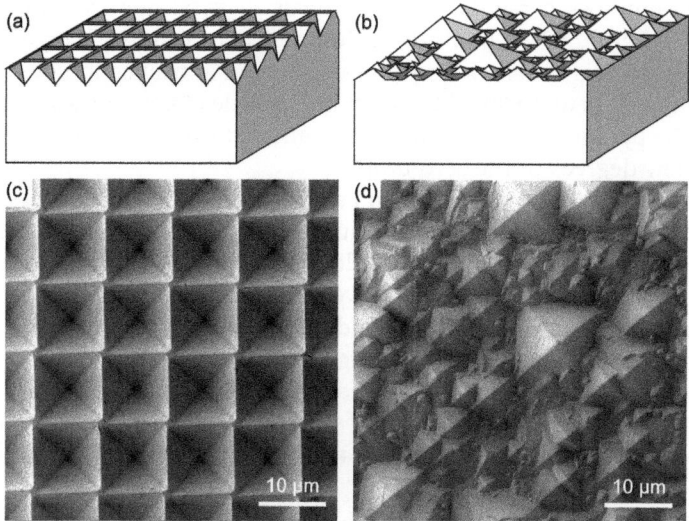

Figure 8.6 (a and b) Schematics of alkali-textured silicon (a) masked, forming a regular array of inverted pyramids, and (b) maskless, forming a random array of pyramids. (c and d) Helium ion microscope images of corresponding examples of: (c) regular inverted pyramids; and (d) random pyramids.

standard terrestrial conditions of 17.2%. The texturing scheme does not require expensive photolithography stages which, combined with the excellent antireflection and light trapping conferred by this approach, led to it becoming the texturing scheme of choice for monocrystalline silicon commercial cells, including the screen printed[44] and buried contact cells.[33]

Potassium hydroxide mixed with isopropanol and heated to 60–70 °C is frequently used for this type of texturing, with mechanical stirring or ultrasonication used to ensure efficient etching at the wafer surface.[45,46] The isopropanol is used to ensure good uniformity of pyramids by reducing the interfacial energy and so improving the wettability of the surface,[47] and by acting as a complexing agent to dissolve the hydrous silica formed during the etch reaction.[48] In an effort to reduce chemical costs, other etchants have been investigated including hydrazine monohydrate ($N_2H_4 \cdot H_2O$),[47] K_2CO_3[49] or Na_2CO_3.[50]

8.6.1.2 Regular Pyramid Arrays

The positions of nucleation sites for the formation of pyramids can be controlled using photolithography to define a mask through which anisotropic etching can proceed. Pyramids can thus be formed in regular arrays and can be upright or inverted depending on the mask geometry. Developed at the University of New South Wales in Australia, the PERL cell—which holds the world record for efficiency of a single junction silicon solar cell (at 25%)—employs a regular array of inverted pyramids as a top surface texturing scheme.[24,51] In addition to low reflectance over a range of wavelengths, pyramidal schemes also confer excellent antireflection over a range of incident angles up to 80°.[52] Diffraction has little effect for these structures because the pyramids are 5–10 μm in size and so visible diffractive orders are within a few degrees of the zero order.

Ray tracing simulations have been used to investigate the effect of varying the arrangement of pyramids, which influences the degree of light trapping conferred by the texture but has little effect on the AR properties of the surface.[10,53] Various tiling schemes have been proposed as alternatives to the close-packed square array of pyramids in an effort to break the pattern symmetry and enhance light trapping. Breaking of the symmetry of each pyramid by slicing the silicon ingot into wafers at a small angle with respect to the ⟨100⟩ crystallographic direction surface has also been investigated for increased light trapping,[54] but the departure from standard wafers is likely to increase costs.

Reflectance of between 10% and 20% in the 400–1000 nm wavelength range are typically achieved using alkali texturing, which is a significant improvement on the 35–40% reflectance exhibited by a planar silicon surface. A further improvement in antireflection can be achieved by combining alkali texturing with thin film ARCs. The minimum reflectance is lower compared with textured bare silicon and reflectance reduction over a broad wavelength range is improved compared with thin film ARCs on planar surfaces. Commercial implementation of the non-random pyramid schemes has been hindered by the extra cost associated with the photolithography stages necessary for defining the patterns before etching. Consequently, regular arrays of pyramids are only employed in high-efficiency, high-cost solar cells, such as the PERL cell, and random arrays still dominate the low-cost bulk market.

Unfortunately, alkali etching is not as effective for light trapping and antireflection on multicrystalline silicon because each grain has a different orientation and so creating deep, dramatic surface relief is not possible. Consequently, this approach does not tend to produce low reflectance surfaces, with most attempts reporting little reduction in reflection compared with planar surfaces.[55,56] Some success using a mixture of NaOH and NaOCl as an etchant for texturing multicrystalline silicon has been reported, with large area (150 mm × 150 mm) cell efficiencies of 14.5%,[57] but in this case, a SiN_x thin film was employed for additional antireflection.

8.6.2   Acid Etching

8.6.2.1   Maskless Acid Texturing

The cheapest, most industry compatible method of acidic wet etching is performed without masking, using the surface damage resulting from wafer sawing as seeds for texturing. Mixtures of HF and HNO_3 have been used, often with catalytic agents such as CH_3COOH or H_3PO_4 to control the etch

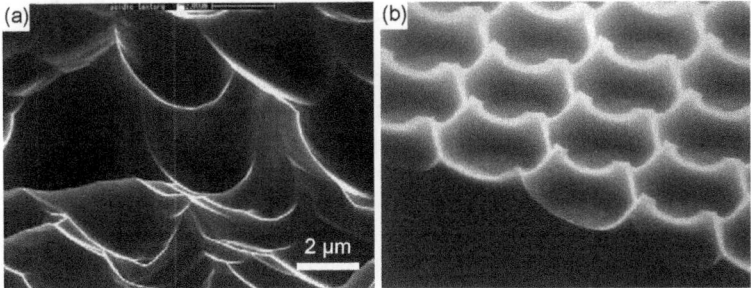

Figure 8.7 SEM images of surfaces textured by acid etching: (a) maskless;[55] and (b) masked (bowl-like features are ~14 μm in diameter).[24] (a) Reprinted from D. H. Macdonald, A. Cuevas, M. J. Kerr, C. Samundsett, D. Ruby, S. Winderbaum and A. Leo, 'Texturing industrial multicrystalline silicon solar cells', *Sol. Energy*, **76**(7), 277–283, Copyright 2004, with permission from Elsevier. (b) Reprinted with permission from J. Zhao, A. Wang, M. A. Green and F. Ferrazza, '19.8% efficient 'honeycomb' textured multicrystalline and 24.4% monocrystalline silicon solar cells', *Appl. Phys. Lett.*, **73**(14), 1991–1993, Copyright 1998, American Institute of Physics.

rate.[55,56] This results in smooth, bowl-like features [Figure 8.7(a)], but the isotropic nature of the etchants means that the resulting surfaces are not textured sufficiently to reduce reflectance to a low level by the multiple reflection mechanism.[55,56] It seems that this technique is only useful for damage removal and is not effective for reflection reduction, with reflectances of between 20 and 30% in the wavelength range from 450 to 1050 nm[56] being demonstrated.

8.6.2.2 Masked Acidic Texturing

By controlling the nucleation of the texture features through masking of the surface, improvements in the performance of acid texturing for antireflection are possible, albeit with the added cost of the photolithographic mask-defining process. In one example, an $HF/HNO_3/H_3PO_4$ isotropic wet etch through an oxide mask was used to create a square array of hemispherical bowls with a diameter of 10 μm on a silicon surface.[58] Although shallower than the inverted pyramid type features achieved with alkali etching, the bowl texture proved effective for encapsulated cells, where it increased total internal reflection from the underside of the glass cover, thus redirecting light back onto the cell.

A similar technique was employed for a multicrystalline cell based on the PERL solar cell design [Figure 8.7(b)].[24] A hexagonal 'honeycomb' array of 4 μm diameter holes, spaced 14 μm apart, was defined in a layer of oxide using photolithography. An isotropic etch ($HNO_3 : HF = 50 : 1$) was then used to create hemispherical bowls in the underlying silicon substrate, through the oxide mask. When combined with a DLAR, the reflectance of the surface was reduced to 3–10% for the 500 nm to 1 μm wavelength range. When compared with the inverted pyramids texturing scheme, which is used on the PERL monocrystalline solar cells, the honeycomb array has ∼8% higher reflectance, which is mainly due to the flat areas at the base of the wells and at the interstices. In terms of light trapping capabilities, however, ray tracing calculations show that the honeycomb structure would outperform the inverted pyramids design, with 85–90% of light remaining after two passes compared with only 65% remaining with the inverted pyramids design. The cell exhibited an efficiency of 19.8%, a world record for multicrystalline silicon cells at the time.

In another example, a patterned photoresist layer was used as an etch mask, eliminating the need for a patterned oxide layer.[59] When etching is complete, the photoresist layer peels off, simplifying the fabrication process. However, the resulting profiles tend to contain more flat areas, with surfaces parallel to the plane of the wafer, than the corresponding anisotropically etched structures, especially in the troughs of the pattern. Light reflected from these flat areas is unlikely to be deflected by a sufficiently large angle to be incident on another part of the cell or undergo total internal reflection (TIR) at the underside of the glass cover and so these areas are deleterious to the antireflective effectiveness of the texturing scheme.

8.6.3 Dry Etching

Reactive ion etching (RIE) employs chemical and physical mechanisms to selectively and directionally remove material. For solar cells, it is used mainly to texture on the sub-micron scale but it can also be used for forming micron-scale features by employing an etch mask. It is mainly applied to multicrystalline cells to achieve larger aspect ratio features than are possible with wet etching, and so to increase the effectiveness of the multiple reflection AR mechanism.

In one example, RIE was performed through a photolithographically patterned nichrome etch mask, leading to a regular array of micron scale pyramids and grooves (Figure 8.8).[60] AR properties were similar to those achieved with anisotropic wet etching of single crystal cells, with the advantage that RIE texturing is equally applicable to multicrystalline cells. Despite the possibility of enhanced surface recombination due to surface damage cause by the etch, cell efficiency improvements of 22% were reported over an untextured cell. Vacuum requirements for RIE cause it to be a relatively expensive technique to implement on a large scale.

In another example, an array of hemispherical bowls was formed in a silicon surface by SF_6 plasma etching through a mask.[61] The dry etch technique enabled the formation of deeper bowls than was possible with acid etching through the same mask by a factor of almost 3 (16 μm compared with 6 μm). Reflectivities similar to those of monocrystalline wafers with inverted pyramids were demonstrated, with corresponding efficiency improvements that indicate that damage due to plasma etching does not hinder the cell performance if a suitable damage removal step is included.

8.6.4 Ablation Techniques

Direct ablation texturing methods including mechanical grooving and laser texturing have also been developed (Figure 8.9). Indeed, the first allusion to using pyramids to reduce reflection from the surface of a solar cell was made in a 1964 patent which suggests a mechanical texturing method.[62] These alternative texturing techniques involve cutting out a pattern using a saw or a laser and then etching in an attempt to remove the damage whilst keeping

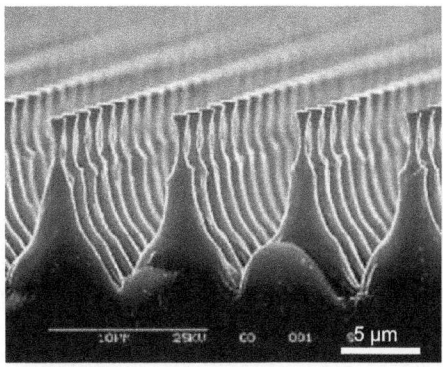

Figure 8.8 Silicon surface textured by RIE through a photolithographically defined nichrome etch mask.[60] Reprinted from S. Winderbaum, O. Reinhold and F. Yun, 'Reactive ion etching (RIE) as a method for texturing polycrystalline silicon solar cells', *Sol. Energy Mater. Sol. Cells*, **46**(3), 239–248, Copyright 1997, with permission from Elsevier.

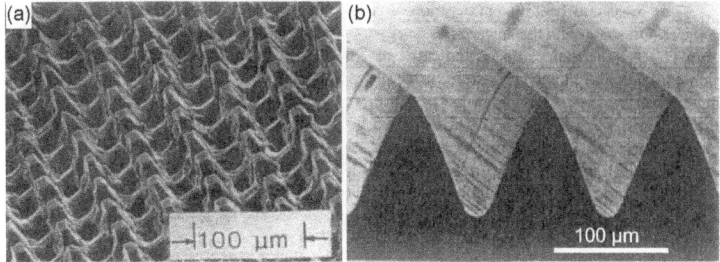

Figure 8.9 SEM images of silicon surfaces textured by: (a) laser scribing;[63] (b) mechanical scribing.[64] (a) Reprinted with permission from J. C. Zolper, S. Narayanan, S. R. Wenham and M. A. Green, '16.7% efficient, laser textured, buried contact polycrystalline silicon solar cell', *Appl. Phys. Lett.*, **55**(22), 2363–2365, Copyright 1989, American Institute of Physics. (b) Reprinted with permission from H. Bender, J. Szlufcik, H. Nussbaumer, G. Palmers, O. Evrard, J. Nijs, R. Mertens, E. Bucher and G. Willeke, 'Polycrystalline silicon solar cells with a mechanically formed texturization', *Appl. Phys. Lett.*, **62**(23), 2941–2363, Copyright 1993, American Institute of Physics.

the desired texture.[63,64] The main problem with these techniques is that considerable damage is caused during texturing and the subsequent etch step is unable to completely remove this. Surface recombination is therefore an issue. Another drawback for laser scribing is the excessive groove depth at the crossover areas where the laser beam has to pass over twice during the texturing process. Effective doping of these deep wells has proven difficult to achieve. Nevertheless, these processes are valid and useful texturing methods for polycrystalline cells, with efficiencies demonstrated at 16.7%[63] and 16.4%[65] for laser and mechanical scribing, respectively. Monocrystalline silicon cells with efficiencies of $\approx 18.5\%$ have also been achieved using laser texturing and several damage removal etches.[66] Reflectivity behaviour similar to alkali-textured monocrystalline cells has been demonstrated.

8.7   Submicron Texturing

If the texture features have dimensions below 1 micron, they are in the subwavelength regime for parts of the solar spectrum. Arrays of sub-wavelength features on a surface can confer an AR effect through the graded index mechanism. One form of this approach is employed by nature in the form of arrays of regular-shaped, tapered protuberances on the eyes and wings of some species of moth and butterfly (which are commonly termed 'moth-eye' structures/arrays)[67-69] (Figure 8.10). Similar structures have been identified on the corneal surfaces of some species of fly and mosquito.[70,71] In nature, it is thought that these arrays have developed to reduce reflections and glare which could reveal the location of the insect to a passing predator.

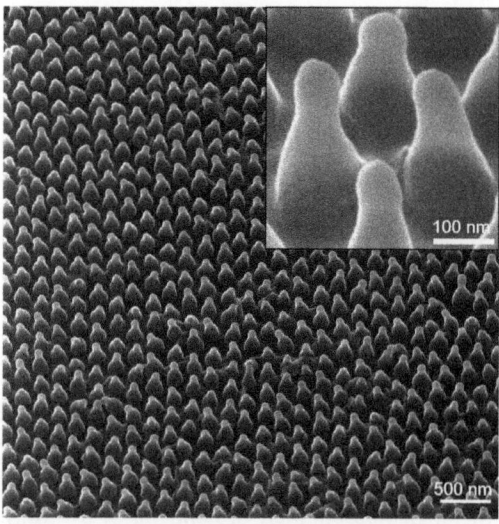

Figure 8.10   Helium ion microscope image of natural moth-eye structures found on the transparent wing of *Cephanodes hylas*.

Another theory is that the increased transmission accompanying the reduction in reflection enhances the sensitivity of the insect's visual system. In any case, natural moth-eye structures have developed to minimize the reflectance of sunlight and therefore have inspired efforts to fabricate artificial versions suitable for photovoltaic applications. Some of these efforts are attempts to directly mimic the array of regularly shaped features found in nature, whereas others realize the graded refractive index profile with irregular subwavelength features.

Subwavelength features are by definition too small to change the direction of propagation of light and so such structures cannot be used for light trapping. If the period of the features approaches the solar wavelength range, however, light trapping is possible in the form of diffraction with the case of regular arrays forming a grating, or scattering in the form of randomly sized features.

8.7.1 Subwavelength Array Theory

The simplest way to treat subwavelength arrays theoretically is to use effective medium theory (EMT) whereby the interface region is modelled as a stack of many layers, each with an effective refractive index in between those of the media either side of the interface.[72] The effective refractive index in each layer is determined by averaging the refractive indices of the two constituent materials, weighted by the amount of each within the layer. In this way, any refractive index profile perpendicular to the surface can be modelled. Studies of how the reflectance varies with the height-to-wavelength ratio all show that, if the graded index region is thicker than approximately half the wavelength, reflectance is below 0.5%.[73-76] For EMT to apply, the features at the interface must be sufficiently small so that they cannot be resolved by the incident light. In other words, the features should not be large enough to scatter light appreciably, either incoherently, as with randomly arranged features, or coherently, as with regular arrays of features such as those found on moth-eye surfaces. By treating the array of subwavelength features as a diffraction grating, this condition translates as the period being sufficiently small so that all orders other than the zeroth are suppressed. This condition is given in Table 8.2 and is calculated using the diffraction grating equation for the cases of reflection and transmission, and normal and grazing incidence.

For applications using infrared light, it is relatively easy to satisfy these conditions and produce antireflective zero order gratings that can be analysed using EMT. Not only are the wavelengths in this region longer but the refractive index of silicon is lower, at $\sim$3.4, than for visible wavelengths. For this reason, EMT is a popular technique for analysing subwavelength AR schemes working at infrared wavelengths.[8,77-79] Fabricating a grating with $d \ll \lambda$ for applications in the visible light range is more challenging and so these subwavelength AR schemes tend to have characteristic lengths closer to the operating wavelengths, *i.e.* $d \rightarrow \lambda$, for which EMT is no longer valid.[79,80] Therefore, most theoretical work on subwavelength AR surfaces has been

Table 8.2   Zero order grating condition from diffraction grating equation

	Normal incidence	*Grazing incidence*
In reflection	$d < \lambda$	$d < \dfrac{\lambda}{2}$
In transmission	$d < \dfrac{\lambda}{n}$	$d < \dfrac{\lambda}{n+1}$

Note: d is the grating period, λ is the wavelength of incident light and n is the refractive index of the substrate (the incident medium is assumed to be air).

carried out using an approach know as rigorous coupled wave analysis (RCWA),[81] which is a differential-based method to calculate efficiencies of orders in diffraction gratings. Various commercial implementations of this theory are available, *e.g.* GD-Calc™ from K. J. Innovation[82] and DiffractMOD™ from RSoft Design Group,[83] which allow subwavelength structures of any profile to be modelled using a staircase approximation. The approach is often used to support reflectance measurements of artificial moth-eye structures,[84–86] but can also be used to explore the parameter space in artificial moth-eye design and perform optimizations for solar applications.[86–89]

Studies using RCWA have shown that feature shape, height and period (or spacing) are all important factors for the AR performance of the subwavelength moth-eye structures.[88] Simulations of the total reflectance of an array of subwavelength features as a function of period and height or aspect ratio for a single wavelength can be used to guide the design of practical moth-eye arrays. An example of the typical results of such simulations is presented in Figure 8.11. For small periods, effective medium behaviour is observed, with a dramatic decrease in reflectance, followed by the appearance of low amplitude interference fringes as the feature height is increased. More complicated behaviour is observed as the period approaches the resonance region (*i.e.* as $d \rightarrow \lambda$). The period region between effective medium and resonance represents an optimum in the design, whereby low reflectances are achieved at modest feature heights. Higher aspect ratio features lead to a lower reflectance but are more difficult to fabricate and incorporate into solar cell designs, a trade-off that is balanced at a height of approximately half of the design wavelength. The feature shape is also key as it specifies the refractive index gradient profile. Unsurprisingly, a smoothly tapered feature profile, similar to those found on natural moth-eyes, is found to out-perform designs based on sharper or flat-topped features.[88]

Theory and experiments have shown that although the moth-eye AR effect is broadband, there is variation with wavelength (Figure 8.12) and the period can be tuned to position a local reflectance minimum in the spectral region important for a particular application.[88] One such optimization for monocrystalline silicon solar cells encapsulated in EVA arrived at a period of 312 nm for the maximum reduction in reflection of the AM1.5 solar spectrum at the silicon–EVA interface.[89] The short-circuit current daily average was

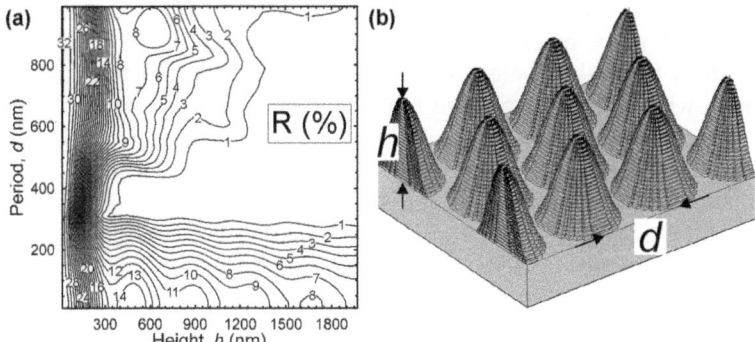

Figure 8.11 (a) Parameter map from RCWA simulations implemented in GD-Calc™ of a silicon moth-eye structure showing reflectance *vs.* height and period at a wavelength of 1000 nm. (b) A three-dimensional rendering of the simulated structure.[88]

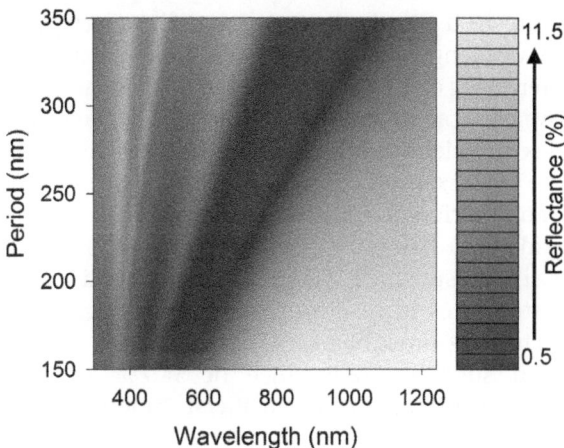

Figure 8.12 Contour plot of reflectance *vs.* period and wavelength of a silicon moth-eye with a feature height of 400 nm, calculated using RCWA implemented in GD-Calc™.[89]

calculated to be only 0.6% less than that of a cell with zero reflectance at the EVA–silicon interface, an improvement of approximately 3% on the best DLAR coating.

8.7.2 Subwavelength Texturing Practical Realization

8.7.2.1 Fabrication of Artificial Moth-eye Structures

Experimental realization of regular arrays of tapered pillars, mimicking the moth-eye arrays found in nature generally involves first defining the pattern in a thin layer of polymeric resist covering a hard substrate using electron beam or interference lithography. These processes consist of chemically modifying the resist by exposure to either an electron beam or the interference pattern formed between two or more laser beams. The resist is then developed whereby the unwanted material (exposed or unexposed, depending on whether the resist is a positive or negative type) is removed, leaving the required subwavelength-scale pattern defined in the resist. For glass substrates, the patterned resist layer is an effective AR coating because of the similar refractive index values of the two materials.[90] However, a pattern defined in photoresist is delicate and so an etch is often employed, using the resist as a mask, to transfer the structures into the harder underlying substrate. Examples of patterning glass/quartz substrates this way include using CHF_3 reactive ion etching (RIE)[91] and SF_6 fast atom beam (FAB) etching[92] through a resist mask. To fabricate taller features, a more robust etch mask is often used. In one example, electron beam lithography (EBL) was used to define a pattern of holes in resist. Chromium was then deposited and the resist lifted off, leaving behind an array of Cr disks which were employed as a mask for a subsequent $C_4F_8/CH_2F_2/O_2$ RIE to form the artificial moth-eye structure in a fused silica substrate.[84] Reflectivities of less than 0.5% across the wavelength range of 400–800 nm were achieved.

Pattern transferral *via* etching is required to form an artificial moth-eye AR array in the active layer of a solar cell (*e.g.* in silicon or GaAs) as polymers with a refractive index sufficiently high to match to the active layer material are not available. In one example, a CHF_3/O_2 RIE was employed to transfer features from an interference lithography patterned resist layer into an intermediate oxide layer before a Cl_2/BCl_3 RIE was employed to transfer the pattern onto a silicon substrate.[93] Another group used a SF_6 FAB to form subwavelength structures in a silicon substrate through an EBL patterned resist mask,[94] achieving broadband AR, down to 0.5% at 400 nm (from an initial untextured substrate reflectance of 54.7% at this wavelength). Figure 8.13 shows an example of an artificial, biomimetic moth-eye array defined in resist by EBL and then transferred into silicon with HBr dry etching along with, for comparison, a natural subwavelength structure found on the transparent wings of the cicada *Cryptotympana aquila*.[88]

In efforts to reduce costs, alternatives to the top–down approaches to moth-eye pattern definition have also been developed involving the self-assembly of an etch mask in the form of polystyrene or silica nanospheres from a colloidal suspensions.[85,95,96] In one example, an SF_6/O_2 RIE was used to pattern a silicon surface through a mask of silica spheres deposited through spin coating a colloidal suspension. Homogeneous monolayer coverage is difficult to achieve with this technique, especially on polycrystalline wafers where the surface roughness hinders the formation of large domains of close-packed spheres. Progress has been made in this area by controlling the viscosity of the nanosphere suspension through the incorporation of a polymer.[97] In a demonstration of this technique, Sun *et al.*[85] coated a 4 inch silicon wafer with a polymer containing 360 nm diameter silica spheres. Removal of the polymer left an array of spheres which was

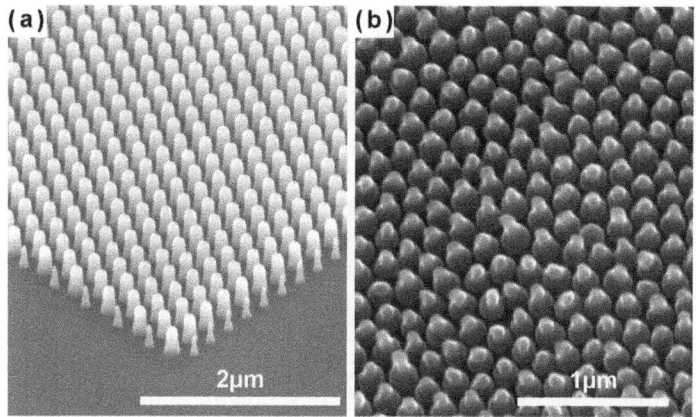

Figure 8.13 SEM images of: (a) artificial, biomimetic moth-eye array in silicon, fabricated by EBL and dry etching; and (b) subwavelength structures found on the transparent wing sections of *Cryptotympana aquila*.[88] Reprinted with permission from S. A. Boden and D. M. Bagnall, 'Tunable reflection minima of nanostructured antireflective surfaces', *Appl. Phys. Lett.*, **93**(13), 133108, Copyright 2008, American Institute of Physics.

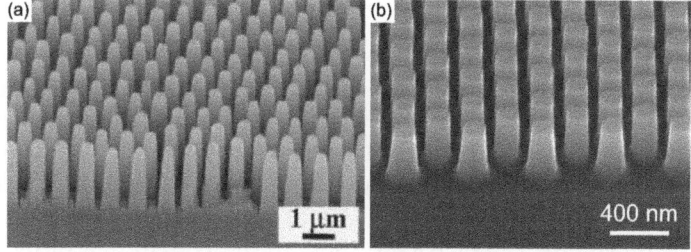

Figure 8.14 SEM images of AR pillar arrays formed in silicon by RIE through: (a) a mask consisting of a monolayer of silica nanospheres deposited from a colloidal suspension;[98] and (b) an aluminium mask patterned by nanoimprint lithography and lift-off.[100] (a) Reprinted with permission from W.-L. Min, B. Jiang and P. Jiang, 'Bioinspired Self-Cleaning Antireflection Coatings', *Adv. Mater.*, **20**(20), 3914–3918, Copyright 2008, John Wiley and Sons. (b) Reprinted with permission from S. A. Boden and D. M. Bagnall, 'Nanostructured biomimetic moth-eye arrays in silicon by nanoimprint lithography', *Proceedings of SPIE*, 7401, Copyright 2009, Society of Photo Optical Instrumentation Engineers.

used as an etch mask for the formation of a moth-eye texture in the silicon by an SF_6 RIE[85,98] [Figure 8.14(a)], achieving a reflectance less than 2.5% in the 350–900 nm wavelength range. This technique has also been applied to GaSb, an important material for thermophotovolatic devices, with similar reflectance reductions.[99]

To enhance the scalability of artificial moth-eye surfaces, embossing or nanoimprinting techniques have been developed whereby a hard stamp or master is fabricated using one of the processes above and then used to replicate the pattern multiple times by imprinting into heated polymer layers.[101–103] A related technique known as step-and-flash imprint lithography employs UV light to cure the polymer through a transparent substrate during the imprint. Using this technique, moth-eye arrays can be replicated in hard wearing polymer layers on glass. In one example, scratch-resistant inorganic–organic co-polymers were imprinted with nickel masters patterned with moth-eye arrays using interference lithography,[104] achieving a ~5% enhancement in transmittance through glass.[105] Again, for moth-eye arrays on higher index substrates, pattern transfer by etching is required. In one approach, an EBL defined stamp is used to imprint holes in a polymer layer on silicon. Then, through a deposition and lift-off process, a 1 cm × 1 cm array of aluminium disks is formed on the silicon surface and used as an etch mask for an SF_6/C_4H_8 dry etch.[100] This is followed by various shaping etches to create the desired feature profile [Figure 8.14(b)]. Large area patterning is possible with nanoimprinting, with nanoimprint masters made by interference lithography on the scale of tens of centimetres available commercially.[106,107] Roller processes have also been developed that use flexible masters and/or substrates for AR moth-eye fabrication on the wafer-scale and beyond.[108–110]

8.7.2.2   Random Subwavelength Texturing

Not all subwavelength texturing is designed to directly mimic moth-eye features; graded index layers can be created with features less regular in size, shape and arrangement. These techniques offer potentially cheaper, more scalable routes to effective AR surfaces because a lithographically defined mask is not required.

Reactive ion etching can be used without masking to form a surface that is randomly textured with subwavelength features. Chlorine- and fluorine-based plasmas have been used to texture silicon wafers in this way,[111–114] with reflectances less than 5% across the wavelength range of 300 nm to 1 μm demonstrated. In the case of maskless SF_6/O_2 etching, deep texturing of silicon can be achieved by etching through a thin oxide layer. The etch perforates the oxide at random points and these areas become more susceptible to the etch. Thus pits appear on the surface and the walls of these pits quickly become covered with absorbed $Si_xO_yF_z$, which protects these areas from further etching. The resulting surface has a random, needle-like texture. When applied in the fabrication of a multicrystalline silicon solar cell, the texturing was shown to increase the device efficiency from 8.5% to 10.9%.[111] The increase in surface area and damage caused by the texturing process leading to increases in surface recombination, and issues related to doping the textured silicon limited the improvement in device performance. A wet etch (e.g. HNO_3, HF) can be performed post-RIE to remove some of the RIE-related defects, with reported improvements of up to 7% over an untextured cell.[115] Similar dry etch process have achieved exceeding low broadband silicon reflectivities, notably Yoo *et al.*,[116] who used a RF multi hollow cathode discharge system for SF_6/O_2 plasma etching, forming a surface texture that exhibited negligible reflectance across the 200–1000 nm wavelength range.

Another related approach is to deposit small amounts of material either prior to or during the dry etch to facilitate the formation of high aspect ratio features. In one example, silicon surfaces were textured in an electron cyclotron resonance reactor using a mixture of silane, methane, argon and hydrogen.[117] Nano-sized silicon carbide clusters are formed in a reaction between silane and methane. These are deposited onto the silicon surface, protecting the silicon directly underneath from etching by the Ar/H_2 plasma,[118] and so leading to the formation of a dense array of conical 'nanotips', uniformly over the surface of a six-inch wafer. The close-packed, high aspect ratio nanotips confer broadband, wide angle antireflection to the surface, but questions remain over the compatibility of such high aspect ratio features and damage-inducing plasma etching processes to successful solar cell fabrication.

Although fewer defects are introduced to a surface by wet etching, the resulting texture is usually insufficient for effective AR. One promising technique circumvents this limitation by firstly depositing a 1–2 nm thick, discontinuous gold film onto a silicon surface before etching with a HF : H_2O_2 mixture.[119] The resulting surface is textured to depths of between 200 and 300 nm with randomly arranged features of widths in the 50–100 nm range. The technique was successfully applied to c-Si, microcrystalline Si (mc-Si) and amorphous Si (a-Si), creating surfaces exhibiting reflectances <5% in the 300–1000 nm wavelength range. A 40% improvement in short-circuit current compared with an untextured reference cell was demonstrated.[120]

Reflectances approaching those of the best micron-scale textures combined with multilayer thin film ARCs are readily achieved with maskless submicron texturing without the need for the addition of thin film ARCs or expensive lithography steps. The challenges for this approach when texturing the active region of a cell are increased surface recombination due to etch damage and surface area increases. Nevertheless, recent progress in this area suggests it holds great promise for future PV applications.

8.7.2.3 Porosity Techniques for Glass AR

Other techniques have been specifically developed for the texturing of glass surfaces in order to reduce PV module top surface reflection. These involve introducing some form of subwavelength-scale porosity into the first few hundred nanometres of the glass surface to form an effective medium and so invoke destructive interference or graded index AR mechanisms. There is considerable overlap here with the principle of thin film ARCs described in Section 8.5. Porosity can be introduced during vacuum deposition of thin films, for example, with magnetron sputtering of porous SiO_2 and TiO_2 films on

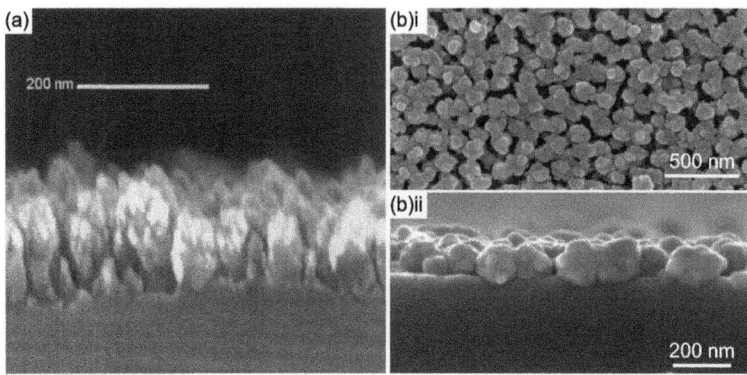

Figure 8.15   SEM images of porous AR schemes for glass surfaces: (a) porous SiO$_2$
ARC by RPECVD;[122] and (b) SiO$_2$–TiO$_2$ colloidal particle coating (i)
top-down view, (ii) cross-section.[128] (a) Reprinted from H. Nagel,
A. Metz and R. Hezel, 'Porous SiO$_2$ films prepared by remote plasma-
enhanced chemical vapour deposition—a novel antireflection
coating technology for photovoltaic modules', *Sol. Energy Mater. Sol.
Cells*, **65**(1–4), 71–77, Copyright 2001, with permission from Elsevier.
(b) Reprinted (adapted) with permission from X.-T. Zhang, O. Sato,
M. Taguchi, Y. Einaga, T. Murakami and A. Fujishima, 'Self-cleaning
particle coating with antireflection properties', *Chem. Mater.*, **17**(3),
696–700, Copyright 2005, American Chemical Society.

glass.[121] Remote plasma-enhanced chemical vapour deposition (RPECVD) has
also been successfully used for porous SiO$_2$ ARC deposition [Figure 8.15(a)].[122]
However, cost is even more important for glass AR as the relative gains are
smaller due to the lower refractive index compared with the silicon and other
active area materials. Therefore, these relatively expensive vacuum techniques
have not seen wide scale commercial implementation. However, combustion
CVD,[123] which does not require a vacuum chamber but instead involves
burning a silicate precursor solution as it is sprayed out of a nozzle onto the
glass surface, has been commercialized for large-scale PV glass AR.

Selective etching or leaching of a component of the glass with an acid
treatment can be used to introduce porosity into a surface layer. This is
achieved in borosilicate glasses by first heating to induce nanoscale phase
separation into high and low silicate components, with the latter being more
susceptible to a subsequent acid etch.[124] Acid etching to produce porous AR
glass surfaces has been shown to increase transmission by ~5% and has
been successfully commercialized by a number of companies including
Sunarc Technologies and CSG Solar Ltd.[125–127]

Various solution-based techniques have also been developed to produce
porous and therefore low effective refractive index layers on glass for AR.
These so-called 'sol–gel' processes[129] generally involve deposition of a solution which forms a gel on a surface, the liquid phase of which is removed
during drying to leave a hard, porous coating. Tetraethyl orthosilicate (TEOS)

is often used as a precursor, resulting in a porous silica layer.[130-132] A non-ionic surfactant is sometimes added to control pore size and improve adhesion.[133] Industrial-scale processes for stable sol–gel glass AR layers have been developed, with the coating of 1.5 m × 2.5 m panels in a single dip process.[134] Other solution-based techniques for glass AR involve colloidal deposition of nanospheres, similar to the nanosphere lithography approaches described in Section 8.7.2.1 but the with the nanosphere coating itself acting as the AR layer because the refractive index of the nanospheres is close to that of the glass substrate. In one example, polymer-silica core-shell particles were dip coated onto a glass substrate. Removal of the polymer core using a high temperature treatment resulted in a durable but highly porous silica layer and a 5% increase in transmittance.[135] In a similar approach, polyelectrolyte multilayers were deposited onto glass to create a positively charged surface onto which negatively charged SiO_2, TiO_2 and polymer colloidal particles were deposited to form an AR coating [Figure 8.15(b)].[128,136,137] In all these examples, no vacuum chamber is needed and expensive patterning processes are avoided, making these techniques attractive to the PV industry for their scalability.

8.8 Metal Nanoparticle Techniques

Metal nanoparticles can interact strongly with UV, visible and NIR photons due to the excitation and decay of localized surface plasmons (LSPs), which are resonant oscillations of conduction electrons.[138] LSPs decay either radiatively or non-radiatively, resulting in far-field scattering or absorption, respectively. The spectral position of the resonance and the dominant decay channel can be modified by changing the composition or morphology of the nanoparticle, and by changing the optical properties of the surrounding medium.[139] Importantly, metal nanoparticles can strongly interact with light despite having dimensions considerably smaller than the wavelength. Therefore metal nanoparticles can be used as highly tuneable subwavelength-sized scattering elements for light trapping in photovoltaics.[12,140] Metal nanoparticles have been shown to improve the efficiency of various types of solar cell, including amorphous,[141] microcrystalline[142] and crystalline[12,143] silicon and GaAs.[144]

The design challenges for metal nanoparticles include the minimization of parasitic absorption (*i.e.* absorption within the nanoparticle rather than within the semiconductor) and maximizing scattering across the wavelength range of interest. These properties can be optimized by appropriate choice of nanoparticle size, shape and position within the solar cell.

8.8.1 Optical Properties of Metal Nanoparticles

Metal nanoparticles both scatter and absorb light. The sum of scattering and absorption is known as extinction and represents the total far-field interaction of light with metal nanoparticles. The extinction spectrum consists of a peak that reaches a maximum at the resonant frequency of the localized surface plasmon. The radiative efficiency is the ratio of scattering to extinction, and varies from 0 for completely absorbing nanoparticles to 1 for completely scattering nanoparticles. The extinction peak magnitude, width, spectral position and radiative efficiency can be tuned by modifying the refractive index of the surrounding medium and by modifying the particle composition, size or shape.

Extinction, scattering and absorption by metal nanoparticles are usually defined as dimensionless efficiencies, which are the ratio of the optical cross-section to the geometric cross-section. The optical cross-section of a metal nanoparticle typically exceeds the geometric cross-section,[145,146] and so a two-dimensional (2D) array of metal nanoparticles with a surface coverage lower than 100% is sufficient to interact with all incident photons. Arranging the nanoparticles in a 2D planar array minimizes multiple interactions and also simplifies fabrication. The design of the nanoparticles within the array must be chosen to maximize the extinction efficiency at wavelengths where light trapping is required, and to maximize radiative efficiency (*i.e.* minimize absorption) for all wavelengths in the incident spectrum.[147] The magnitude of the extinction efficiency determines the minimum surface coverage of nanoparticles required to scatter all incident photons: the higher the extinction efficiency the lower the required surface coverage.

The optical properties of metal nanoparticles can be modelled using analytical or numerical simulation techniques. Analytical solutions to Maxwell's equations are available for the case of a spherical or spheroidal metal nanoparticle in a non-absorbing homogeneous medium.[146] More complicated particle geometries and environments require the use of numerical methods such as the finite-difference time-domain (FDTD) technique,[148] the finite-element method (FEM),[149] or the discrete dipole approximation (DDA).[150] In the following sections, Mie theory and DDA simulation results are used to investigate the influence of size, shape and surrounding medium on the optical properties of metal nanoparticles, with the aim of achieving high scattering efficiencies in the NIR.

8.8.1.1   Position within the Solar Cell

Metal nanoparticles can be integrated into a solar cell as a planar 2D array positioned at the front or rear of the solar cell, or within the semiconductor. Nanoparticles positioned within the semiconductor are likely to degrade the electrical properties of the layer due to the introduction of defects and the diffusion of metal impurities. Situating the nanoparticles on the front or rear of the solar cell ensures that they have minimal impact on the electrical properties of the semiconductor. In this case the array can be embedded within one of the transparent layers already present in the solar cell structure, for example, a front or rear transparent conducting oxide (TCO) or a dielectric passivation layer. Metal nanoparticles situated on the front surface can both reduce reflection and increase light trapping,[12] but can in some cases increase reflectance due to back-scattering.[151,152] Nanoparticles positioned at the rear of the device only interact with photons that are not absorbed within a single pass of the semiconductor, which reduces parasitic absorption losses, and the inclusion of a planar reflector eliminates out-coupling losses.[153]

8.8.1.2   Size and Surrounding Medium

Small metal nanoparticles exhibit a single narrow extinction peak, which corresponds to the dipolar LSP resonance.[146] As the diameter is increased the dipolar peak is broadened, attenuated and shifted to longer wavelengths (Figure 8.16). Additional peaks occur in the spectra of large nanoparticles due to the excitation of higher order modes, with the overall spectra being superpositions of each individual mode.[146,151]

Spherical nanoparticles in air ($n = 1$) require large diameters to achieve extinction peaks in the NIR. However, metal nanoparticles embedded within the solar cell structure will have a higher surrounding refractive index ($n \geq 1.5$). Increasing the refractive index of the surrounding medium shifts the peaks to longer wavelengths, with a linear relationship between peak position and the surrounding refractive index (Figure 8.17). Increasing the refractive index of the surrounding medium reduces the minimum nano-particle size required to achieve scattering in the NIR. The radiative efficiency of the extinction peak rapidly increases with particle size until a saturation region is reached. This is achieved at smaller diameters as the refractive index is increased. The refractive index has negligible effect on the radiative efficiency of metal nanoparticles with diameters of 150 nm or larger, and so in this size range only its influence on the peak position need be considered. Spherical nanoparticles with diameters of 100 nm or smaller cannot be tuned to the NIR and can suffer from high parasitic absorption.

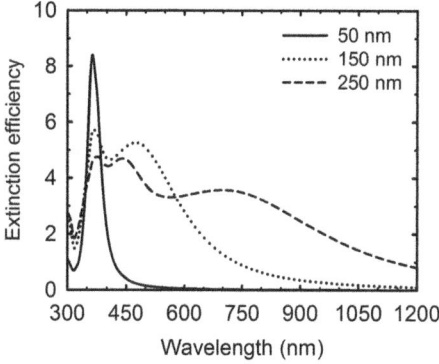

Figure 8.16   Simulated optical extinction spectra in vacuum of Ag spheres with diameters of 50, 150 and 250 nm.

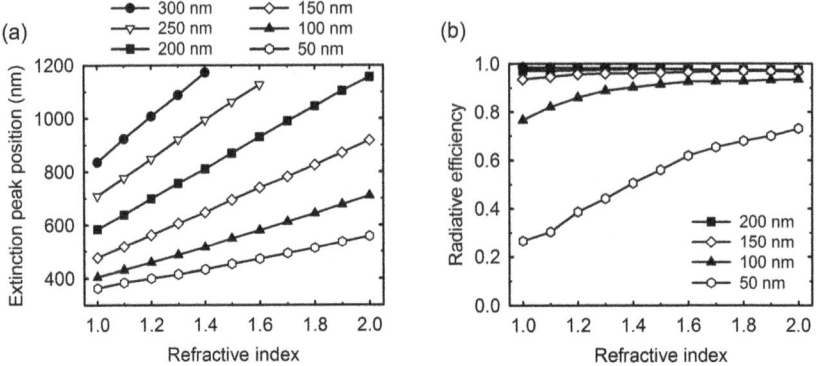

Figure 8.17 (a) Extinction peak position and (b) radiative efficiency of spherical Ag nanoparticles as a function of the refractive index of the surrounding medium for diameters ranging from 50 to 300 nm.

8.8.1.3 Shape: Spheroids and Prisms

The optical properties of metal nanoparticles are highly sensitive to the particle shape because this influences the collective electron oscillation dynamics and hence the resonance frequency. Therefore it is possible to tune the optical properties of metal nanoparticles while keeping the particle volume constant. For example, stretching one axis of a sphere results in a prolate spheroid, which supports distinct resonances along the longitudinal and transverse axes.[154] The result is a polarization-sensitive extinction spectrum, with a weak short wavelength peak for polarization aligned along the transverse axes, and a strong long wavelength peak for polarization aligned along the longitudinal axis. As the aspect ratio (long axis to short axis) is increased, the longitudinal mode becomes strongly shifted to longer wavelengths, while the transverse mode is attenuated and weakly blue-shifted (Figure 8.18). The response to unpolarized light is the average of both modes but is dominated by the longitudinal mode.[155] Tuning the peak position by increasing the aspect ratio rather than the size means that the peak is not attenuated or broadened.

Figure 8.19 demonstrates that a wide range of extinction spectra can be achieved by varying both the aspect ratio and the particle size. There is a nearly linear relationship between extinction peak position and the aspect ratio, and larger particles exhibit a stronger shift for a given change in aspect ratio. The relationship between aspect ratio and radiative efficiency is more complicated, with small particles generally exhibiting an improved radiative efficiency as the aspect ratio is increased, but larger particles exhibit a modest decrease for aspect ratios larger than 2.

Metal nanoparticles fabricated by lithography have a prism shape, with a constant 2D cross-section extruded along an axis. The optical properties of prisms can be changed by modifying either the height or the cross-sectional size and shape. The cross-sectional shape has a strong influence on the

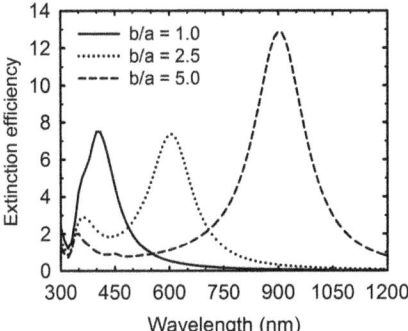

Figure 8.18 Simulated optical extinction spectra of Ag spheroids with aspect ratios of 1, 2.5 and 5, and volume equal to that of a 100 nm diameter sphere, illuminated by unpolarized light.

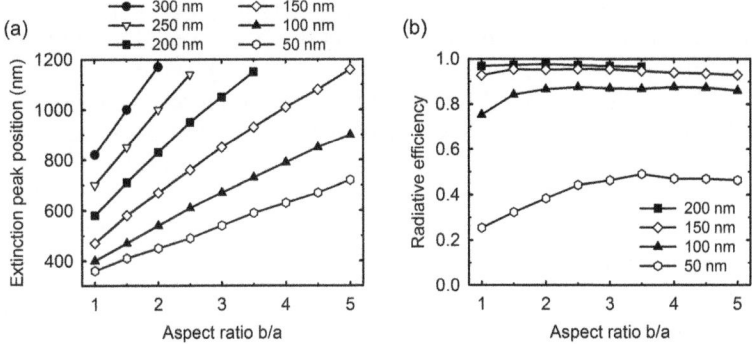

Figure 8.19 (a) Extinction peak position and (b) radiative efficiency of spheroidal Ag nanoparticles as a function of the aspect ratio for volumes corresponding to spheres with diameter ranging from 50 to 300 nm.

optical properties, even if the cross-sectional area is kept constant.[155] The extinction spectra and peak radiative efficiencies of Ag nanoparticles with circular, square, triangular and rectangular cross-sections are shown in Figure 8.20. Rectangular nanoparticles exhibit a similar extinction spectrum to the spheroids discussed above, and have the highest extinction efficiency and longest wavelength peak position of the four shapes. For the other three shapes there is a correlation between the peak position and the maximum radius of curvature of the shape vertices, with sharper corners resulting in a more red-shifted peak position.[156] However, sharp features also result in a large drop in radiative efficiency and so should be avoided for photovoltaic applications.[155]

Decreasing the height of the prism generally shifts the peak position to longer wavelengths, with the magnitude of the shift strongly increasing for heights less than 10 nm [Figure 8.21(a)]. The effect of height on extinction

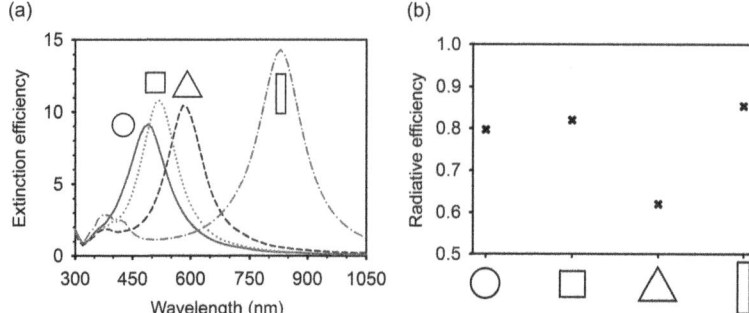

Figure 8.20   (a) Extinction spectra and (b) peak radiative efficiency of Ag prisms with
height 50 nm and cross-sections of circle with diameter 115 nm, square
width side length 100 nm, equilateral triangle with side length 150 nm,
and rectangle 50 nm × 200 nm.

peak position becomes very weak for heights greater than approximately 40
nm. The radiative efficiency is decreased as the prism height is reduced and
so very thin prisms are predominantly absorbing [Figure 8.21(b)]. For
example, decreasing the height of a 150 nm square nanoparticle from 10 to 5
nm shifts the extinction peak from 860 to 1350 nm, but also decreases the
radiative efficiency from 0.71 to 0.15. Therefore reducing the nanoparticle
height is not a suitable tuning method for photovoltaic applications because
the beneficial red-shifting of the extinction peak position is accompanied by
a detrimental reduction in radiative efficiency.

8.8.1.4   Constituent Metal

So far in this section we have only considered Ag nanoparticles, but other
metals also support LSPs in the visible and NIR range.[157-159] LSPs can only be
excited efficiently at wavelengths where the optical properties of the metal are
predominantly due to the behaviour of electrons in the conduction band (*i.e.*
free electrons). Interband transitions act as additional non-radiative decay
channels and so either damp or prohibit the excitation of LSPs. The metals
with predominantly free-electron behaviour in the visible and NIR are the
noble metals (Ag, Au, Cu), the alkali metals (K, Na, Li) and Al. The alkali
metals are too chemically unstable to use in practical applications. The noble
metals (Ag, Au, Cu) are characterized by a threshold below which the optical
properties are dominated by interband transitions, which occurs at wavelengths of approximately 327, 517 and 590 nm for Ag, Au and Cu, respectively.[160] Al has a weak interband region centred near 827 nm, but supports
free-electron behaviour above and below this region.[159]

Free-electron behaviour is a requirement for both high reflectivity of bulk
metal and the excitation of LSPs in metal nanoparticles. Therefore the
reflectance spectrum of bulk metal can be used to gauge the suitability of the
metal for plasmonic applications at a given wavelength. Figure 8.22(a) shows

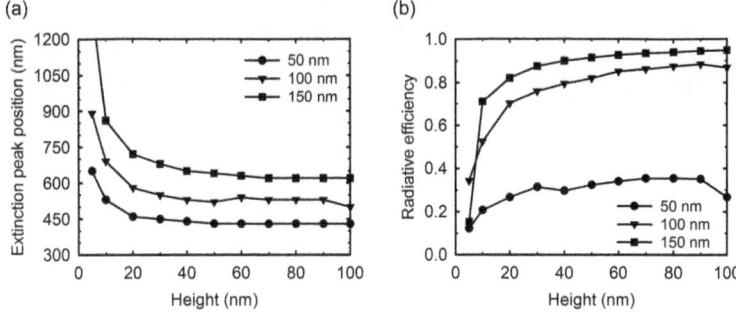

Figure 8.21 (a) Extinction peak position and (b) radiative efficiency of Ag nanoparticle as a function of the height of the prism for a square cross-section with side length of 50, 100 and 150 nm.

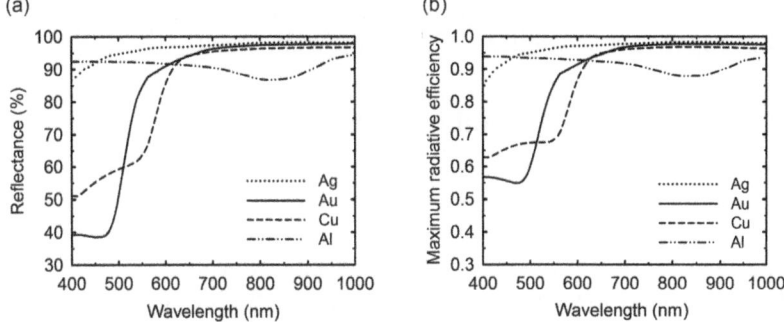

Figure 8.22 (a) Reflectance spectra of bulk Ag, Au, Cu and Al calculated using Fresnel's equations. (b) The maximum radiative efficiency found for Ag, Au, Cu and Al spherical nanoparticles with diameters ranging from 5 to 300 nm.

the reflectance spectra of bulk Ag, Au, Cu and Al calculated using eqn (8.3). The interband regions of Au and Cu can be identified as regions of low reflectivity in the UV and visible, which gives rise to the well-known colouration of these metals in bulk form. Al has the highest reflectance of all the metals at 400 nm, but has the lowest in the NIR due to the presence of interband transitions. The maximum radiative efficiency that can be attained at each wavelength by spherical particles as the diameter is varied from 5 to 300 nm is shown in Figure 8.22(b), and clearly correlates with the reflectance spectrum. The presence of interband transitions results in low reflectivity of bulk metal and poor radiative efficiency in metal nanoparticles.

The Au and Cu interband regions are more detrimental to reflectance and radiative efficiency than the Al interband region, but the latter occurs within the NIR where light trapping is required. In general, Ag gives the highest radiative efficiency, although the values of all of the noble metals are similar

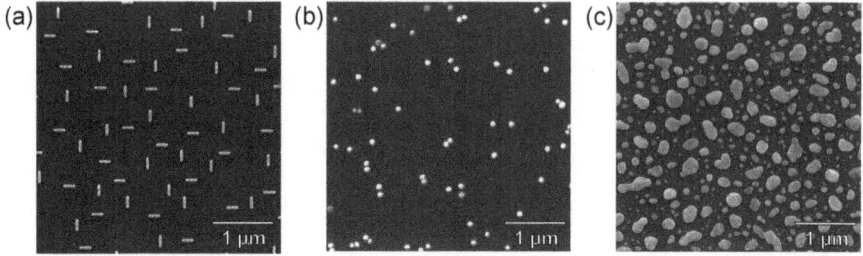

Figure 8.23   SEM images showing arrays of metal nanoparticles fabricated using (a) lithography; (b) chemical synthesis; and (c) thermal recrystallization.

in the NIR. In general, Ag can be used for applications in the visible and NIR, Au and Cu can be used for applications in the NIR, and Al can be used for applications in the UV and visible.[147] It is interesting to note that the maximum radiative efficiency of a metal nanoparticle is nearly equal to the reflectance, *i.e.* the minimum achievable absorption losses within an array of metal nanoparticles is similar to that of the bulk metal.

8.8.2   Fabrication of Metal Nanoparticles

Metal nanoparticles can be fabricated at low cost over large areas and so are a viable option for the PV industry. For PV applications there is a considerable difference in the industrial and laboratorial requirements of a metal nanoparticle fabrication technique. Industrial applications require a technique that is fast, low-cost and large area, and does not require complex equipment. These criteria are less important for laboratory use, and instead the level of control over particle size, shape and arrangement are prioritized such that the parameter space available for experiments is maximized.

Metal nanoparticles can be fabricated using a wide variety of techniques, which can broadly be split into three categories: lithography; chemical synthesis; and thermal recrystallization. Each category of technique results in considerably different arrays of metal nanoparticles in terms of the arrangement of the particles and the distribution of size and shape (Figure 8.23). For example, lithographically defined arrays of nanoparticles generally consist of nearly identical nanoparticles arranged periodically, while arrays made by thermal recrystallization have a broad distribution of size and shape and are randomly arranged.

8.8.2.1   *Lithography*

Lithographic techniques utilize a resist layer to mask the deposition or etching of a planar metal film, resulting in an array of planar metal nanoparticles. The primary difference between lithographic techniques is the method used to pattern the resist layer. The most flexible option is electron beam lithography (EBL), where an electron beam is used to sequentially write the pattern of each nanoparticle into the resist. This allows complete freedom of the lateral design of the array: each nanoparticle can have an arbitrarily chosen size, shape and position, with feature sizes down to tens of nanometres.[161,162] The drawback is that EBL equipment is expensive and complex, and requires many hours to write a 1 mm^2 array. Interference lithography, as introduced in Section 8.7.2.1 for the fabrication of sub-wavelength AR arrays, can be used to define large areas in a single exposure by using the interference of multiple beams,[163] but still requires relatively complex equipment and is restricted to the fabrication of periodic arrays. Nanoimprint lithography (NIL) uses physical compression of a reusable stamp into a polymer layer to transfer the pattern[164,165] and so can be used to replicate patterns fabricated by EBL or interference lithography for metal particle arrays, in addition to the subwavelength AR structures described in Section 8.7.2.1.

A range of low-cost lithographic techniques are based on spherical latex particles. One of the first of these techniques to be developed is known as nanosphere lithography (NSL).[166] In NSL, latex nanospheres are self-assembled into a close-packed monolayer onto a substrate and metal is then deposited into the gaps between the spheres to form a hexagonal array of triangular (or, more accurately, truncated-tetrahedral) nanoparticles. NSL is a very simple technique but the geometry and arrangement of the nano-particles is limited by the array structure, and any defects in the nanosphere array are transferred to the substrate as large areas of metal. Hole-mask colloidal lithography (HCL) also utilises latex nanospheres, but these are randomly dispersed and used to pattern an intermediate resist layer.[167] The resulting metal nanoparticles have a disc-like structure with the same diameter as the latex spheres and are randomly arranged on the surface.[158] HCL has also been used to create other nanoparticle geometries by altering the angle of the substrate during evaporation, resulting in rod-shaped and cone-shaped nanoparticles, and disc dimers.[167]

There is no restriction to the constituent metal that can be used in nanoparticle arrays fabricated using lithography, provided it can be deposited in a smooth film that adheres to the substrate. The metal deposition stage can even be modified to produce nanoparticles that comprise multiple metal and dielectric layers.[168] Conventional lift-off procedures are restricted to arrays of nanoparticles that have identical height and composition, but the lateral size and shape are only limited by the technique used to pattern the resist.

8.8.2.2 Chemical Synthesis

Chemical synthesis can be used to produce metal nanoparticle colloids, which can subsequently be precipitated onto a substrate in 2D or 3D arrays. Synthesis is typically achieved by the reduction of metal salts, and involves separate stages to nucleate, grow and sort the nanoparticles within solution. Simple one-pot recipes can be used to produce spherical nanoparticles in gram scale quantities.[169-171] In contrast to other fabrication methods, chemical synthesis also allows a large degree of control over the three-dimensional (3D) shape of the nanoparticles, as the growth of specific crystal facets can be enhanced or suppressed by changing the reaction conditions.[172] There are an ever increasing number of published recipes for a wide variety of geometries, including spheres, rods, platelets, cages, core-shells and multi-branched nanoparticles. In general, the larger or more complex the nanoparticle is, the lower the yield will be. Each metal type requires a distinct synthesis recipe, with Au being by far the most commonly investigated material. Nanoparticles of other metals can also be synthesized, but have a lower number of available recipes, *i.e.* a more limited choice of geometries.

After synthesis the size distribution of nanoparticles within the colloid can be reduced by various techniques such as filtering and centrifuging.[173] The final stage is to precipitate the nanoparticles from solution onto the substrate, ideally as a submonolayer with minimal aggregation. One method is to simply drop-cast the colloid onto the substrate, but this can result in uneven surface coverage and aggregation due to the surface tension of the evaporating drop. Improved results can be achieved by dunk-coating, where the substrate is prepared with an adhesion layer and immersed in the colloid. Adhesion is conventionally achieved either by electrostatic attraction to a charged polymer layer,[174] or by covalent bonding to a linker layer such as an organosilane monolayer.[175] Aggregation is limited by mutual repulsion of nanoparticles due to the surface charge of capping layers. Depositions of different colloids on one substrate can be used to form an array with more than one type of nanoparticle and the surface coverage of each can be controlled by varying the immersion time.

8.8.2.3 Thermal Recrystallization

Thermal recrystallization is the process of forming or modifying metal nanoparticles from larger or smaller starting materials using heat.[176] Temperatures well below the melting point can still result in dramatic reshaping of metal films and nanoparticles due to the diffusion of metal atoms along grain boundaries and the substrate. Annealing semi-continuous or continuous metal thin films at temperatures on the order of 100 °C or higher can result in the formation of nanoparticle arrays, known as metal island films (MIFs). These are dense arrays of nanoparticles with a wide distribution of nanoparticle size and shape, which is a complex function of both the properties of the starting layer and of the annealing conditions. Increasing the amount of deposited material in the starting layer increases the average particle size but results in more irregular particle shapes, while increasing the anneal temperature results in smaller, more rounded particles.[177] The metal deposition speed and substrate properties also affect the size and shape of nanoparticles.[178,179] Critically, there is very little control over the overall surface coverage of the nanoparticles.

Other thermal methods have been developed to achieve more control of nanoparticle size and shape. In vapour coalescence the nanoparticles are formed by evaporating metal atoms into a flowing gas, with the ambient temperature controlled to achieve nucleation of nanoparticles and reshaping or aggregation.[180] Various methods can be used to select a narrow range of particle sizes to reach the substrate and the surface coverage can be controlled by varying the deposition time.

8.8.2.4 Summary of Metal Nanoparticle Fabrication

Despite the large variety of fabrication methods, metal island films are currently the most common type of nanoparticle used in laboratory solar cell experiments. This is because the technique requires no additional equipment or expertise than is already available in a typical solar cell fabrication laboratory, needing only a single metal evaporation and a low temperature anneal. Most other metal nanoparticle fabrication techniques require additional equipment or extensive process development. The MIF technique is very simple, but the resulting arrays are highly complex, often being comprised of closely spaced, irregularly shaped nanoparticles with a wide distribution of sizes.

Low cost, ease of process integration and large area coverage make metal island films attractive for industrial applications, but they are far from ideal for research purposes. In contrast, EBL offers the highest level of control over particle shape and position and so is highly suited to research studies, but cost and complexity prohibit its use in industry. Simpler lithography methods such as NSL and HCL are useful for research studies when EBL is unavailable or otherwise unsuitable. Chemical synthesis offers an interesting middle ground, with a high level of control and the possibility for large area processing, but requires extensive expertise to implement and develop. Fabrication techniques can also be combined, for example, annealing can be used to reshape any nanoparticle array, and lithography can be used for site-selective adhesion of chemically synthesised nanoparticles.

8.8.3 Integration of Metal Nanoparticles into Silicon Solar Cells

Metal nanoparticles have been integrated into many different types of solar cell and photodetectors. In this section we review the experimental results for metal nanoparticles integrated into crystalline (wafer-based), polycrystalline and amorphous silicon solar cells.

Nanoparticles deposited onto the front surface of thick, wafer-based silicon solar cells primarily increase photocurrent by reducing reflection. Lim *et al.*[143] deposited a sparse array of 100 nm Au spheres onto the front surface of a silicon photodiode, and found that photocurrent was increased at short wavelengths and decreased at long wavelengths. The decrease was attributed to interference between scattered and unscattered photons. Pillai *et al.*[12]

reported a 19% increase in photocurrent after depositing a Ag island film on the front surface of a planar PERL cell.

The light-trapping effect of metal nanoparticles becomes more important when the thickness of the semiconductor is reduced. Beck et al.[153] deposited a Ag island film at the rear of a 20 μm thick bifacial solar cell, which was used in combination with a conventional thin-film ARC on the front surface. The nanoparticles increased the photocurrent by 10% when used alone and by 13% when used in conjunction with a detached mirror. The thinnest single-crystal devices are made using silicon-on-insulator (SOI) wafers, which are not economically viable for solar cells but provide an interesting experimental platform. Improving on earlier results by Stuart and Hall,[181] Pillai et al.[12] demonstrated a 33% increase in photocurrent after depositing a Ag island film onto the front surface of an SOI solar cell with an active layer thickness of 1.25 μm. They also found that the enhancement was strongly sensitive to the particle size.

Thin film solar cells have a semiconductor layer thickness of a few microns or less and so can strongly benefit from light trapping. Ouyang et al.[182] studied the influence of a Ag island film on the efficiency of a 2 μm thick evaporated silicon solar cell, and reported photocurrent enhancements of 27% and 44% for rear-mounted nanoparticles with and without a white paint reflector, respectively. Amorphous silicon has higher absorption but poorer carrier lifetimes than crystalline silicon, and so is limited to active layers of a few hundred nanometres. Eminian et al.[141] integrated a dense array of silver nanoparticles into the rear reflector of an amorphous silicon solar cell with an intrinsic layer thickness of 150 nm. They reported a 33% increase in total absorbance, which led to a 20% increase in short-circuit current. However, this was accompanied by a reduction of fill factor from 58% to 42%, which was attributed to reduced quality semiconductor growth due to the roughness of the nanoparticle array.

Despite the impressive results already published there is still considerable scope for the improvement of antireflection and light trapping by metal nanoparticles. Only a small section of the available parameter space has been explored and there have been few studies that attempt to experimentally optimize the size, shape and arrangement of metal nanoparticles for solar cell applications.

8.9 Summary

A wide variety of techniques for capturing light in PV devices have been developed in an effort to reduce optical losses in solar cells. These are broadly divided into techniques to reduce reflection and so increase transmission of light into the active region of a cell, and techniques to enhance absorption through light trapping within the active region. Antireflection technologies include single or multilayer thin film approaches exploiting destructive interference mechanisms, either standalone or combined with texturing on the micron scale to enhance transmission by causing light to undergo multiple reflections. More recently, subwavelength texturing schemes, inspired by natural 'moth-eye' structures, have been developed which exploit graded refractive index mechanisms to confer AR to a surface. Efficient broadband and wide angle AR can be achieved with these approaches, but the costs involved in fabricating dense arrays of such small features have limited industrial adoption of these techniques. Nevertheless, the development of scalable processes based on nanoimprinting, interference lithography and colloidal techniques is paving the way towards commercial viability.

Micron scale texturing can also scatter or refract light, changing its direction of propagation through the active layer of a cell and so increasing the optical path length and therefore the amount absorbed. Photolithography and controlled etching can be used to accurately define arrays of features and this has been used to good effect in the form of inverted pyramids and bowl-like features for high efficiency laboratory type cells. Cheaper techniques for forming random textures are used in commercial devices. A new light trapping approach that is gaining increasing popularity involves the use of metal particle arrays. Here, the excitation of localized surface plasmons is exploited to scatter light leading to enhanced optical path lengths and therefore increased absorption. The scattering properties of such arrays can be tuned by varying the particle size, shape, density, material and surrounding medium to maximize the extinction efficiency and radiative efficiency (minimizing absorption by the particles) across the spectral range important for PV devices. Theoretical studies are promising and various fabrication techniques have been developed, but it has yet to be seen whether the balance between extra cost and performance enhancement can be tipped towards enabling commercial implementation of such approaches.

Light capture strategies are essential for maximizing the absorption of light within a solar cell and hence to achieve high conversion efficiencies. The design of antireflection and light trapping schemes is complicated by the need for them to function over a broad wavelength range and the requirement that they do not substantially increase the cost or complexity of the solar cell fabrication process. New developments in light capture are needed to further increase optical absorption, particularly in thin film solar cells.

References

1. M. Born and E. Wolf, *Principles of Optics: Electromagnetic Theory of Propagation, Interference and Diffraction of Light*, Cambridge University Press, Cambridge, 7th edn, 1999, p. 43.
2. S. A. Boden, *Biomimetic Nanostructured Surfaces for Antireflection in Photovoltaics*, PhD thesis, University of Southampton, 2009.
3. M. A. Green, *High Efficiency Silicon Solar Cells*, Trans Tech Publications, Aedermannsdorf, Switzerland, 1987.

4. H. Nagel, A. G. Aberle and R. Hezel, Optimised antireflection coatings for planar silicon solar cells using remote PECVD silicon nitride and porous silicon dioxide, *Prog. Photovoltaic,* 1999, **7**(4), 245–260.

5. E. Yablonovitch and G. Cody, Intensity enhancement in textured optical sheets for solar cells, *IEEE Trans. Electron Devices,* 1982, **29**(2), 300–305.

6. D. M. Callahan, J. N. Munday and H. A. Atwater, Solar cell light trapping beyond the ray optic limit, *Nano Lett.,* 2012, **12**(1), 214–218.

7. D. M. Bagnall and S. A. Boden, in *Energy Harvesting for Autonomous Systems,* ed. S. Beeby and N. White, Artech House, Norwood, MA, 2010, ch. 3, pp. 45–85.

8. D. H. Raguin and G. M. Morris, Antireflection structured surfaces for the infrared spectral region, *Appl. Opt.,* 1993, **32**(7), 1154–1167.

9. E. Hecht, in *Optics,* Addison Wesley, San Francisco, 4th edn, 2002, ch. 10, p. 476.

10. P. Campbell and M. A. Green, Light trapping properties of pyramidally textured surfaces, *J. Appl. Phys.,* 1987, **62**(1), 243–249.

11. J. H. Lambert, *Lamberts Photometrie: Photometria, sive De mensure et gradibus luminis, colorum et umbrae,* W. Englemann, Augsburg and Leipzig, Germany, 1760.

12. S. Pillai, K. R. Catchpole, T. Trupke and M. A. Green, Surface plasmon enhanced silicon solar cells, *J. Appl. Phys.,* 2007, **101**(9), 093105.

13. H. R. Stuart and D. G. Hall, Enhanced dipole–dipole interaction between elementary radiators near a surface, *Phys. Rev. Lett.,* 1998, **80**(25), 5663–5666.

14. T. L. Chu and S. S. Chu, Thin film II–VI photovoltaics, *Solid-State Electron.,* 1995, **38**(3), 533–549.

15. J. Britt and C. Ferekides, Thin-film CdS/CdTe solar cell with 15.8% efficiency, *Appl. Phys. Lett.,* 1993, **62**(22), 2851.

16. H.-H. Yang and G.-C. Park, A study on the properties of MgF_2 antireflection film for solar cells, *Trans. Electr. Electron. Mater.,* 2010, **11**(1), 33–36.

17. J. Rancourt, M. Kamerling and D. Monaco, 'New antireflection coating for solar cell covers,' in *Proceedings of the 22nd IEEE Photovoltaic Specialists Conference,* 1991, vol. 95407, no. 707, pp. 1518–1520.

18. J. Shimada and M. Hoshino, Surface fluorination of transparent polymer film, *J. Appl. Polym. Sci.,* 1975, **19**(5), 1439–1448.

19. G. Jorgensen and P. Schissel, Effective antireflection coatings of transparent polymeric materials by gas-phase surface fluorination, *Sol. Energy Mater.,* 1985, **12**(6), 491–500.

20. L. J. Hayes and D. D. Dixon, Surface tension versus barrier property for fluorinated surfaces, *J. Appl. Polym. Sci.,* 1978, **22**(4), 1007–1013.

21. H. Shinohara, M. Iwasaki, S. Tsujimura, K. Watanabe and S. Okazaki, Fluorination of polyhydrofluoroethylenes. I. Direct fluorination of poly(vinyl fluoride) film, *J. Polym. Sci., Part A-1: Polym. Chem.,* 1972, **10**(7), 2129–2137.

22. J. Zhao and M. a. Green, Optimized antireflection coatings for high-efficiency silicon solar cells, *IEEE Trans. Electron Devices,* 1991, **38**(8), 1925–1934.

23. I. Lee, The effects of a double layer anti-reflection coating for a buried contact solar cell application, *Surf. Coat. Technol.,* 2001, **137**(1), 86–91.

24. J. Zhao, A. Wang, M. A. Green, F. Ferrazza and V. A. D. Andrea, 19.8% efficient 'honeycomb' textured multicrystalline and 24.4% monocrystalline silicon solar cells, *Appl. Phys. Lett.,* 1998, **73**(14), 1991–1993.

25. A. B. Djurisic and E. H. Li, Modeling the index of refraction of insulating solids with a modified lorentz oscillator model, *Appl. Opt.,* 1998, **37**(22), 5291–5297.

26. T. Bruton, N. Mason, S. Roberts, O. N. Hartley, J. Fernandez, R. Russell, W. Warta, S. Glunz, O. Shultz, M. Hermle and G. Willeke, in *Proceedings of the 3rd World Conference on Photovoltaic Energy Conversion,* 2003, pp. 899–902.

27. P. Doshi, G. E. Jellison and A. Rohatgi, Characterization and optimization of absorbing plasma-enhanced chemical vapor deposited antireflection coatings for silicon photovoltaics, *Appl. Opt.,* 1997, **36**(30), 7826–7837.

28. S. Winderbaum, Application of plasma enhanced chemical vapor deposition silicon nitride as a double layer antireflection coating and passivation layer for polysilicon solar cells, *J. Vac. Sci. Technol., A,* 1997, **15**(3), 1020–1025.

29. B. Richards, Single-material TiO_2 double-layer antireflection coatings, *Sol. Energy Mater. Sol. Cells,* 2003, **79**(3), 369–390.

30. G. Hass, Preparation, properties and optical applications of thin films of titanium dioxide, *Vacuum,* 1952, **2**(4), 331–345.

31. M. W. Ribarsky, in *Handbook of Optical Constants,* ed. E. Palik, Academic Press, Orlando, FL, 1985, vol. 1, pp. 795–804.

32. B. Richards, J. Cotter, C. Honsberg and S. Wenham, in *Proceedings of the 28th IEEE Photovoltaic Specialists Conference,* 2000, pp. 375–378.

33. S. Wenham, Buried-contact silicon solar cells, *Prog. Photovoltaic,* 1993, **1**(1), 3–10.

34. A. G. Aberle, Surface passivation of crystalline silicon solar cells: a review, *Prog. Photovoltaic,* 2000, **8**(5), 473–487.

35. A. Rohatgi, P. Doshi, J. Moschner, T. Lauinger, A. G. Aberle and D. S. Ruby, Comprehensive study of rapid, low-cost silicon surface passivation technologies, *IEEE Trans. Electron Devices,* 2000, **47**(5), 987–993.

36. S. Narayanan, J. Creager, S. Roncin, A. Rohatgi and Z. Chen, in *Proceedings of 1994 IEEE 1st World Conference on Photovoltaic Energy Conversion,* 1994, vol. 2, pp. 1319–1322.

37. A. W. Stephens, A. G. Aberle and M. A. Green, Surface recombination velocity measurements at the silicon–silicon dioxide interface by microwave-detected photoconductance decay, *J. Appl. Phys.,* 1994, **76**(1), 363.

38. J. Schmidt, T. Lauinger, A. G. Abe and R. Hezel, in *Proceedings of the 25th IEEE Photovoltaic Specialist Conference*, 1996, pp. 17–20.
39. J. Schmidt, M. Kerr and A. Cuevas, Surface passivation of silicon solar cells using plasma-enhanced chemical-vapour-deposited SiN films and thin thermal SiO_2/plasma SiN stacks, *Semicond. Sci. Technol.*, 2001, **16**, 164–170.
40. Z. Chen, P. Sana, J. Salami and A. Rohatgi, A novel and effective PECVD SiO/sub 2//SiN antireflection coating for Si solar cells, *IEEE Trans. Electron Devices*, 1993, **40**(6), 1161–1165.
41. B. S. Richards, Comparison of TiO_2 and other dielectric coatings for buried-contact solar cells: a review, *Prog. Photovoltaic*, 2004, **12**(4), 253–281.
42. A. W. Blakers and M. A. Green, 20% efficiency silicon solar cells, *Appl. Phys. Lett.*, 1986, **48**(3), 215–217.
43. J. Haynos, J. Allison, R. Arndt and A. Meulenburg, in *Proceedings of International Conference on Photovoltaic Power Generation*, 1974, p. 487.
44. E. L. Ralph, in *Proceedings of the 11th IEEE Photovoltaic Specialist Conference*, 1975, pp. 315–316.
45. D. L. King and M. E. Buck, in *Proceedings of the 22nd IEEE Photovoltaic Specialists Conference*, 1991, pp. 303–308.
46. J. M. Kim and Y. K. Kim, The enhancement of homogeneity in the textured structure of silicon crystal by using ultrasonic wave in the caustic etching process, *Sol. Energy Mater. Sol. Cells*, 2004, **81**(2), 239–247.
47. U. Gangopadhyay, K. Kim, A. Kandol, J. Yi and H. Saha, Role of hydrazine monohydrate during texturization of large-area crystalline silicon solar cell fabrication, *Sol. Energy Mater. Sol. Cells*, 2006, **90**(18–19), 3094–3101.
48. M. J. Declercq, L. Gerzberg and J. D. Meindl, Optimization of the hydrazine-water solution for anisotropic etching of silicon in integrated circuit technology, *J. Electrochem. Soc.*, 1975, **122**, 545–552.
49. R. Chaoui, M. Lachab, F. Chiheub and N. Seddiki, in *Proceedings of the 14th European Photovoltaic Solar Energy Conference*, 1997, pp. 812–814.
50. Y. Nishimoto and K. Namba, Investigation of texturization for crystalline silicon solar cells with sodium carbonate solutions, *Sol. Energy Mater. Sol. Cells*, 2000, **61**(4), 393–402.
51. M. A. Green, K. Emery, Y. Hishikawa, W. Warta and E. D. Dunlop, Solar cell efficiency tables (version 39), *Prog. Photovoltaic*, 2012, **20**(1), 12–20.
52. A. Parretta, A. Sarno, P. Tortora, H. Yakubu, P. Maddalena, J. Zhao and A. Wang, Angle-dependent reflectance measurements on photovoltaic materials and solar cells, *Opt. Commun.*, 1999, **172**(1–6), 139–151.
53. P. Campbell and M. A. Green, High performance light trapping textures for monocrystalline silicon solar cells, *Sol. Energy Mater. Sol. Cells*, 2001, **65**(1–4), 369–375.
54. P. Campbell, S. Wenham and M. Green, in *Proceedings of the 20th IEEE Photovoltaic Specialists Conference*, 1988, pp. 713–716.

55. D. H. Macdonald, A. Cuevas, M. J. Kerr, C. Samundsett, D. Ruby, S. Winderbaum and A. Leo, Texturing industrial multicrystalline silicon solar cells, *Sol. Energy,* 2004, vol. **76**(1–3), 277–283.
56. Y. Nishimoto, T. Ishihara and K. Namba, Investigation of acidic texturization for multicrystalline silicon solar cells, *J. Electrochem. Soc.,* 1999, **146**(2), 457–461.
57. U. Gangopadhyay, S. K. Dhungel, K. Kim, U. Manna, P. K. Basu, H. J. Kim, B. Karunagaran, K. S. Lee, J. S. Yoo and J. Yi, Novel low cost chemical texturing for very large area industrial multi-crystalline silicon solar cells, *Semicond. Sci. Technol.,* 2005, **20**(9), 938–946.
58. M. Stocks, A. Carr and A. Blakers, Texturing of polycrystalline silicon, *Sol. Energy Mater. Sol. Cells,* 1996, **40**(1), 33–42.
59. Y. Bai, A. Barnett and J. Rand, Light-trapping and back surface structures for polycrystalline silicon solar cells, *Prog. Photovoltaic,* 1999, **361**, 353–361.
60. S. Winderbaum, O. Reinhold and F. Yun, Reactive ion etching (RIE) as a method for texturing polycrystalline silicon solar cells, *Sol. Energy Mater. Sol. Cells,* 1997, **46**(3), 239–248.
61. O. Schultz, G. Emanuel, W. Glunz and G. Willeke, in *Proceedings of 3rd World Conference on Photovoltaic Energy Conversion,* 2003, vol. 2, pp. 1360–1363.
62. H. G. Rudenberg and B. Dale, 'Radiant energy transducer,' *US Pat.,* 315099929, 1964.
63. J. C. Zolper, S. Narayanan, S. R. Wenham and M. a. Green, 16.7% efficient, laser textured, buried contact polycrystalline silicon solar cell, *Appl. Phys. Lett.,* 1989, **55**(22), 2363.
64. H. Bender, J. Szlufcik, H. Nussbaumer, G. Palmers, O. Evrard, J. Nijs, R. Mertens, E. Bucher and G. Willeke, Polycrystalline silicon solar cells with a mechanically formed texturization, *Appl. Phys. Lett.,* 1993, **62**(23), 2941.
65. T. Machida, K. Nakajima, Y. Takeda, S. Tanaka, N. Shibuya, K. Okamoto, T. Nammori, T. Nunoi and T. Tsuji, in *Proceedings of the 20th IEEE Photovoltaic Specialists Conference,* 1991, pp. 1033–1034.
66. M. Abbott and J. Cotter, Optical and electrical properties of laser texturing for high-efficiency solar cells, *Prog. Photovoltaic,* 2006, **14**(3), 225–235.
67. C. G. Bernhard, Structural and functional adaption in a visual system, *Endeavour,* 1967, **26**, 79–84.
68. A. R. Parker and H. E. Townley, Biomimetics of photonic nanostructures, *Nat. Nanotechnol.,* 2007, **2**(6), 347–353.
69. D. G. Stavenga, S. Foletti, G. Palasantzas and K. Arikawa, Light on the moth-eye corneal nipple array of butterflies, *Proc. R. Soc. B,* 2006, **273**(1587), 661–667.
70. A. R. Parker, Z. Hegedus and R. A. Watts, Solar-absorber antireflector on the eye of an Eocene fly (45 Ma), *Proc. R. Soc. B,* 1998, **265**(1398), 811–815.

71. X. Gao, X. Yan, X. Yao, L. Xu, K. Zhang, J. Zhang, B. Yang and L. Jiang, The dry-style antifogging properties of mosquito compound eyes and artificial analogues prepared by soft lithography, *Adv. Mater.*, 2007, **19**(17), 2213–2217.

72. P. Lalanne and D. Lemercier-lalanne, On the effective medium theory of subwavelength periodic structures, *J. Mod. Opt.*, 1996, **43**(10), 2063–2085.

73. L. Rayleigh, On reflection of vibrations at the confines of two media between which the transition is gradual, *Proc. London Mater. Soc.*, 1879, **1**(1), 51–56.

74. R. B. Stephens and G. D. Cody, Optical reflectance and transmission of a textured surface, *Thin Solid Films*, 1977, **45**(1), 19–29.

75. W. Lowdermilk and D. Milam, Graded-index antireflection surfaces for high-power laser applications, *Appl. Phys. Lett.*, 1980, **36**(11), 891–893.

76. S. J. Wilson and M. C. Hutley, The optical properties of 'moth eye' antireflection surfaces, *Opt. Acta*, 1982, **29**(7), 993–1009.

77. Y. Ono, Y. Kimura, Y. Ohta and N. Nishida, Antireflection effect in ultrahigh spatial-frequency holographic relief gratings, *Appl. Opt.*, 1987, **26**(6), 1142–1146.

78. L. Escoubas, An antireflective silicon grating working in the resonance domain for the near infrared spectral region, *Opt. Commun.*, 2003, **226**(1–6), 81–88.

79. E. B. Grann, M. G. Moharam and D. A. Pommet, Artificial uniaxial and biaxial dielectrics with use of two-dimensional subwavelength binary gratings, *J. Opt. Soc. Am. A*, 1994, **11**(10), 2695–2703.

80. C. W. Haggans, L. Li and R. K. Kostuk, Effective-medium theory of zeroth-order lamellar gratings in conical mountings, *J. Opt. Soc. Am. A*, 1993, **10**(10), 2217.

81. M. Neviere and E. Popov, *Light Propagation in Periodic Media*, Marcel Dekker, New York, 2003.

82. K. J. Johnson, Grating diffraction calculator (GD-Calc®) [online], 2008, available at http://software.kjinnovation.com/GD-Calc.html.

83. RSoft: DiffractMOD [online], 2012, available at http://optics.synopsys.com/rsoft/.

84. H. Toyota, K. Takahara, M. Okano, T. Yotsuya and H. Kikuta, Fabrication of microcone array for antireflection structured surface using metal dotted pattern, *Jpn. J. Appl. Phys., Part 2*, 2001, **40**(7B), 747–749.

85. C.-H. Sun, P. Jiang and B. Jiang, Broadband moth-eye antireflection coatings on silicon, *Appl. Phys. Lett.*, 2008, **92**(6), 061112.

86. H. Sai, H. Fujii, Y. Kanamori, K. Arafune, Y. Ohshita, H. Yugami and M. Yamaguchi, in *Proceedings of the 4th IEEE World Conference on Photovoltaic Energy Conversion*, 2006, vol. 1, pp. 1191–1194.

87. H. Sai, Y. Kanamori, K. Arafune, Y. Ohshita and M. Yamaguchi, Light trapping effect of submicron surface textures in crystalline Si solar cells, *Prog. Photovoltaic*, 2007, **15**, 415–423.

88. S. A. Boden and D. M. Bagnall, Tunable reflection minima of nanostructured antireflective surfaces, *Appl. Phys. Lett.,* 2008, **93**(13), 133108.

89. S. A. Boden and D. M. Bagnall, Optimization of moth-eye antireflection schemes for silicon solar cells, *Prog. Photovoltaic,* 2010, **18**(3), 195–203.

90. P. Clapham and M. C. Hutley, Reduction of lens reflexion by the 'moth eye' principle, *Nature,* 1973, **244**, 281–282.

91. R. C. Enger and S. K. Case, Optical elements with ultrahigh spatial-frequency surface corrugations, *Appl. Opt.,* 1983, **22**(20), 3220.

92. Y. Kanamori, H. Kikuta and K. Hane, Broadband antireflection gratings for glass substrates fabricated by fast atom beam etching, *Jpn. J. Appl. Phys.,* 2000, **39**(7), 735.

93. P. Lalanne and G. M. Morris, Antireflection behavior of silicon subwavelength periodic structures for visible light, *Nanotechnology,* 1997, **8**, 53–56.

94. Y. Kanamori, M. Sasaki and K. Hane, Broadband antireflection gratings fabricated upon silicon substrates, *Opt. Lett.,* 1999, **24**(20), 1422–1424.

95. C. L. Cheung, R. J. Nikolić, C. E. Reinhardt and T. F. Wang, Fabrication of nanopillars by nanosphere lithography, *Nanotechnology,* 2006, **17**(5), 1339–1343.

96. W. A. Nositschka, C. Beneking, O. Voigt and H. Kurz, Texturisation of multicrystalline silicon wafers for solar cells by reactive ion etching through colloidal masks, *Sol. Energy Mater. Sol. Cells,* 2003, **76**, 155–166.

97. P. Jiang, T. Prasad, M. J. McFarland and V. L. Colvin, Two-dimensional nonclose-packed colloidal crystals formed by spincoating, *Appl. Phys. Lett.,* 2006, **89**(1), 011908.

98. W.-L. Min, B. Jiang and P. Jiang, Bioinspired self-cleaning antireflection coatings, *Adv. Mater.,* 2008, **20**(20), 3914–3918.

99. W.-L. Min, A. P. Betancourt, P. Jiang and B. Jiang, Bioinspired broadband antireflection coatings on GaSb, *Appl. Phys. Lett.,* 2008, **92**(14), 141109.

100. S. A. Boden and D. M. Bagnall, Nanostructured biomimetic moth-eye arrays in silicon by nanoimprint lithography, *Proc. SPIE,* 2009, **7401**, 74010J-1–74010J-12.

101. C. M. Sotomayor Torres, S. Zankovych, J. Seekamp, A. P. Kam, C. Clavijo Cedeño, T. Hoffmann, J. Ahopelto, F. Reuther, K. Pfeiffer, G. Bleidiessel, G. Gruetzner, M. V. Maximov and B. Heidari, Nanoimprint lithography: an alternative nanofabrication approach, *Mater. Sci. Eng. C,* 2003, **23**(1), 23–31.

102. S. Y. Chou, P. R. Krauss and P. J. Renstrom, Imprint of sub-25 nm vias and trenches in polymers, *Appl. Phys. Lett.,* 1995, **67**, 3114.

103. L. J. Guo, Recent progress in nanoimprint technology and its applications, *J. Phys. D: Appl. Phys.,* 2004, **37**(11), R123–R141.

104. A. Gombert, W. Glaubitt, K. Rose, J. Dreibholz, B. Bläsi, A. Heinzel, D. Sporn, W. Döll and V. Wittwer, Subwavelength-structured antireflective surfaces on glass, *Thin Solid Films,* 1999, **351**(1–2), 73–78.

105. A. Gombert, K. Rose, A. Heinzel, W. Horbelt, C. Zanke, B. Bläsi and V. Wittwer, Antireflective submicrometer surface-relief gratings for solar applications, *Sol. Energy Mater. Sol. Cells*, 1998, **54**(1–4), 333–342.

106. Holotools GmbH, *HT Standard Models* [online], available at http://www.holotools.de/english/standardmolds.html [Accessed October 2011].

107. NIL Technology, *NILT® Imprint Stamps* [online], available at http://www.nilt.com [Accessed October 2011].

108. Q. Chen, G. Hubbard, P. A. Shields, C. Liu, D. W. E. Allsopp, W. N. Wang and S. Abbott, Broadband moth-eye antireflection coatings fabricated by low-cost nanoimprinting, *Appl. Phys. Lett.*, 2009, **94**(26), 263118.

109. C.-J. Ting, F.-Y. Chang, C.-F. Chen and C. P. Chou, Fabrication of an antireflective polymer optical film with subwavelength structures using a roll-to-roll micro-replication process, *J. Micromech. Microeng.*, 2008, **18**(7), 075001.

110. J. Lee, S. Park, K. Choi and G. Kim, Nano-scale patterning using the roll typed UV-nanoimprint lithography tool, *Microelectron. Eng.*, 2008, **85**(5–6), 861–865.

111. M. Schnell, R. Ludemann and S. Schaefer, in *Proceedings of the 28th IEEE Photovoltaic Specialists Conference*, 2000, pp. 367–370.

112. Y. Inomata, K. Fukui and K. Shirasawa, Surface texturing of large area multicrystalline silicon solar cells using reactive ion etching method, *Sol. Energy Mater. Sol. Cells*, 1997, **48**(1–4), 237–242.

113. H. Craighead, R. Howard and D. Tennant, Textured thin-film Si solar selective absorbers using reactive ion etching, *Appl. Phys. Lett.*, 1980, **37**(7), 653–655.

114. J. Gittleman, E. Sichel, H. Lehmann and R. Widmer, Textured silicon: a selective absorber for solar thermal conversion, *Appl. Phys. Lett.*, 1979, **35**(10), 742–744.

115. D. Ruby, S. Zaidi, S. Narayanan, B. M. Damiani and A. Rohatgi, RIE-texturing of multicrystalline silicon solar cells, *Sol. Energy Mater. Sol. Cells*, 2002, **74**(1), 133–137.

116. J. S. Yoo, I. O. Parm, U. Gangopadhyay, K. Kim, S. K. Dhungel, D. Mangalaraj and J. Yi, Black silicon layer formation for application in solar cells, *Sol. Energy Mater. Sol. Cells*, 2006, **90**(18–19), 3085–3093.

117. Y.-F. Huang, S. Chattopadhyay, Y.-J. Jen, C.-Y. Peng, T.-A. Liu, Y.-K. Hsu, C.-L. Pan, H.-C. Lo, C.-H. Hsu, Y.-H. Chang, C.-S. Lee, K.-H. Chen and L.-C. Chen, Improved broadband and quasi-omnidirectional anti-reflection properties with biomimetic silicon nanostructures, *Nat. Nanotechnol.*, 2007, **2**(12), 770–774.

118. C.-H. Hsu, H.-C. Lo, C.-F. Chen, C. T. Wu, J.-S. Hwang, D. Das, J. Tsai, L.-C. Chen and K.-H. Chen, Generally applicable self-masked dry etching technique for nanotip array fabrication, *Nano Lett.*, 2004, **4**(3), 471–475.

119. S. Koynov, M. S. Brandt and M. Stutzmann, Black nonreflecting silicon surfaces for solar cells, *Appl. Phys. Lett.*, 2006, **88**(20), 203107.

120. S. Koynov, M. S. Brandt and M. Stutzmann, Black multi-crystalline silicon solar cells, *Phys. Status Solidi RRL,* 2007, **1**(2), R53–R55.
121. M. Vergöhl, N. Malkomes, T. Staedler, T. Matthee and U. Richter, Ex situ and in situ spectroscopic ellipsometry of MF and DC-sputtered TiO$_2$ and SiO$_2$ films for process control, *Thin Solid Films,* 1999, **351**(1–2), 42–47.
122. H. Nagel, A. Metz and R. Hezel, Porous SiO$_2$ films prepared by remote plasma-enhanced chemical vapour deposition-a novel antireflection coating technology for photovoltaic modules, *Sol. Energy Mater. Sol. Cells,* 2001, **65**(1–4), 71–77.
123. A. Hunt, W. Carter and J. Cochran, Combustion chemical vapor deposition: a novel thin-film deposition technique, *Appl. Phys. Lett.,* 1993, **63**(2), 266–268.
124. M. J. Minot, Single-layer, gradient refractive index antireflection films effective from 035 to 25 μ, *J. Opt. Soc. Am.,* 1976, **66**(6), 515–519.
125. M. Brogren, P. Nostell and B. Karlsson, Optical efficiency of a PV-thermal hybrid CPC module for high latitudes, *Sol. Energy,* 2001, **69**, 173–185.
126. G. K. Chinyama, A. Roos and B. Karlsson, Stability of antireflection coatings for large area glazing, *Sol. Energy,* 1993, **50**, 105–111.
127. M. J. Keevers, T. L. Young, U. Schubert and M. A. Green, in *Proceedings 22nd European Photovoltaic Solar Energy Conference,* 2007, pp. 1783–1791.
128. X.-T. Zhang, O. Sato, M. Taguchi, Y. Einaga, T. Murakami and A. Fujishima, Self-cleaning particle coating with antireflection properties, *Chem. Mater.,* 2005, **17**(3), 696–700.
129. D. Chen, Anti-reflection (AR) coatings made by sol-gel processes: a review, *Sol. Energy Mater. Sol. Cells,* 2001, **68**(3–4), 313–336.
130. M. Bautista, Silica antireflective films on glass produced by the sol–gel method, *Sol. Energy Mater. Sol. Cells,* 2003, **80**(2), 217–225.
131. Y. Xiao, J. Shen, Z. Xie, B. Zhou and G. Wu, Microstructure control of nanoporous silica thin film prepared by sol–gel process, *J. Mater. Sci. Technol.,* 2007, **23**(4), 504–508.
132. I. M. Thomas, High laser damage threshold porous silica antireflective coating, *Appl. Opt.,* 1986, **25**(9), 1481–1483.
133. G. S. Vicente, R. Bayo+n and A. Morales, Effect of additives on the durability and properties of antireflective films for solar glass covers, *J. Sol. Energy Eng.,* 2008, **130**(1), 011007.
134. C. Ballif, J. Dicker, D. Borchert and T. Hofmann, Solar glass with industrial porous SiO antireflection coating: measurements of photovoltaic module properties improvement and modelling of yearly energy yield gain, *Sol. Energy Mater. Sol. Cells,* 2004, **82**(3), 331–344.
135. P. Buskens, N. Arfsten, R. Habets, H. Langermans, A. Overbeek, J. Scheerder, J. Thies and N. Viets, 'Innovation at DSM: state of the art single layer anti-reflective coatings for solar cell cover glass,' in *Glass Performance Days 2009,* 2009, pp. 505–507.

136. H. Hattori, Anti-reflection surface with particle coating deposited by electrostatic attraction, *Adv. Mater.,* 2001, **13**(1), 51–54.

137. H. Y. Koo, D. K. Yi, S. J. Yoo and D.-Y. Kim, A snowman-like array of colloidal dimers for antireflecting surfaces, *Adv. Mater.,* 2004, **16**(3), 274–277.

138. S. A. Maier, *Plasmonics: Fundamentals and Applications*, Springer, New York, 2003.

139. K. Kelly, E. Coronado, L. Zhao and G. C. Schatz, The optical properties of metal nanoparticles: the influence of size, shape, and dielectric environment, *J. Phys. Chem. B,* 2003, **107**(3), 668–677.

140. H. A. Atwater and A. Polman, Plasmonics for improved photovoltaic devices, *Nat. Mater.,* 2010, **9**(3), 205–213.

141. C. Eminian, F.-J. Haug, O. Cubero, X. Niquille and C. Ballif, Photocurrent enhancement in thin film amorphous silicon solar cells with silver nanoparticles, *Prog. Photovoltaic,* 2011, **19**(3), 260–265.

142. E. Moulin, J. Sukmanowski, M. Schulte, A. Gordijn, F. X. Royer and H. Stiebig, Thin-film silicon solar cells with integrated silver nanoparticles, *Thin Solid Films,* 2008, **516**(20), 6813–6817.

143. S. H. Lim, W. Mar, P. Matheu, D. Derkacs and E. T. Yu, Photocurrent spectroscopy of optical absorption enhancement in silicon photodiodes *via* scattering from surface plasmon polaritons in gold nanoparticles, *J. Appl. Phys.,* 2007, **101**(10), 104309.

144. K. Nakayama, K. Tanabe and H. A. Atwater, Plasmonic nanoparticle enhanced light absorption in GaAs solar cells, *Appl. Phys. Lett.,* 2008, **93**, 121904.

145. C. F. Bohren, How can a particle absorb more than the light incident on it?, *Am. J. Phys.,* 1983, **51**(4), 323.

146. C. F. Bohren and D. R. Huffman, *Absorption and Scattering of Light by Small Particles*, Wiley VCH, 1998.

147. T. L. Temple and D. M. Bagnall, Broadband scattering of the solar spectrum by spherical metal nanoparticles, *Prog. Photovoltaic,* 2013, **21**(4), 600–611.

148. A. Taflove and S. C. Hagness, *Computational Electrodynamics: The Finite-Difference Time-Domain Method*, Artech House, Norwood, MA, 3rd edn, 2005.

149. J. L. Volakis, A. Chatterjee and L. C. Kempel, Review of the finite-element method for three-dimensional electromagnetic scattering, *J. Opt. Soc. Am. A,* 1994, **11**(4), 1422–1433.

150. B. T. Draine and P. Flatau, Discrete-dipole approximation for periodic targets: theory and tests, *J. Opt. Soc. Am. A,* 2008, **25**(11), 2693–2703.

151. T. L. Temple, G. D. K. Mahanama, H. S. Reehal and D. M. Bagnall, Influence of localized surface plasmon excitation in silver nanoparticles on the performance of silicon solar cells, *Sol. Energy Mater. Sol. Cells,* 2009, **93**(11), 1978–1985.

152. K. R. Catchpole and A. Polman, Plasmonic solar cells, *Opt. Express,* 2008, **16**, 21793–21800.

153. F. J. Beck, S. Mokkapati and K. R. Catchpole, Plasmonic light-trapping for Si solar cells using self-assembled, Ag nanoparticles, *Prog. Photovoltaic,* 2010, **18**(7), 500–504.

154. K.-S. Lee and M. A. El-Sayed, Dependence of the enhanced optical scattering efficiency relative to that of absorption for gold metal nanorods on aspect ratio, size, end-cap shape, and medium refractive index, *J. Phys. Chem. B,* 2005, **109**(43), 20331–20338.

155. T. L. Temple and D. M. Bagnall, Optical properties of gold and aluminium nanoparticles for silicon solar cell applications, *J. Appl. Phys.,* 2011, **109**(8), 084343.

156. A. J. Haes, C. L. Haynes, A. D. McFarland, G. C. Schatz, R. P. Van Duyne and S. Zou, Plasmonic materials, *MRS Bull.,* 2005, **30**(5), 368–375.

157. M. G. Blaber, M. D. Arnold, N. Harris, M. J. Ford and M. B. Cortie, Plasmon absorption in nanospheres: A comparison of sodium, potassium, aluminium, silver and gold, *Phys. B: Condens. Matter,* 2007, **394**(2), 184–187.

158. C. Langhammer, B. Kasemo and I. Zorić, Absorption and scattering of light by Pt, Pd, Ag, and Au nanodisks: absolute cross sections and branching ratios, *J. Chem. Phys.,* 2007, **126**(19), 194702.

159. C. Langhammer, M. Schwind, B. Kasemo and I. Zorić, Localized surface plasmon resonances in aluminum nanodisks, *Nano Lett.,* 2008, **8**(5), 1461–1471.

160. P. B. Johnson and R. W. Christy, Optical constants of the noble metals, *Phys. Rev. B,* 1972, **6**, 4370–4379.

161. J. Grand, *et al.*, Optical extinction spectroscopy of oblate, prolate and ellipsoid shaped gold nanoparticles: Experiments and theory, *Plasmonics,* 2006, **1**(2–4), 135–140.

162. I. Zoric, M. Zach, B. Kasemo and C. Langhammer, Gold, platinum and aluminum nanodisk plasmons: Material independence, subradiance and damping mechanisms, *ACS Nano,* 2011, (4), 2535–2546.

163. Y. Ekinci, H. H. Solak and J. F. Löffler, Plasmon resonances of aluminum nanoparticles and nanorods, *J. Appl. Phys.,* 2008, **104**(8), 083107.

164. S. Y. Chou and P. R. Krauss, Imprint lithography with sub-10 nm feature size and high throughput, *Microelectron. Eng.,* 1997, **35**(1), 237–240.

165. B. D. Lucas, J.-S. Kim, C. Chin and L. J. Guo, Nanoimprint lithography based approach for the fabrication of large-area, uniformly-oriented plasmonic arrays, *Adv. Mater.,* 2008, **20**(6), 1129–1134.

166. J. C. Hulteen and R. P. V. Duyne, Nanosphere lithography: A materials general fabrication process for periodic particle array surfaces, *J. Vac. Sci. Technol., A,* 1995, **13**(3), 1553–1558.

167. H. Fredriksson, Y. Alaverdyan, A. Dmitriev, C. Langhammer, D. S. Sutherland, M. Zäch and B. Kasemo, Hole–mask colloidal lithography, *Adv. Mater.,* 2007, **19**(23), 4297–4302.

168. A. Dmitriev, T. Pakizeh, M. Käll and D. S. Sutherland, Gold–silica–gold nanosandwiches: tunable bimodal plasmonic resonators, *Small*, 2007, **3**(2), 294–299.

169. J. Turkevich, P. C. Stevenson and J. Hillier, A study of the nucleation and growth processes in the synthesis of colloidal gold, *Discuss. Faraday Soc.*, 1951, **11**(c), 55.

170. S. Stoeva, K. J. Klabunde, C. M. Sorensen and I. Dragieva, Gram-scale synthesis of monodisperse gold colloids by the solvated metal atom dispersion method and digestive ripening and their organization into two- and three-dimensional structures, *J. Am. Chem. Soc.*, 2002, **124**(10), 2305–2311.

171. N. R. Jana and X. Peng, Single-phase and gram-scale routes toward nearly monodisperse Au and other noble metal nanocrystals, *J. Am. Chem. Soc.*, 2003, **125**(47), 14280–14281.

172. M. M. Alvarez, J. T. Khoury, T. G. Schaaff, M. N. Shafigullin, I. Vezmar and R. L. Whetten, Optical absorption spectra of nanocrystal gold molecules, *J. Phys. Chem. B*, 1997, **101**(19), 3706–3712.

173. D. D. Evanoff and G. Chumanov, Size-controlled synthesis of nanoparticles. 1. 'Silver-only' aqueous suspensions *via* hydrogen reduction, *J. Phys. Chem. B*, 2004, **108**(37), 13948–13956.

174. D. M. Schaadt, B. Feng and E. T. Yu, Enhanced semiconductor optical absorption *via* surface plasmon excitation in metal nanoparticles, *Appl. Phys. Lett.*, 2005, **86**(6), 063106.

175. K. C. Grabar, *et al.*, Two-dimensional arrays of colloidal gold particles: A flexible approach to macroscopic metal surfaces, *Langmuir*, 1996, **12**(10), 2353–2361.

176. A. Heilmann, *Polymer Films with Embedded Metal Nanoparticles*, Springer, Berlin, 2003.

177. Y. Suzuki, Y. Ojima, Y. Fukui, H. Fazyia and K. Sagisaka, Post-annealing temperature dependence of infrared absorption enhancement of polymer on evaporated silver films, *Thin Solid Films*, 2007, **515**(5), 3073–3078.

178. R. Gupta, M. J. Dyer and W. A. Weimer, Preparation and characterization of surface plasmon resonance tunable gold and silver films, *J. Appl. Phys.*, 2002, **92**(9), 5264.

179. V. L. Schlegel and T. M. Cotton, Silver-island films as substrates for enhanced Raman scattering: effect of deposition rate on intensity, *Anal. Chem.*, 1991, **63**(3), 241–247.

180. G. Biskos, V. Vons, C. U. Yurteri and A. Schmidt-Ott, Generation and Sizing of Particles for Aerosol-Based Nanotechnology, *KONA Powder Part. J.*, 2008, **26**(26), 13–35.

181. H. R. Stuart and D. G. Hall, Absorption enhancement in silicon-on-insulator waveguides using metal island films, *Appl. Phys. Lett.*, 1996, **69**(16), 2327.

182. Z. Ouyang, X. Zhao, S. Varlamov, Y. Tao, J. Wong and S. Pillai, Nanoparticle-enhanced light trapping in thin-film silicon solar cells, *Prog. Photovoltaic*, 2011, **19**, 917–926.

CHAPTER 9

Photon Frequency Management Materials for Efficient Solar Energy Collection

LEFTERIS DANOS[*†a], THOMAS J. J. MEYER[b],
PATTAREEYA KITTIDACHACHAN[c], LIPING FANG[a],
THOMAS S. PAREL[a], NAZILA SOLEIMANI[a], AND
TOMAS MARKVART[a]

[a]Solar Energy Laboratory, Engineering Materials, School of Engineering
Sciences, University of Southampton SO17 1BJ, UK; [b]Teknova Renewable
Energy Research Group, Teknova, Gimlemoen 19, 4630, Kristiansand,
Norway; [c]Department of Physics, Faculty of Science, King Mongkut's Institute
of Technology Ladkrabang, Bangkok 10520, Thailand
*E-mail: l.danos@lancaster.ac.uk

9.1 Introduction

Photon management in solar cells usually refers to processes that aim to
enhance light capture as the first step in photovoltaic solar energy conversion
(Figure 9.1). Proposed methods include sub-wavelength and nanoscale
techniques,[1] embracing at times near-field interaction[2,3] and plasmonics.[4]
This chapter focuses on photon management where the frequency is
changed between absorption of the incident light and its re-emission as
fluorescence.[5] We show that, within the realm of classical optics, current

---

†Present address: Department of Chemistry, Lancaster University, Lancaster LA1 4YB, UK.

RSC Energy and Environment Series No. 12
Materials Challenges: Inorganic Photovoltaic Solar Energy
Edited by Stuart J C Irvine
© The Royal Society of Chemistry 2015
Published by the Royal Society of Chemistry, www.rsc.org

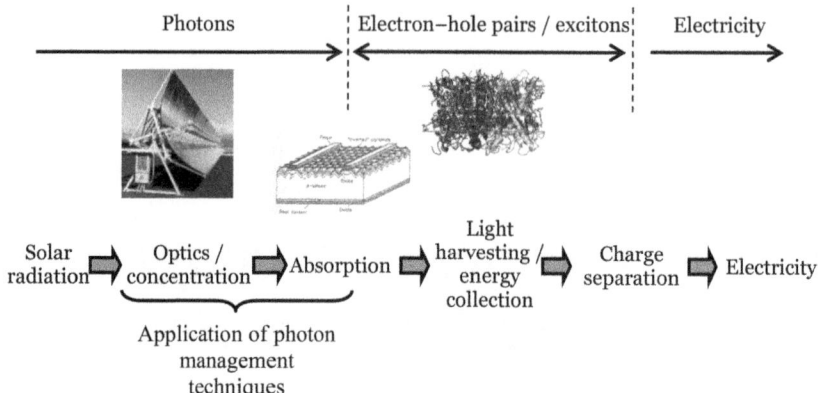

Figure 9.1 Examples of photon management structures for improved solar energy collection in PV. From left to right: a parabolic concentrator, a University of New South Wales passivated emitter, rear locally diffused (PERL) cell where light trapping is achieved *via* surface texturing, and a schematic of a light harvesting unit. Copyright © 1999 John Wiley & Sons, Ltd.

results and potential applications outline a landscape that offers substantial scope to increase the efficiencies and reduce material requirements in the manufacture of solar cells. A particularly useful tool for frequency management is Förster resonance energy transfer[6] which provides a powerful platform to optimise independently the absorption and emission spectra in structured dye mixtures using, for example, zeolites,[7-10] Langmuir Blodgett (LB) films[11,12] or biological complexes.[13]

This type of photon management has been successfully applied to cadmium telluride (CdTe) and copper indium gallium (di)selenide (CIGS) solar cells which employ a relatively thick CdS layer and results in a low quantum efficiency in the blue region of the solar spectrum.[14,15] Luminescent solar concentrators (or fluorescent solar collectors), have been studied for some time[16-21] with the aim of reducing the size of the solar cell and thus the cost of the overall system. We discuss a novel form of fluorescent collector which directs light effectively onto the edge of a solar cell (and not in the usual way on the front surface), making it possible to reduce the amount of crystalline silicon by some two orders of magnitude. A similar idea is to employ photon frequency management to maximise the photon path length inside the solar cell and obtain a novel form of optical confinement (or light trapping) compared with the traditional scheme based on surface texturing.

At the fundamental level, photon frequency management can be pictured as a transformation of a high temperature solar light beam into a room temperature beam at a lower frequency. In this formulation, the principal constraints are revealed as thermodynamic in nature. Unlike geometric concentrators, the energy exchange with the absorbing/fluorescent medium allows the entropy of the captured photon gas to be reduced, making it possible for the resulting beam to be emitted with lower ètendue and thus containing smaller number of quantum states than the incident beam.[22,23]

9.2 Fundamentals

9.2.1 Introduction

In contrast to geometric concentrators, the opportunity to change frequency (invariably implemented through absorption and re-emission as luminescence) introduces a new degree of freedom which opens up new avenues to enhance the capture of sunlight. The stochastic nature of these processes, however, must be taken fully on board for a satisfactory understanding and description of device operation. Viewed in a more fundamental setting, the ergodic features of re-absorption (also known as photon recycling) introduce a fundamentally new facet to optical devices and move optics into the novel arena of the thermodynamics of light.[24]

Detailed numerical models of photon transport in fluorescent collectors have emerged in recent years. A model based on the radiative transfer of all fluxes between mesh points in the concentrator plate has been proposed.[25] Also popular is a generic approach using ray tracing.[26,27] In this chapter, we adopt a simpler but more instructive two-flux model, developed originally in our group.[28]

9.2.2 Re-absorption

Viewed quite generally, the probability r_λ that light emitted in a volume V at wavelength λ is re-absorbed is given by:

$$1 - r_\lambda = \int dV\, p(\underline{r}) \int \frac{d\Omega}{4\pi} e^{-\ell\alpha_\lambda} \tag{9.1}$$

where ℓ denotes the optical path length of a ray inside the volume, α_λ is the absorption coefficient at wavelength λ, and $p(\underline{r})$ describes the probability distribution of emission events inside V. Eqn (9.1) assumes that emission occurs isotropically over all elements of solid angle $d\Omega$. For simplicity we have also assumed that a ray is either completely transmitted or completely reflected at the edges of the volume, for example, by total internal reflection. Specific examples of the re-absorption probability [eqn (9.1)] are considered below.

In terms of r_λ, the total re-absorption probability for light emitted with spectrum $f_1(\lambda)$ is equal to:

$$R = \int f_1(\lambda) r_\lambda d\lambda \tag{9.2}$$

The spectrum $f_1(\lambda)$—the first generation spectrum—is the 'usual' fluorescence spectrum of the dye without re-absorption, normalised to a unit total emission probability, $\int f_1(\lambda)d\lambda = 1$.

The re-absorption probability R in the framework of fluorescent concentrators was first analysed by Weber and Lambe[16] on the example of infinite strip geometry. By symmetry, such geometry also applies to a rectangular collector where a solar cell is attached to one edge and the other three edges are covered by perfect mirrors.[27] In this case, noting that for fluorescent collectors with a reasonably high gain, it is sufficient to take the probability $p(r) = 1/V$, where V is the volume of the collector. The volume integral in eqn (9.1) then reduces to an integral over a single coordinate which can be evaluated analytically to give:

$$\{1 - r(\lambda)\}_{\text{WL}} = \frac{1}{4\pi(1-P)} \int\limits_{0}^{\pi} d\phi \int\limits_{\theta_c}^{\pi-\theta_c} \frac{\sin\theta\sin\phi}{\alpha_e(\lambda)L} \left\{1 - \exp\left[\frac{-2\alpha_e(\lambda)L}{\sin\theta\sin\phi}\right]\right\} \sin\theta d\theta$$

$$(9.3)$$

where $P = \Omega/4\pi = 1 - \cos\theta_c$, $\theta_c = \sin^{-1}(1/n)$ is a half apex angle of the 'escape cone' subtending a solid angle Ω, L is the length of the collector and n is the refractive index of the medium. The integral, eqn (9.3), has to be evaluated numerically and the result is shown in Figure 9.2(b).

The validity of the Weber and Lambe model has been verified by means of ray tracing simulations.[27] Simulations of flat-plate, liquid and thin film collectors have demonstrated the validity and versatility of the Weber and Lambe model for re-absorption at high gain ratios. Ray tracing simulations have also shown that collector geometry has little or no influence on collector operation.[26,27] These results confirm experiments by Roncali and Garnier[29] in which no efficiency improvement was recorded with various collector geometries. In addition, it has been shown that thin film collector structures perform no better than standard homogeneous collectors.[26,30] The optimal configuration is reached when the refractive index of the dye film layer is the same as that of the substrate, which is usually the case for spin coated polymethyl methacrylate (PMMA) films on glass substrates. In this case a thin film device can be treated like a homogeneous device, with the resulting absorption efficiency identical to a plain collector.

Eqn (9.3) describes the re-absorption of rays outside the solid angle of the escape cone shown in Figure 9.2(a), and is appropriate for rays which enter a solar cell of a higher refractive index than the collector. If we wish to describe the photon flux emitted from the edge of the collector—a useful characterisation technique, as we shall see in Section 9.4.2—a modification of Weber and Lambe's theory is required [Figure 9.2(b)] since only photons emitted in escape cone at the edge can be collected by the detector

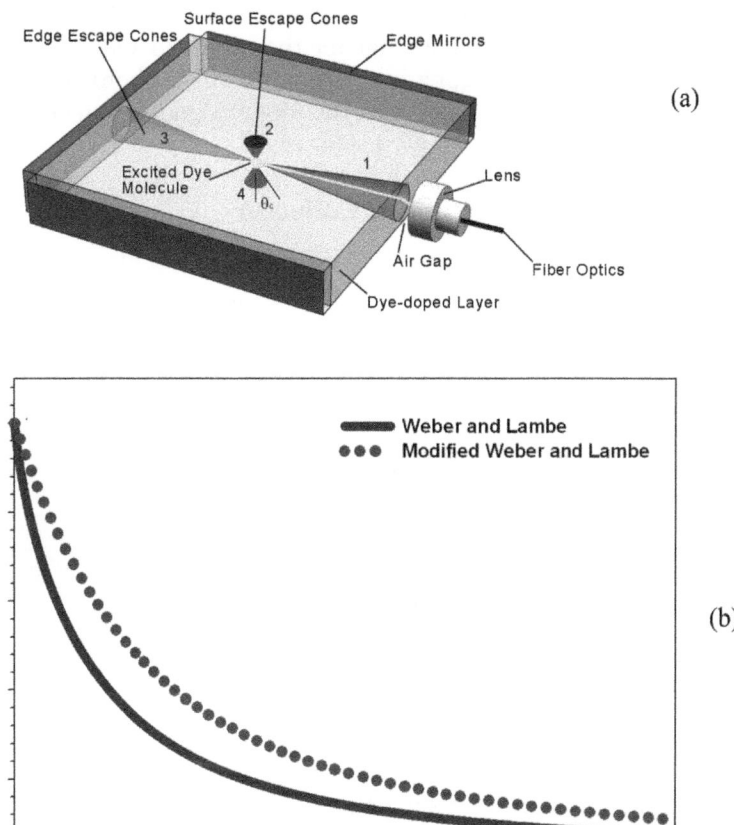

Figure 9.2   (a) Schematic of a collector structure and definition of the light escape cones. θ_c is the critical angle for total internal reflection. (b) Comparison of re-absorption probabilities calculated from Weber and Lambe's theory (solid line) and modified Weber and Lambe's theory (dashed line). Reprinted with permission from ref. 31, Copyright 2012, AIP Publishing LLC.

[Figure 9.2(a), cone 1 and cone 3]. Thus, using now the angle from the normal to the edge of the collector as the integration variable:

$$\{1 - r(\lambda)\}_{\mathrm{MWL}} = \frac{1}{2P\alpha_e(\lambda)L} \int_0^{\theta_c} \left\{ 1 - \exp\left[\frac{-2\alpha_e(\lambda)L}{\cos\theta} \right] \right\} \cos\theta \sin\theta \, d\theta \qquad (9.4)$$

where $\{1 - r(\lambda)\}_{\mathrm{MWL}}$ is the spectral re-absorption probability for the modified Weber and Lambe's theory applicable to edge fluorescence experiments.[31]

The results show that the spectral re-absorption probability for Weber and Lambe's theory is higher than that for the modified Weber and Lambe's theory [Figure 9.2(b)]. The apparent reason is that the photons considered to be collected in Weber and Lambe's theory propagate a longer path length before reaching the edge mounted solar cell.

9.2.3 Photon Balance in the Collector

The performance of the collector is usually assessed by means of optical efficiency η_{opt}, equal to the ratio of photon flux $\dot{N}_{work}$ emitted from the collector and reaching the solar cell and the photon flux I_o incident on the collector:

$$\eta_{opt} = \frac{\dot{N}_{work}}{I_o} = \frac{\dot{N}_{work}}{\dot{N}_{abs}} \frac{\dot{N}_{abs}}{I_o} = Q_A Q_C \tag{9.5}$$

Here, we also define the absorption efficiency:

$$Q_A = \frac{\dot{N}_{abs}}{I_o} \tag{9.6}$$

and the photon collection efficiency:

$$Q_C = \frac{\dot{N}_{work}}{\dot{N}_{abs}} \tag{9.7}$$

In terms of the incident photon flux $\Phi(\lambda)$ and absorbance $A(\lambda) = \varepsilon(\lambda)Cd$ (where C is the concentration, d is the thickness of the collector and $\varepsilon(\lambda)$ is the decadic extinction coefficient) the absorption efficiency can be written as:

$$Q_A = \frac{\int_{\lambda_{min}}^{\lambda_g} \Phi(\lambda) \cdot \left[1 - 10^{-A(\lambda)}\right] d\lambda}{\int_{\lambda_{min}}^{\lambda_g} \Phi(\lambda) d\lambda} \tag{9.8}$$

where λ_{min} and λ_g are the minimum and maximum (near bandgap) wavelength in solar spectrum that are taken into consideration. In our analysis below we take $\lambda_{min} = 300$ nm, unless otherwise stated.

An expression for Q_c has been obtained by different methods by Zewail and co-workers[20] and Kittidachachan *et al.*[28]

$$Q_C = \frac{\dot{N}_{work}}{\dot{N}_{abs}} = \varphi_f \frac{(1 - P)(1 - R)}{1 - R_{tot}\varphi_f} \tag{9.9}$$

where φ_f is the quantum yield of fluorescence and R_{tot} is the probability of re-absorption, averaged over all directions.

Let us suppose that the fluorescence emitted by the structure is in the form of a sharp single sharp line at wavelength λ_{em}. The useful current collected by the collector can then be written as:

$$J_{col} = q\eta_{opt}QE_{cell}(\lambda_{em}) \int_{\lambda_{min}}^{\lambda_g} \Phi(\lambda) d\lambda = Q_A Q_C QE_{cell}(\lambda_{em}) J_{ideal} \tag{9.10}$$

where $J_{ideal} = q \int_{\lambda_0}^{\lambda_g} \Phi(\lambda) d\lambda$ is the maximum current available for conversion. For a broader emission spectrum, $QE_{cell}(\lambda_{em})$ should be replaced by the average of the quantum efficiency of the cell weighted over the fluorescence spectrum $I_f(\lambda)$:

$$\langle QE_{cell} \rangle = \frac{\int_{\lambda_{min}}^{\lambda_g} QE_{cell} \cdot I_f(\lambda) d\lambda}{\int_{\lambda_{min}}^{\lambda_g} I_f(\lambda) d\lambda} \tag{9.11}$$

In luminescence down-shifting layers where the solar cell is positioned behind the collector, the current collected by the down-shifting structure should be supplemented by the current absorbed directly by the solar cell:

$$J_{dir} = q \cdot \int_{\lambda_{min}}^{\lambda_g} \Phi(\lambda) \cdot 10^{-A(\lambda)} \cdot QE_{cell}(\lambda) d\lambda \tag{9.12}$$

For down-shifting structures that rely on energy transfer between dyes, the photon collection efficiency can be written as:

$$Q_C = \phi_f^{(A)} \eta_{ET} \eta_{LDS} \tag{9.13}$$

where $\phi_f^{(A)}$, is the fluorescence quantum yield of the acceptor dye, η_{ET} is the energy transfer efficiency (see Section 9.3) and η_{LDS} ($\leq 1 - R$) is the down-shifting efficiency which takes into account escape cone losses and re-absorption. The total current (sum of the collector current [eqn (9.10) and (9.12)] should, of course, be greater than the current that would be produced by the cell without the down-shifting structure:

$$J_{cell} = q \cdot \int_{\lambda_0}^{\lambda} \Phi(\lambda) \cdot QE_{cell}(\lambda) d\lambda \tag{9.14}$$

We apply these expressions to fluorescent collectors and down-shifting structures in Sections 9.4 and 9.5.

9.3   Förster Resonance Energy Transfer

9.3.1   Introduction

Förster Resonance energy transfer (FRET)[6,32] is a powerful tool for efficient photon management. An example of such process is photosynthesis which makes use of efficient energy transfer in the antenna system to increase the absorption efficiency of energy conversion. Light is absorbed by chlorophylls and other accessory pigments that surround the reaction centre and the molecular excitation energy created is transported to the reaction centre. Borrowing concepts from light harvesting in photosynthesis suitable antenna pigment structures can be envisaged to improve the capture of solar radiation in artificial structures used in photovoltaic and photochemical conversion.[13]

Fluorescent collectors and down-shifting structures can benefit from a similar process but for a subtly different reason. Energy transfer in photosynthesis is a spatial phenomenon which increases the cross-section for optical absorption. At the same time, however, the change of spectrum on energy transfer between different molecules can be used to advantage and improve the efficiency of solar cells or concentrate light, as in fluorescent collectors. We discuss these aspects in Sections 9.3.5 and 9.4.3.

9.3.2 Basic Theory

Excitation energy transfer results from the resonance interaction between the electronic transition dipole moments of a donor molecule (D) in the excited state and an acceptor molecule (A) in the ground state (Figure 9.3). It is a near field interaction which occurs during the excitation lifetime of the donor molecule and the electronic excitation energy transfer is radiationless. The FRET mechanism is effective over distances ranging from <1 nm to about 10 nm. The pioneering theoretical work was carried out half a century ago by Theodor Förster who showed that the excitation energy transfer rate varies with the separation distance r between their transition dipole centres as:[32,33]

$$k_{DA}(r) = \frac{1}{\tau_D}\left(\frac{R_0}{r}\right)^6 \tag{9.15}$$

where τ_D is the lifetime of the excited state for the donor molecule in the absence of the acceptor molecule and R_0 is known as the Förster critical radius given by:

$$R_0 = 0.2108\left[\kappa^2\Phi_D^0 n^{-4}\int_0^\infty I_D\varepsilon_A(\lambda)\lambda^4 d\lambda\right]^{1/6} \tag{9.16}$$

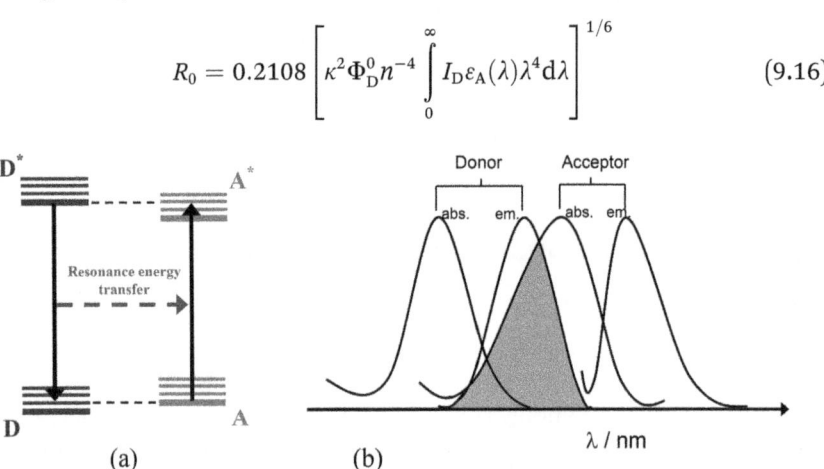

(a) (b)

Figure 9.3 (a) Energy level scheme of a donor–acceptor dye molecular system showing the resonance coupled transitions and (b) illustration of the overlap integral (shaded area) between the emission spectrum of the donor and the absorption of the acceptor. Reproduced from ref. 33, Copyright © 1999 John Wiley & Sons, Ltd.

where κ^2 is an orientational factor which takes into account the orientation of the electronic transition dipole moments of the donor and acceptor and can take the values from 0 (perpendicular transition moments) to 4. Φ_D^0 is the fluorescence quantum yield of the donor in the absence of the acceptor, n is the refractive index of the medium, $I_D(\lambda)$ is the fluorescence spectrum of the donor normalised so that $\int_0^\infty I_D(\lambda)\mathrm{d}\lambda = 1$, $\varepsilon_A(\lambda)$ is the molar absorption coefficient of the acceptor (in $\mathrm{dm}^3\ \mathrm{mol}^{-1}\ \mathrm{cm}^{-1}$), and λ is the wavelength in nanometres.

The excitation energy transfer efficiency is defined by:

$$\mathrm{ET}_{\mathrm{eff}} = \frac{k_{\mathrm{DA}}(r)}{1/\tau_D + k_{\mathrm{DA}}(r)} = \frac{1}{1 + (r/R_0)^6} \tag{9.17}$$

The excitation energy transfer efficiency in dye mixture layers can be estimated by comparing the absorption spectrum of the collector layer with the excitation spectrum of the acceptor molecule in the layer. The energy transfer efficiency between a donor and acceptor molecule can be calculated:[33]

$$\mathrm{ET}_{\mathrm{eff}} = \frac{A_A(\lambda_A)}{A_D(\lambda_D)}\left[\frac{I_A(\lambda_D, \lambda_A^{\mathrm{em}})}{I_A(\lambda_A, \lambda_A^{\mathrm{em}})} - \frac{A_A(\lambda_D)}{A_A(\lambda_A)}\right] \tag{9.18}$$

where λ_A and λ_D are the absorption maxima of the donor $A_D(\lambda_D)$ and acceptor $A_A(\lambda_A)$ in the fluorescent layer respectively, I_A and I_D are the excitation fluorescence spectra observed at a wavelength λ_A^{em} where there is no absorption from the donor.

In the case of time-resolved measurements the fluorescence decay of the donor or acceptor can be used directly to estimate the excitation transfer rate. If the fluorescence decay lifetime of the donor is not a single exponential, the transfer efficiency can be estimated from the average decay times of the donor in the presence and absence of the acceptor:

$$\mathrm{ET}_{\mathrm{eff}} = 1 - \frac{\langle \tau_D \rangle}{\langle \tau_D^0 \rangle} \tag{9.19}$$

where the average decay lifetime of the donor can be empirically modelled as a sum of exponentials:[34]

$$i_D(t) = \sum_i \alpha_i \exp(-t/\tau_i) \tag{9.20}$$

where α_i is the fractional amplitude and τ_i is the decay lifetimes for the ith component.

Clearly, it is important to distinguish between radiative and non-radiative transfer. Radiative transfer or re-absorption occurs when absorption by molecule A of a photon emitted by a molecule D is observed. Such interaction does not require close proximity between the molecules and depends on both the spectral overlap and concentration. In contrast, non-radiative transfer occurs without the emission of photons and occurs over distances much less than the wavelength of emission.

9.3.3 Materials for Improved Photon Energy Collection

Over the past years a variety of organic dyes have been extensively employed in fluorescent collectors because of their high absorption coefficients combined with high fluorescence quantum yields. However, they are not photostable and the few dyes with high quantum efficiency emit in the green–yellow spectral region which introduces a severe constraint on the absorption efficiency in application with commercial solar cells.[35]

The BASF laboratories have developed organic fluorophores based on perylene and naphthalimide dyes for application in fluorescent collectors.[24,36] The Lumogen F series,[37] in particular, have been used extensively in almost every fabricated collector since 2007.[38-40] All Lumogen dyes show very high quantum yields (near 100%) in plastic plates[41] and strong absorption of light, are non-toxic, and have photostability guaranteed over 10 years if shielded from ultraviolet (UV) radiation. The absorption and emission spectra of the Lumogen F dyes embedded in PMMA are shown in Figure 9.4. It is evident that by using a mixture of the dyes it is possible to absorb all solar radiation from 350 nm up to 620 nm. In addition, their absorption and emission spectra exhibit good overlap between them as shown in Figure 9.5 and will be good candidates for efficient energy transfer.

9.3.4 Estimation of Quantum Yield

The fluorescence spectra and intensities of fluorescent samples are dependent on the optical density of the samples and the geometry of fluorescence detection.[34] Of particular importance is fluorescence quenching at high dye concentrations, and the fluorescence emitted by the collectors provides a good vehicle how this quenching can be quantified. A simple method is to observe the fluorescence intensities at the red end of the spectrum where we can ensure that the re-absorption is negligible.[42] At that wavelength range the fluorescence intensity is proportional to concentration except for any excitation energy loss due to quenching. Thus, a plot of the fluorescence intensity per absorption efficiency as a function of concentration should be constant, with any deviation observed due to quenching effects. Examples of these plots are shown in Figure 9.6.

Figure 9.6 shows fluorescence quenching plots for Violet (V570), Orange (O240) and Rhodamine 101 dyes. In the case of Rhodamine 101 there is a rapid drop in the fluorescence efficiency for concentrations higher than 7×10^{-4} M. For the BASF dyes V570 and O240, the plots show a decrease in fluorescence efficiency when the concentration reaches around 1000 parts per million (ppm). The decrease in fluorescence efficiency occurs because, as the dye molecules come closer together with increased concentration, they

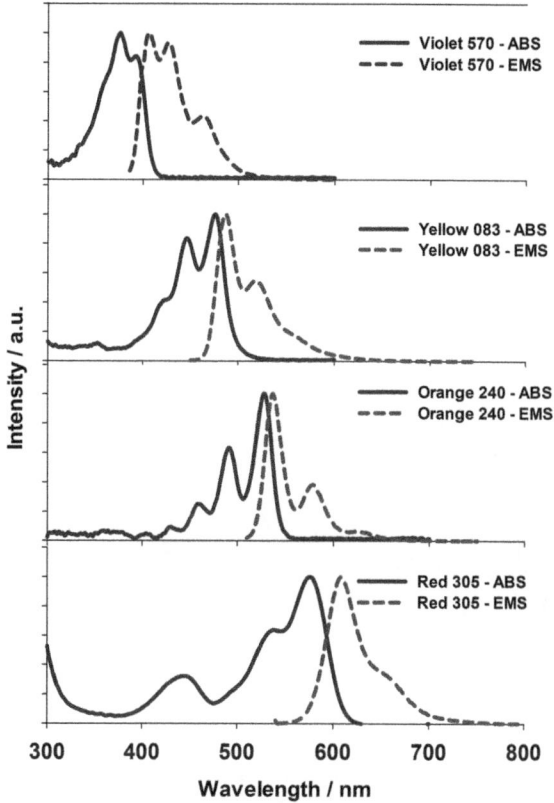

Figure 9.4 Normalised absorption and emission spectra of selected BASF Lumogen organic dyes embedded in PMMA.

can transfer their excitation energy to each other which greatly increase non-radiative relaxation losses. In a recent study the Lumogen F dyes have shown to have very high fluorescent yields (near 100%) in plastic plate luminescent collectors and with no significant fluorescence concentration quenching, even at concentrations as high as 1000 ppm.[41] In some instances, however, thin film collectors (fabricated, for example, by spin coating) require high dye doping in order to achieve high absorbance values which inevitably results in fluorescence quenching.

Increasing further the concentration of the dye in the polymer film can result in the presence of aggregates. The fluorescence spectra of the yellow dye (Y083) at high concentration show a significant change on their spectral shape which is indicative of the formation of aggregates (Figure 9.7). This is typical behaviour of excimer (excited state dimer) formation observed in perylene dyes.[43] The emission spectrum of the Y083 dye changes drastically above 800 ppm signifying the onset of aggregate formation in the layer. Excimer emission is red-shifted with respect to the absorption spectrum of

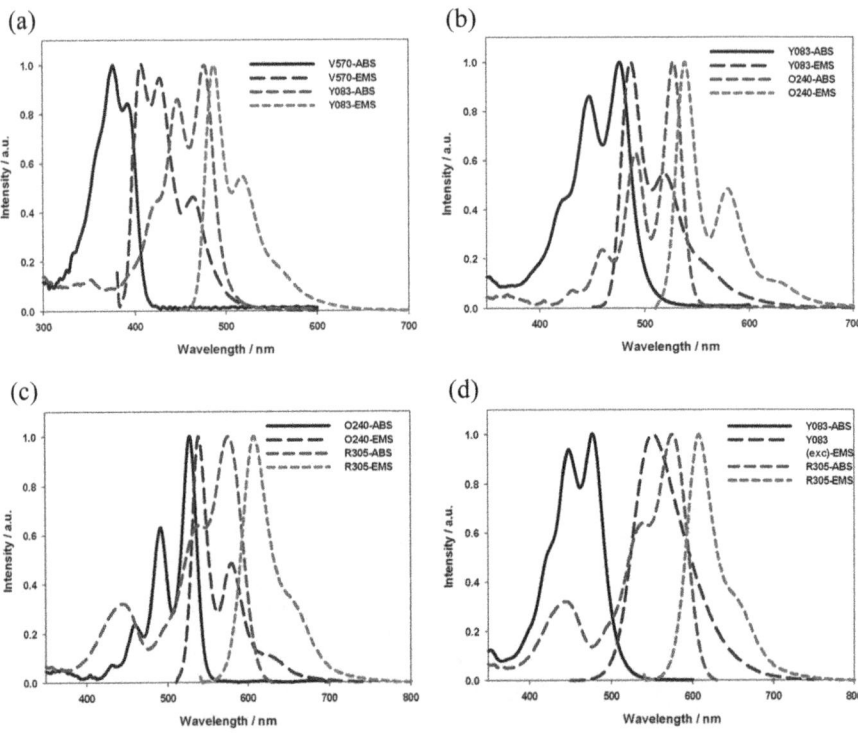

Figure 9.5 Normalised absorption and emission spectra of BASF dyes (donor : acceptor) spin coated PMMA films: (a) V570 and Y083; (b) Y083 and O240; (c) O240 and R305; and (d) Y083 (excimer emission) and R305.

the dye leading to a large Stokes shift and hence a decrease in re-absorption. Excimers can have high fluorescence quantum yields, which makes them potentially a useful candidate for application in fluorescent collectors.[44]

9.3.5 Examples of Energy Transfer for Efficient Photon Management

The good overlap of the absorption and emission spectra of the BASF dyes shown in Figure 9.5 forecasts efficient energy transfer in a mixture of these dyes. This will lead to an increase of the absorption efficiency in the dye layer and, at the same time, the emission wavelength can be shifted towards the red end of the spectrum, helping improve the operation of fluorescent collectors and down-shifting.

Examples of combined absorption and fluorescence spectra for different BASF dye doped PMMA mixtures fabricated by spin coating on glass substrates are shown in Figure 9.8. These mixtures of dyes can be used in fluorescent or down-shifting collectors. The normalised absorption and

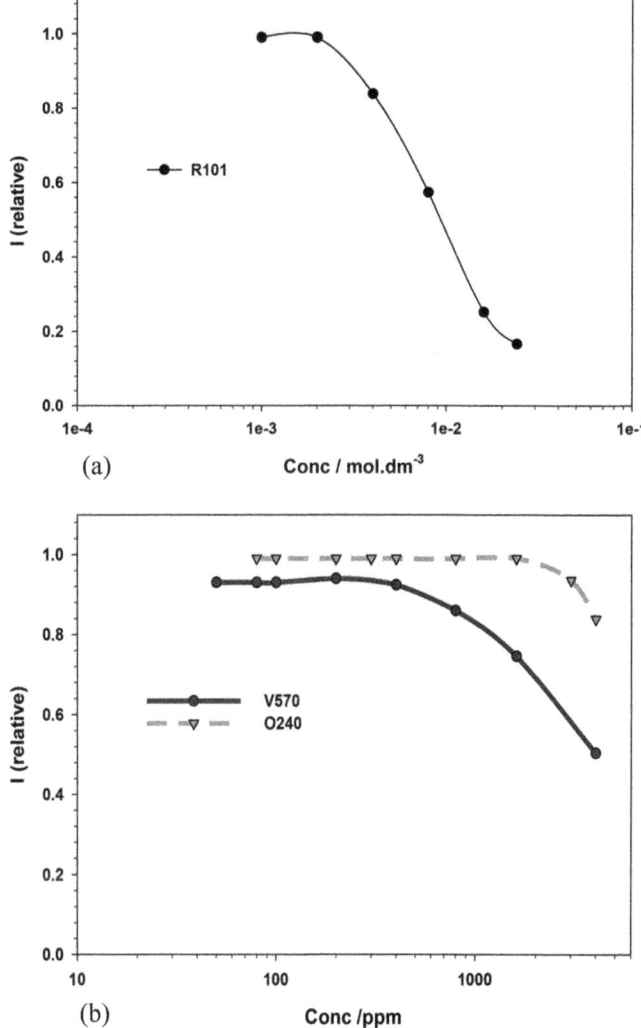

Figure 9.6 Semi-log plots of the dependence of fluorescence efficiency on the concentration of the dye in the fluorescent layer for: (a) Rhodamine 101; and for (b) Violet-V570 and Yellow-Y083.

emission spectra show efficient energy transfer between two and three dye systems, and the ability to shift to the red end of the spectrum the absorbed energy. Nearly all emission in these dye mixtures occurs from the acceptor dye in the layer with estimated energy transfer efficiency in the order of 60–80%. In collectors fabricated by spin coating, the dye concentrations are kept high in order to achieve high absorbance. Thus, the fluorescence efficiency is reduced somewhat due to quenching and possible aggregate formation. In thicker PMMA cast collectors the concentrations can be kept

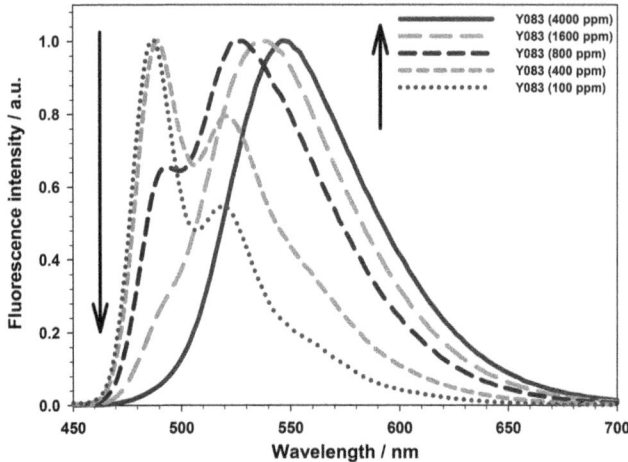

Figure 9.7   Fluorescence spectra for Yellow-Y083 dye in the fluorescent layer at
different dye concentrations (ppm). The monomer fluorescence
spectrum is reduced with increasing concentration (arrow pointing
down). The onset of the excimer emission is observed for dye layer
concentration above 800 ppm and the fluorescence spectrum is
shifted to the red and the fluorescence increases (arrow pointing up).

low, less than 1000 ppm, for mixture of dyes without sacrifice in fluorescence
efficiency.[41] Keeping the concentration low to avoid fluorescence quenching
but at the same time increasing the proximity of the molecules together to
achieve efficient energy transfer is a challenge in standard solution method
approaches and different methods are required.

In Figure 9.9 a different system is shown consisting of two LB dye monolayers which consist of donor and acceptor molecules. The energy transfer
efficiency of the dye monolayer, measured by time-resolved fluorescence,
gives an efficiency of 80%.

Similar systems have been developed during the past years using LB films[45]
and dye-loaded zeolites[10] and are finding their way in applications to
collectors. Figure 9.10(a) shows a cylindrical nanochannel structure in
zeolites which can accommodate individual dyes and create an artificial
photonic system with efficient energy transfer between the dyes. Calzaferri[7,10]
has pioneered the research field in artificial light harvesting by developing
hierarchically organised structures based on one-dimensional channel
materials such as zeolites and mesoporous silicas. His research has
produced artificially photonic antenna systems which can be used as
building blocks for solar energy conversion devices such as fluorescent
collectors.[46]

Figure 9.10(b) illustrates a schematic based on the arrangement of dye
molecules (D-donor) in J-aggregates packed in a brickstone work arrangement using LB films. The optical properties of J-aggregates are dramatically

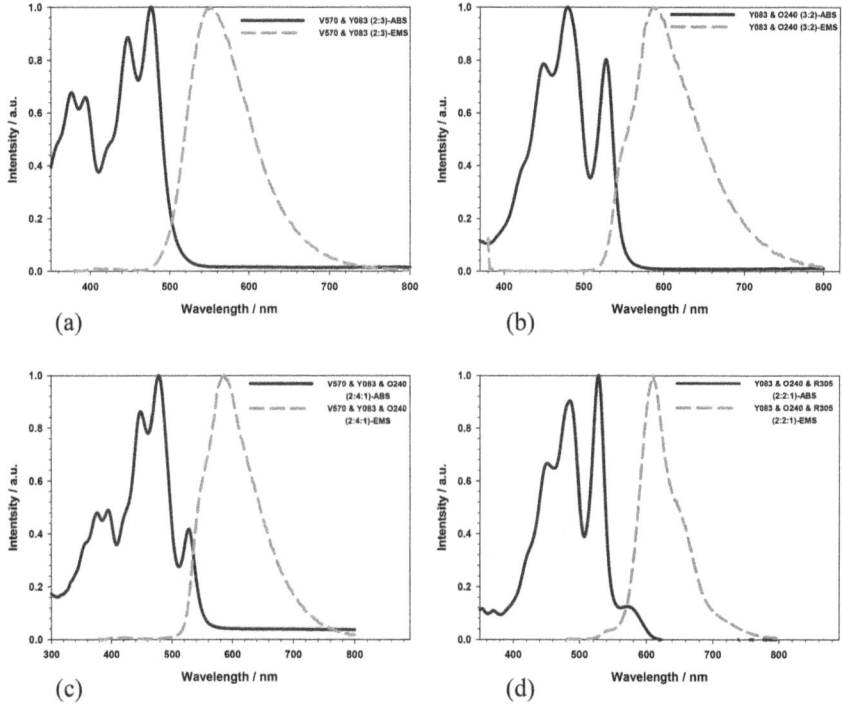

Figure 9.8   (a) Normalised absorption and emission spectra of different mixture of dyes with efficient energy transfer at different donor : acceptor ratios deposited as a thin polymer film *via* spin coating: (a) V570 and Y083; (b) Y083 and O240; (c) V570, Y083 and O240; and (d) Y083, O240 and R305. The excitation wavelength was 370 nm in (a), (b) and (c) and 440 nm in (d).

different from those of their individual molecules. Because the molecules in the J-aggregate are tightly packed, their oscillator strengths are strongly coupled and as a result coherent excitons are created within the monolayer. An acceptor dye (A-acceptor) can be incorporated in the aggregate monolayer which can act as an energy acceptor. J-aggregates can mimic light harvesting arrays[45] and appear to manifest efficient quenching in acceptor : donor mixing ratios as high as (1 : 10,000).[47]

9.4   Luminescent Solar Collectors

9.4.1   Introduction

Luminescence solar collectors (LSCs) or concentrators were first introduced in the late 1970s (see, for example ref. 16–20). Following intense research activity in the 1980s, the area has received renewed interest in the past decade or so due to the availability of new materials and the advent of photonics, leading to more optimistic theoretical predictions.[48,49]

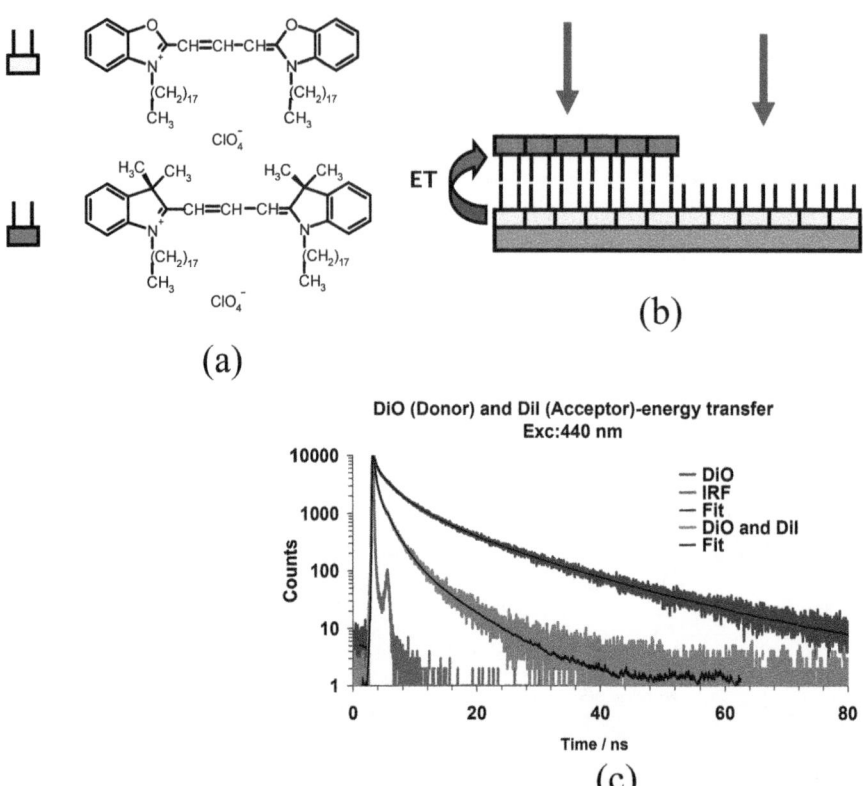

Figure 9.9   (a) Chemical structure of donor (DiO) and acceptor (DiI) carbocyanine dyes used for (b) energy transfer between dye monolayers. (c) Fluorescence decay curves for the DiO in the absence and presence of the acceptor dye (DiI) showing the significant shortening of the decay curve. The decay curve has been fitted with a multi-exponential and the energy transfer was calculated to be 80%. The excitation wavelength was 440 nm.

A fluorescent solar collector usually consists of a flat plate, doped with a luminescent species, which absorbs the incident sunlight (direct or diffuse). A large fraction of the emitted light is then trapped within the collector by total internal reflection (TIR) and is directed to a solar cell at the edge of the collector where the remained edges of the collector can be covered by mirrors. A schematic of the operation of the luminescent solar collector is shown in Figure 9.11.

An increase in the photon flux reaching the solar cell is achieved by virtue of the large area difference between the front face of the collector and the edge area covered by solar cells. The ultimate aim is to produce a sizeable concentration gain ratio A_{ent} to A_{exit}, where A_{ent} and A_{exit} are the areas of the top and edge surfaces, respectively (Figure 9.12).

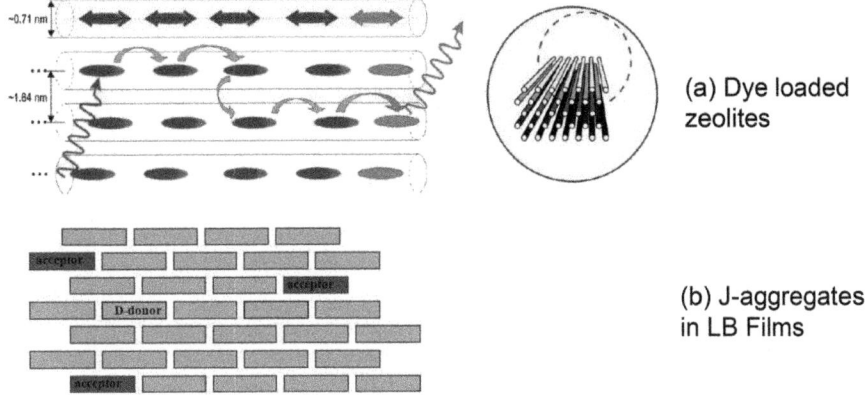

(a) Dye loaded zeolites

(b) J-aggregates in LB Films

Figure 9.10   (a) Schematic overview of an artificial photonic antenna system. The image on the left shows the chromophores being embedded in the channels of the host material (zeolite). The dyes act as donor molecules that absorb the incoming light and transport the electronic excitation energy *via* resonance energy transfer to the acceptors shown at the ends of the channels on the right. The process can be analysed by measuring the emission of the acceptors and comparing it with that of the donors. The double arrows indicate the orientation of the electronic transition dipole moment (ETDM). The image on the right shows a bunch of such strictly parallel channels: a schematic view of some channels in a hexagonal zeolite crystal with cylindrical morphology. Reproduced from ref. 10 with permission from the European Society for Photobiology, the European Photochemistry Association, and the Royal Society of Chemistry. (b) Schematic of brickstone arrangement of dyes to form an aggregate. Reproduced with permission from ref. 45.

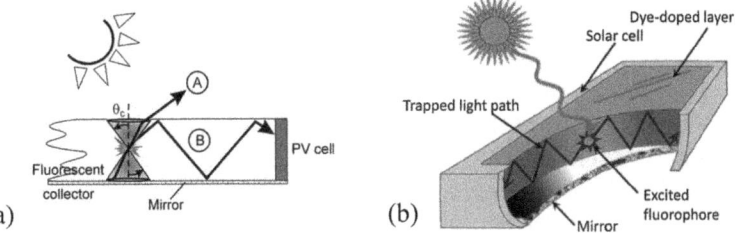

Figure 9.11   (a) Schematic diagram of fluorescent collector, A and B. llustrate the path length of rays emitted with an angle $\theta \leq \theta_c$ and $\theta > \theta_c$ respectively. Reprinted with permission from ref. 28, Copyright © Swiss Chemical Society: CHIMIA. (b) Perspective view of a fluorescent solar collector. Reprinted with permission from ref. 24, Copyright 2009, AIP Publishing LLC.

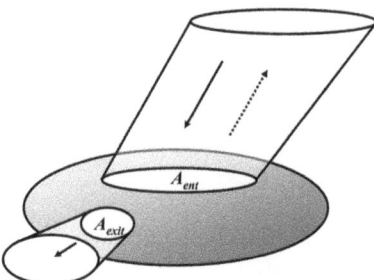

Figure 9.12 Schematic diagram of a generic fluorescent collector defining the areas A_{ent} and A_{exit}. Reprinted with permission from ref. 49, Copyright 2006, AIP Publishing LLC.

The collector may be composed of a transparent matrix such as PMMA,[50] glass[51] or liquid[27] which is doped with a mixture of dyes,[35] quantum dots[52] or rare earth ions.[53] The choice of the luminescent species can be modified to suit the bandgap of the solar cell. For efficient operation, the collector has to absorb a substantial part of the incident light and this usually necessitates the use of several different dyes. Efficient absorption, however, implies losses through re-absorption of the emitted light and a careful understanding of the re-absorption losses in the collector is therefore paramount.[44]

9.4.2 Spectroscopic Characterisation of LSCs

In this section we introduce a simple characterisation technique based on the absorption and emission spectra of fluorescent collectors used for a spectral-based analysis of the performance of the collectors using the two-flux model outlined in Section 9.2.3. The analysis is based on careful measurements of the edge fluorescence of the collector and comparing that with the 'first' generation fluorescence spectra obtained from samples with no re-absorption at low dye concentrations. The re-absorption loss that occurs from a partial overlap of the absorption and emission bands is evaluated by scaling the measured edge fluorescence to the first generation fluorescence.

The photon flux emitted from the edge of the collector is usually observed with a fibre optic or an integrating sphere. These measurements, however, imply photon collection from a limited angular range (restricted to the edge escape cone) and observation of a different photon flux than the one received by the solar cell with perfect optical coupling to the edge, as described by the Weber and Lambe theory[16] (see Section 9.2.2).

The experimental spectral re-absorption probability of different dye concentration collector samples obtained in Section 9.2 are compared with Weber and Lambe's theory and the modified Weber and Lambe's theory in Figure 9.13(a), where the first generation fluorescence spectrum was normalised to the edge fluorescence of BASF Red 305 collector samples in the

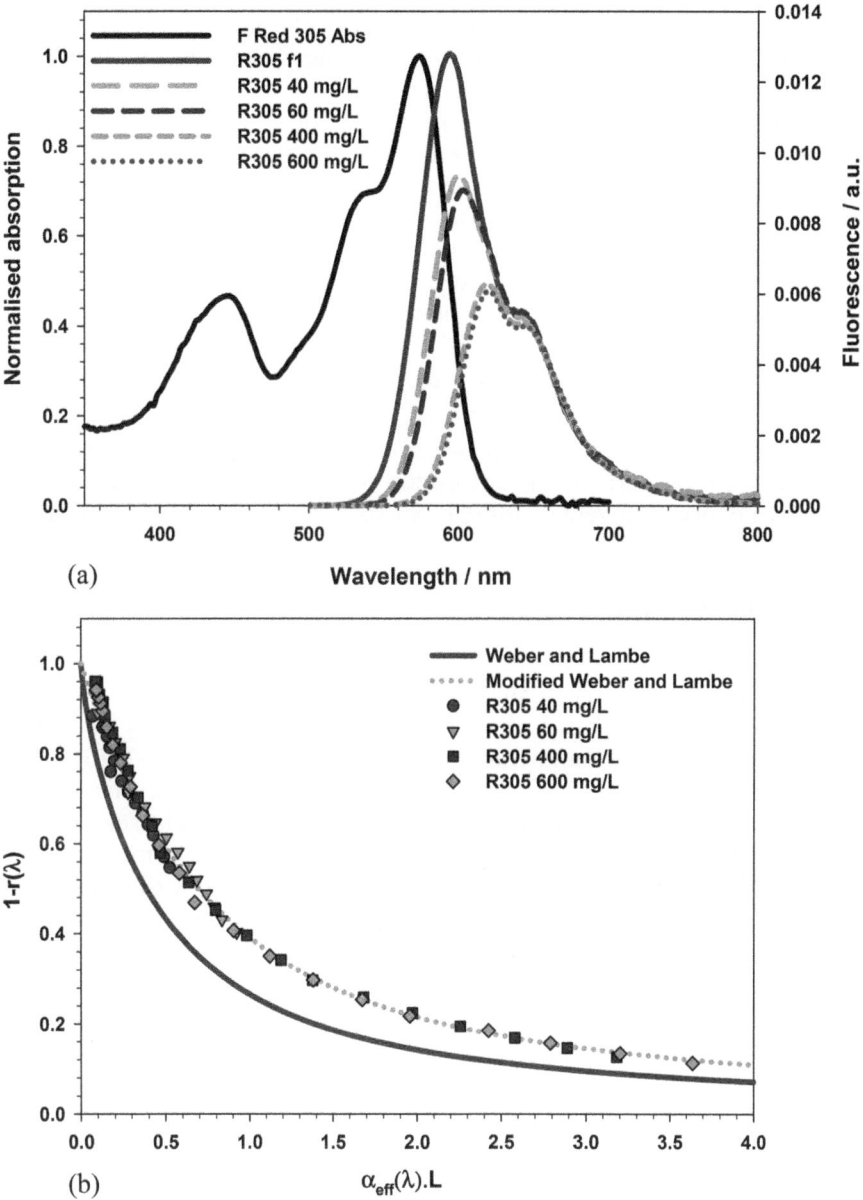

Figure 9.13 (a) Normalised absorption, first generation fluorescence (f1) and fitted edge fluorescence spectra. (b) Experimental re-absorption probability of different dye concentration collector samples compared with theory. Reprinted with permission from ref. 31, Copyright 2012, AIP Publishing LLC.

wavelength range from 650 to 850 nm, where the absorbance of all the collector samples was negligible. The probability of re-absorption $r(\lambda)$ for each collector was estimated from the ratio of the two scaled spectra. Figure 9.13(b) shows that the experimental re-absorption probability fits well with the modified Weber and Lambe's theory.

Through this analysis, it is possible to estimate the performance of any fluorescent collector based on knowledge of its absorption and emission spectra. The analysis is not restricted to single dye-doped collectors. This simple model estimates the re-absorption probability within the collector and can set an upper limit to the achievable collection efficiency Q_C assuming no other losses are present except re-absorption and escape cone losses. The effective absorption coefficient, introduced in ref. 28 and defined by:

$$\alpha_{\text{eff}} = \alpha \frac{l_\alpha}{l_\alpha + l_s} \qquad (9.21)$$

where l_α and l_s are the thicknesses of the absorbing layer and substrate glass, respectively, and α is the absorption coefficient of the absorbing layer, provides a convenient vehicle for comparison with uniform block collectors of the same absorbance.

9.4.3 LSC Examples

The characterisation techniques and principal results can be conveniently illustrated on the example of single-dye collectors. Studies carried out so far on the effect of mirrors on the fluorescent collectors indicate that a small air gap between the mirror and the edge of the collector is needed for better collector efficiency[39] and any attempt on optical coupling disturbs the TIR structure and limits the efficiency. The effect of re-absorption losses also puts a limit on the concentration gain factor, which cannot assume very high values, and so most collectors to date have been fabricated with gain factors up to 50, and is much lower than the high gain ratios near the thousands initially reported.[29] This restriction applies only to TIR configurations and is lifted when a photonic band stop is used together with a near unity fluorescence efficiency dye[49] as seen in Section 9.6. Also, differences between a fluorescent solar concentrator surrounded by four edge mounted solar cells *versus* a single solar cell configuration have shown that the re-absorption probability of trapped photons in a four solar cell configuration is improved.[54]

Despite the improvements of fluorescent collectors over the past 30 years, the overall experimental power conversion efficiencies for fabricated devices (collector and solar cell) remain well below 10% (Table 9.1). Higher efficiencies of fluorescent collectors using GaAs solar cells (Table 9.2) are reported due to a better match of the dye emission spectra to the GaAs bandgap.

The highest power conversion efficiency for a single-dye collector for silicon solar cells reported is 2.4%, increasing to 2.7% (optical efficiency of 14.5%) if combined with a second dye. Recently an efficiency of 2.8% has been reported with CdSe quantum dots.[52] Using the same two dye mixture,

Table 9.1 Efficiencies of fluorescent collectors coupled to a c-Si cell, as reported in the literature for different collector gain; the highest efficiency for a single plate collector is about 3%

Year	Gain	Number of dyes	Efficiency/%	Comments/dyes	References
1981	133	1	1.1		75
1983	68	1	1.3	DCM	19
1983	92	1	0.9		19
1985	62	1	2.3		76
1981	28	1	1		77
1983	23	2	1.9		78
1984	20	1	1.4		29
2008	12.5	1	2.4	R305	39
2008	12.5	2	2.7	R305 and CRS040	39
2009	20	2	1.9	R305 and DCM	27
2011	12.3	1	2.8	CdSe	52
2012	5	1	2.9	R305	79
2012	2.5	2	4.2	Two stage, R305 and perylene perinone	79

Table 9.2 Efficiencies of fluorescent collectors coupled to GaAs cells for different collector gain, as reported in the literature

Year	Gain	Number of dyes	Efficiency/%	Dyes	References
1984	35	1	2.5		29
1985	33	1	2.6		76
1985	33	1	4		76
1984	17	1	4		29
2008	10	2	4.6	R305 and CRS040	38
2008	2.5	2	7.1	R305 and CRS040	38

the power conversion efficiency using a single GaAs solar cell increases to 4.6% and increases further to 7.1% when used with four GaAs solar cells connected parallel, albeit with a reduced collector gain. The best results so far have been obtained with collectors gain in the region of 10–15 due to severe restrictions posed by re-absorption.

The possibility of using many luminescent species together to extend the absorption range in the collector was proposed originally by Zewail and co-workers.[19,20] Making use of Förster resonance transfer (Section 9.3) the light absorption is maximised while the acceptor concentration is kept to a minimum to reduce re-absorption. This approach has been employed successfully in LB films[45] and dye-loaded zeolite channels[55] (Section 9.3), which can lead to collectors absorbing a substantial part of the incoming radiation spectrum (high Q_A) and an emission close to the bandgap of the solar cell (high Q_C). The spectral characteristics of such a collector are shown schematically in Figure 9.14(a), where ΔE denotes the width of the photon

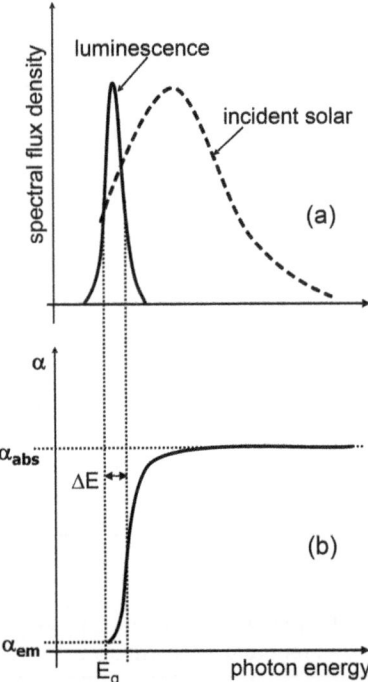

Figure 9.14   (a) Schematic diagram of the incident and emitted fluxes and (b) the absorption spectrum for an optimum fluorescent collector with light trapping by TIR. Reprinted with permission from ref. 31, Copyright 2012, AIP Publishing LLC.

transport channel in energy units. Broad absorption of the incident light is achieved by the use of an appropriate mixture of dyes. For optimal absorption of the solar radiation, the emission region should be spectrally narrow and close to the semiconductor wavelength bandgap. At the same time, it is important to ensure that this emission region is absorption-free as shown in Figure 9.14(b). The absorption coefficients for absorption and emission should therefore satisfy:

$$\alpha_{\text{abs}} \geq 1/d, \ \alpha_{\text{em}} \ll 1/L \tag{9.22}$$

where d and L are the thickness and length of the collector, respectively.

Recently, an LSC scheme has been proposed that mimics a four-level laser design, making use of Förster energy transfer and phosphorescence in thin film organic coatings on glass substrates.[56] Collector optical efficiencies near 50% were measured and device efficiencies of the order of 6.8% when used in tandem have been claimed but were not experimentally verified. Other multi-dye fluorescent collector studies included three dyes with a near-unit efficient energy transfer resulting in high optical efficiencies.[57] The above studies show that it is possible to create multi-dye mixtures in fluorescent collectors

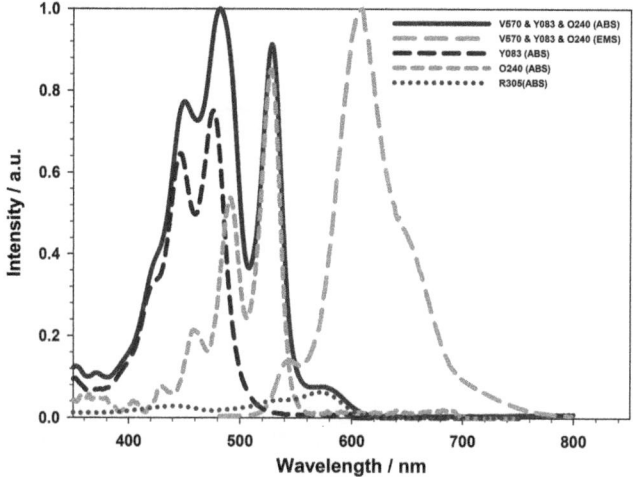

Figure 9.15 Normalised absorption and emission spectra for a three dye collector (Y083, O240 and R305) with donor : acceptor ratios (30 : 20 : 1). The individual dye absorption bands have been deconvoluted in the spectra.

that absorb a significant part of the incident solar spectrum and exhibit efficient photon management.

We have fabricated a three dye collector based on the BASF Lumogen F dyes yellow (Y083), orange (O240) and red (R305). The dyes were dissolved in PMMA and spin coated on glass substrates. In this mixture the Y083 and O240 act as the donor dyes and the R305 acts as the acceptor dye to which all the excitation energy is transferred and emission occurs. By increasing the donor to acceptor ratio we can reduce the re-absorption losses in the collector. An example of the absorption and emission characteristics of a collector is shown in Figure 9.15. The combined absorption spectrum has been deconvoluted to the absorption of the individual dyes. A significant portion of the incident photon flux is absorbed in the donor dyes (Y083 and O240) while the absorption of the acceptor dye (R305) remains low, which is necessary to reduce re-absorption losses. We expect fluorescent collectors with these spectral characteristic to be able to reach significant optical efficiencies.

9.5 Luminescence Down-Shifting (LDS)

9.5.1 Introduction

Luminescence down-shifting (LDS) of the incident solar spectrum was originally proposed by Hovel *et al.*[58] in order to overcome the low spectral response in the blue region of the solar spectrum in some types of solar cells.

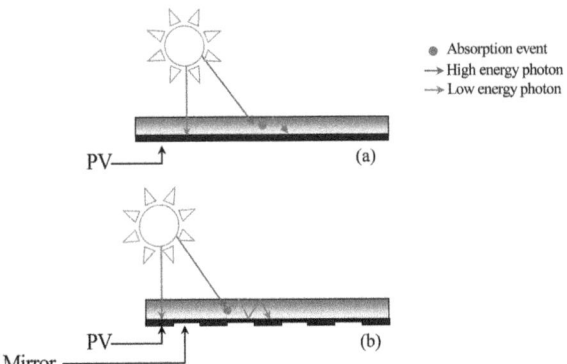

Figure 9.16   (a) A typical down-shifting structure for a CdTe solar cell. (b) A similar structure as a fluorescent concentrator. Reproduced from ref. 27.

Proposed during the same period as luminescent solar collectors, both technologies share many similarities; a down-shifting structure for CdTe solar cells Figure 9.16(a) can be compared with a similar structure acting as a fluorescent collector in Figure 9.16(b). The solar cells in Figure 9.16(b), which cover only a part of the rear of the collecting structure, receive light that travels along the light-guiding collector by virtue of their higher refractive index.[59] A recent review for luminescence down-shifting can be found in ref. 60.

Luminescent down-shifting, of course, is beneficial only if any losses are compensated by gains incurred through the frequency shift in the incoming light. The losses in the LDS layer are similar to those in the LSCs: (i) absorption, reflection and scattering losses in the matrix material; (ii) lower than unity fluorescence quantum yield of the luminescent species; (iii) escape cone losses from the top and the edge of the LDS layer; and (iv) re-absorption losses, although they play less significant role because of the shorter distances traversed by the emitted light.

There are two types of solar cells where luminescent down-shifting can be proven beneficial: solar cells with low quantum efficiencies at short wavelengths due to high minority carrier losses near the surface such as crystalline silicon (c-Si);[61] and gallium arsenide (GaAs)[60] and solar cells with a cut-off in the blue region due to absorption of the incoming light by the presence of window layers such as CdTe[14] and CIGS.[15] In principle, these shortcomings can usually be addressed by improvements in solar cell design, but it is frequently easier and cheaper to make use of LDS where light capture can be optimised independently of the solar cell fabrication process. Due to the availability of a wide range of dyes with high fluorescence efficiencies in the blue region of the spectrum, luminescence down-shifting offers the potential for improved light collection leading to increase of efficiencies in commercial solar cells.

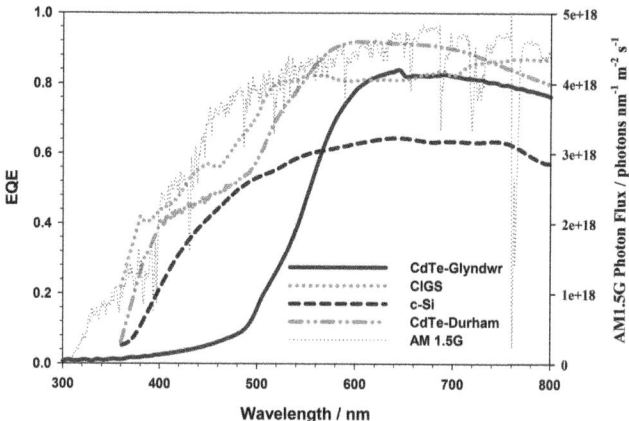

Figure 9.17   Typical EQEs of thin film solar cells fabricated by the PV21 Supergen consortium.

9.5.2   LDS Examples

Typical external quantum efficiency (EQE) curves for thin film solar cells fabricated in the PV21 consortium are shown in Figure 9.17 for three different platforms—amorphous silicon (a-Si), CdTe and CIGS—together with the AM1.5G solar spectrum photon flux. The short wavelength region (300–500 nm), which is of interest in the down-shifting process, shows a reduced spectral response.

From an inspection of the EQE curves at wavelengths below 500 nm it can be seen that the most promising improvements in relative efficiency increase could be achieved with CdTe solar cells. Recent down-shifting studies have also been carried out on c-Si cells but with only small increments observed in efficiencies.[61] The two CdTe EQE curves shown in Figure 9.17 are fabricated from different methods. One contains a thick (240 nm) CdS layer fabricated by metal organic chemical vapour deposition (MOCVD),[62] and the other one has a normal thickness CdS and the cell was fabricated *via* close space sublimation.[63] The thicker CdS layer was grown to demonstrate the feasibility of the LDS layer and also due to the similarities of the spectral response with commercial produced PV modules. The higher efficiency CdTe spectral response can be used to investigate potential efficiency improvements due to luminescence down-shifting in state-of-the-art laboratory CdTe solar cells.

Examples of absorption and emission spectra of LDS layers spin coated on glass containing one, two and three BASF Lumogen F dyes are shown in Figure 9.18. The mixing dye ratios were optimised in the layer for efficient energy transfer. The combination of two (V560 and Y083 or Y083 and O240) and three (V570, Y083 and O240) dyes together can increase the absorption range of the LDS layer and utilising efficient energy transfer between the

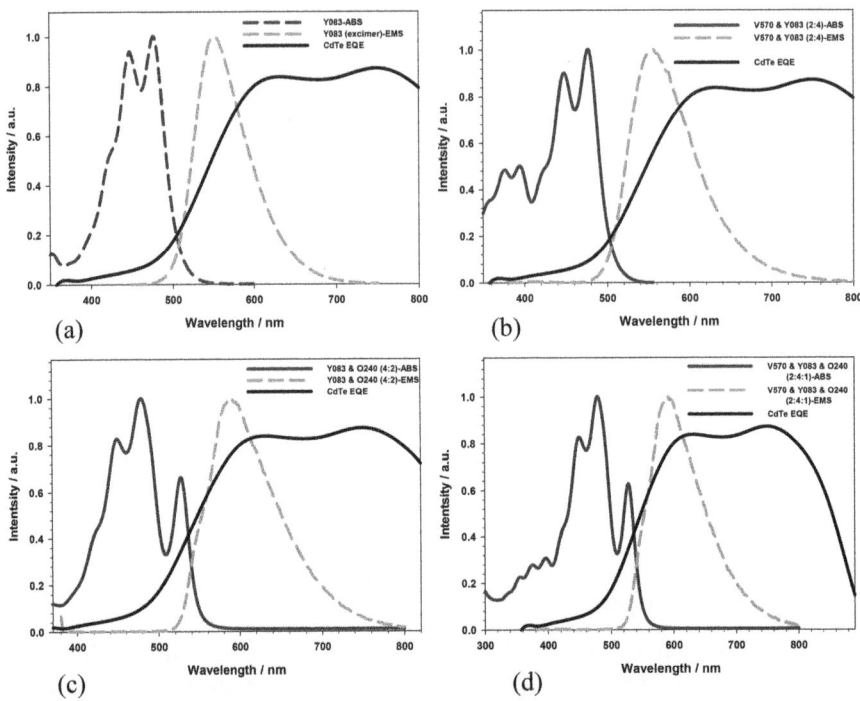

Figure 9.18 Normalised absorption and emission spectra of LDS dye mixtures containing: (a) one (Y083); (b) two (V570 and Y083); (c) (Y083 and O240); and (d) three (V570, Y083 and O240) dyes. A typical EQE curve of a CdTe solar is also shown.

dyes emit the absorbed light in a wavelength region where the EQE response of the solar cell is high. The two dye (V570 and Y083) and one dye (Y083) layers have the same emission profile, but the two dye layer increases the absorption range of the LDS layer into the UV region (350–400 nm). Using the O240 dye as the acceptor dye in a two dye (Y083 and O240) LDS layer, the emission can be shifted further into the red end of the spectrum shown in Figure 9.18(c). If we add the R305 dye, the effect of absorption of the dye becomes detrimental to the efficiency of the cell. Lastly, the LDS layer containing three dyes shown in Figure 9.18(d) absorb almost all the incident light below 550 nm and emission occurs in a region of the solar spectrum with improved spectral response; in this case the maximum of the emission peak is at 584 nm.

These LDS layers have been tested with CdTe solar cells to estimate the increase in EQE from the action of the LDS layer.[64] A blank PMMA layer of the same thickness as the dye-doped LDS layer was used to measure the reference EQE. An example of an improved spectral response of a CdTe solar cell with LDS dye layers on top when compared with a blank PMMA layer is

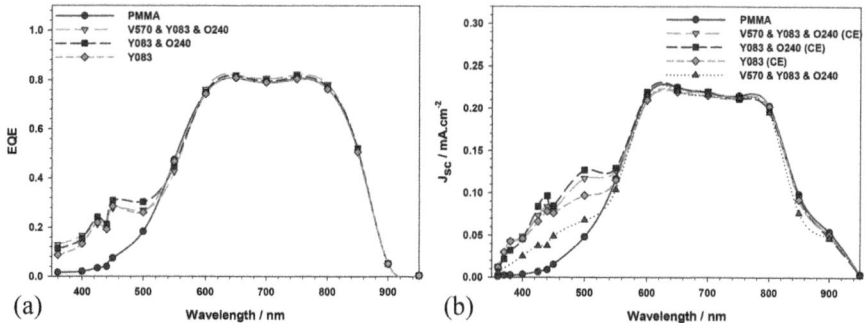

Figure 9.19 (a) EQE comparison of CdTe solar cells (Glyndŵr) with the application of fluorescent down-shifting structures. A single, two and three dye LDS layer was applied and the blank PMMA EQE curve is also shown for comparison. (b) Comparison of short circuit current density (J_{sc}) of a CdTe solar cell with and without concentration of light using fluorescent structures (CE: concentration enhancement).

shown in Figure 9.19(a). The external quantum efficiency (EQE) of the cell is improved in the blue region of the spectrum due to the action of the LDS layer. The EQE values are improved for wavelengths 400–500 nm mirroring the absorption spectrum of the dye in the LDS layer. By taking advantage of the wave guiding structure of Figure 9.16(b), a further increase in current output can be observed when the down-shifting structure is used as a fluorescent collector. The short-circuit current density output is shown in Figure 9.19(b) with and without the effect of concentration. Although further work is needed to optimise the LDS structures, the examples given here give an indication of the efficiency increase that can be achieved.

9.6 Advanced Photonic Concepts

Photon frequency management can benefit from recent advances in photonics and this section provides a brief outline of different ways to increase the efficiency of collectors beyond the TIR efficiency limit.

Spectrally selective filters have been proposed for application to luminescent solar collectors in order to reduce the light losses through the TIR escape cone. The filters can be fabricated from photonic crystals. A photonic crystal has a spatial periodic variation in its dielectric constant and prevents light of certain energy propagating in certain directions.[65] The top face of the collector can be covered by a photonic structure (a band stop) that reflects the fluorescent light (Figure 9.20) blocking much of the escaping light and reducing photon transport losses to a minimum. Simulations carried out by Goldschmidt and co-workers[66,67] have shown 20% increases in efficiency with band edge filters. An increase in edge emission with the use of photonic opal crystals was also reported by Knabe *et al.*[68] Debije *et al.*[69,70] observed up to 12% more light collected on the edge of the collector by using

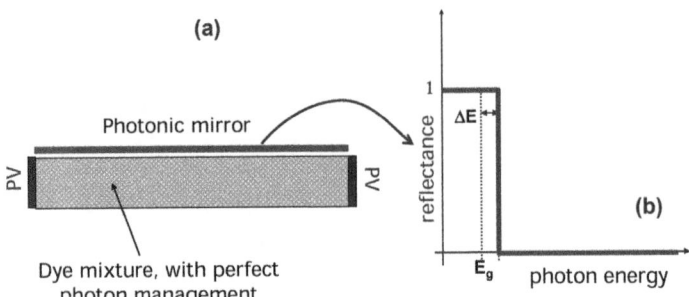

Figure 9.20 (a) Schematic of an advanced fluorescent collector where a photonic band stop covers the entrance aperture, with the reflectance profile shown in (b). Reproduced from ref. 5.

wavelength-selective mirrors consisting of chiral nematic (cholesteric) liquid crystals applied on the top of the collector using an air gap and when used in conjunction with a separate white scatterer layer on the back of the collector.

Recently researchers have started attempting to improve photon collection efficiency by restricting the molecular orientation of the dye inside the collector. For example, a planar aligned dye with its absorbing axis parallel to the edge of collection will have an increased light output and a respective reduction edge light output in a direction perpendicular to the absorption axis. Using liquid crystal hosts to hold common dye molecules such as Coumarin 6 and DCM[71] aligned, Debije *et al.*[70] carried out measurements under polarised light and showed improvements up to 30% in collected light with planar dye alignment with respect to no dye alignment (isotropic). Baldo and co-workers[71a] in a similar experimental set-up saw improvement in collection efficiencies of 23% with respect to isotropic dye alignment. These results clearly demonstrate the potential of increasing collection efficiencies in fluorescent collectors by controlling the orientation of the dye molecules. However, doubts remain about losses introduced by a reduction in absorption[72] and more work is clearly required to substantiate the potential for improvement using this approach.

Fluorescent collectors provide an elegant technique for exceeding the Shockley–Queisser limit by spectral splitting. Stacked collectors were introduced by Goetzberger and Greubel[17] in the 1970s (Figure 9.21). Each plate absorbs part of the solar spectrum and re-emits it onto a small solar cell. Theoretical conversion efficiencies have been estimated to exceed 30%, but practical results remain well below this limit.

A radically new solution to enhance the photoexcitation of silicon by directing the illumination onto the edge of the solar cell by means of fluorescence energy collection has recently been proposed by our group.[73] The generic structure of the new device is shown in Figure 9.22(a). A fluorescent collector, adjacent to a thin crystalline silicon solar cell, captures all or part of

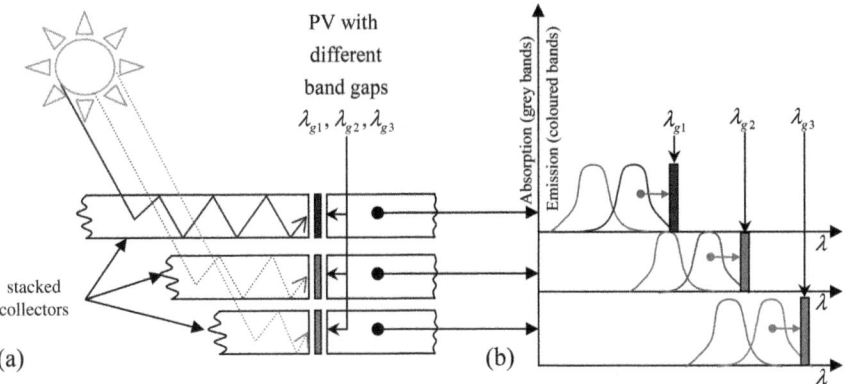

Figure 9.21   (a) Schematic of a cross-section of a fluorescent collector stack. (b) Example of absorption and emission spectra of the stacked layers. Reproduced from ref. 27.

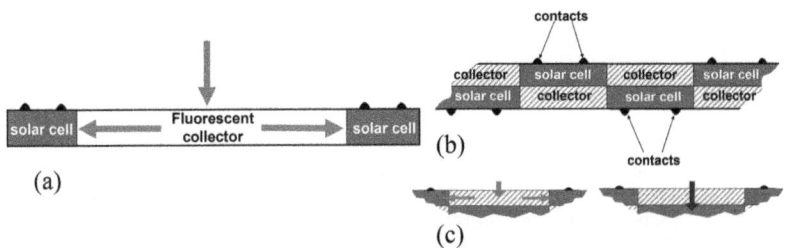

Figure 9.22   (a) Solar cell illuminated from the edge by a fluorescent collector. (b) An alternative 'checkerboard' arrangement, making use of strong silicon absorption at short wavelengths. (c) The absorption and re-emission paths of the red and blue parts of the spectrum. Reproduced from ref. 73.

the incident solar radiation and emits this energy in the form of fluorescence with frequency near the silicon bandgap. In an optimised structure shown in Figures 9.22(b) and 9.22(c), the collector absorbs only the red/near infrared part of the spectrum but the short wavelength radiation is absorbed directly by the thin silicon cell.

If attached to a 1 μm thick crystalline silicon solar cell with nominal 20% efficiency and illuminated from the edge at standard testing conditions, the edge illuminated ultra-thin silicon solar cells using fluorescent collectors can produce conversion efficiencies close to conventional c-Si solar cell but with greatly reduced material requirements as shown in Figure 9.23, depicting the quantum and total efficiencies. Higher efficiencies can still be achieved with thicker cells, different emission wavelengths or a different photon

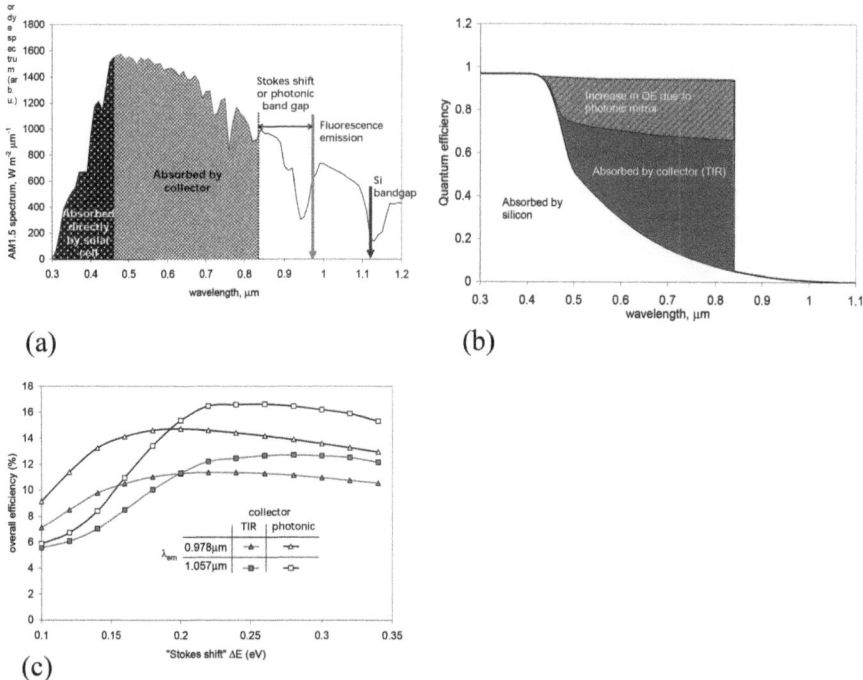

(a)   **(b)**

(c)

Figure 9.23 (a) The spectral management of device operation illustrating the absorption and fluorescence channels and a possible absorption spectrum of the dye. AM1.5 spectrum, W m^{-2} μm^{-1} or dye spectrum (arb. μ.). (b) Calculated quantum efficiencies, showing contributions due to photons absorbed in different parts of the device. (c) Overall efficiencies (cell + collector) for 1 μm c-Si solar cell operating with a collector based on a Nd^{+3} or Yb^{+3} emission channel. Reproduced from ref. 73.

management strategy.[7] In a practical setting, current developments in fluorescent collectors bear the promise of inexpensive practical devices with efficiencies in excess of 10%.

As suggested in Section 9.2.1, frequency management can be used not only to concentrate light but also to enhance light absorption by increasing the path length of light in the solar cell. Figure 9.24(b) shows a structure taking advantage of frequency shift to trap light inside a thin weakly absorbing c-Si solar cell, to be contrasted with a conventional light trapping schemes based on a textured rear surface as shown in Figure 9.24(a).[5,74] We have shown that this photonic scheme has the potential to increases the photon path length by a factor proportional to the Boltzmann factor of the frequency shift.[23] The photonic band stop now takes over as the principal bandgap that governs the operation of the solar cell. Somewhat surprisingly we find that a 1 μm thick c-Si solar cell with a photonic bandgap is not only highly effective in trapping light but can exceed slightly the efficiency of a standard c-Si solar cell![23]

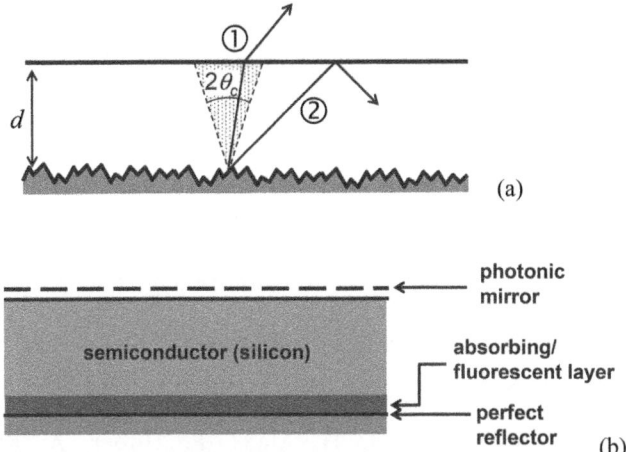

(a)

(b)

Figure 9.24   (a) Light trapping scheme with a textured rear surface, showing external rays ① within the 'escape cone' and trapped rays ②. (b) A photonic scheme where the absorbing/fluorescent layer at the back surface introduces a frequency shift. Reprinted with permission from ref. 23, Copyright 2011, AIP Publishing LLC.

9.7   Conclusions

In this chapter we have presented a unified overview of fluorescent collectors and down-shifting structures in application to solar cells. We have shown how different photon frequency management methods can be used to increase solar cell efficiencies and lower the cost significantly by reducing the size as well as the thickness of the solar cell. A simple two-flux model has been presented which can be used to describe the re-absorption losses and collection efficiency of the collector using as input only the absorption spectrum and edge fluorescence measurements. The application of photon frequency management materials that employ efficient Förster resonance energy transfer can be applied in the near term for increasing the efficiencies of commercial CdTe solar cells *via* down-shifting/light guiding or for the fabrication of fluorescent collectors with reasonable efficiencies approaching 10%. Looking further into the future, the employment of photonic structures with fluorescent collectors or into light trapping structures offers the promise for substantial increase in solar cell efficiencies even beyond the theoretical limits for single-junction solar cells with a significant lower material demand than for current conventional methods.

Acknowledgements

The authors would like to thank the Engineering and Physical Sciences Research Council for financial support through the PV21 Supergen consortium.

References

1. M. Peters, J. C. Goldschmidt and B. Bläsi, *Sol. Energy Mater. Sol. Cells,* 2010, **94**, 1393–1398.
2. P. Bharadwaj, R. Beams and L. Novotny, *Chem. Sci.,* 2011, **2**, 136.
3. W. L. Barnes, *Components,* 1998, **45**, 661–699.
4. H. A. Atwater and A. Polman, *Nat. Mater.,* 2010, **9**, 205–213.
5. T. Markvart, L. Danos, L. Fang, T. Parel and N. Soleimani, *RSC Adv.,* 2012, **2**, 3173.
6. D. L. Andrews and A. A. Demidov, *Resonance Energy Transfer,* John Wiley & Sons, Chichester, UK, 1999.
7. G. Calzaferri, *Top. Catal.,* 2009, **53**, 130–140.
8. D. Brühwiler, L.-Q. Dieu and G. Calzaferri, *Chim. Int. J. Chem.,* 2007, **61**, 820–822.
9. G. Calzaferri, M. Pauchard, H. Maas, S. Huber, A. Khatyr and T. Schaafsma, *J. Mater. Chem.,* 2002, **12**, 1–13.
10. G. Calzaferri and K. Lutkouskaya, *Photochem. Photobiol. Sci.,* 2008, **7**, 879–910.
11. G. Roberts, *Langmuir-Blodgett Films,* Plenum Press, New York, 1990.
12. H. Kuhn, D. Mobius and H. Bucher, in *Techniques of Chemistry,* ed. A. Weisberger and B. Rossiter, Wiley, New York, 1972, pp. 577–702.
13. D. L. Andrews, *Energy Harvesting Materials,* World Scientific Publishing, Singapore, 2005.
14. R. Kitamura and T. Maruyama, *Sol. Energy Mater. Sol. Cells,* 2001, **69**, 61–68.
15. G. Glaeser and U. Rau, *Thin Solid Films,* 2007, **515**, 5964–5967.
16. W. H. Weber and J. Lambe, *Appl. Opt.,* 1976, **15**, 2299–2300.
17. A. Goetzberger and W. Greubel, *Appl. Phys.,* 1977, **14**, 123–139.
18. J. A. Levitt and W. H. Weber, *Appl. Opt.,* 1977, **16**, 2684–2689.
19. J. S. Batchelder, A. H. Zewail and T. Cole, *Appl. Opt.,* 1981, **20**, 3733–3754.
20. J. S. Batchelder, A. H. Zewail and T. Cole, *Appl. Opt.,* 1979, **18**, 3090–3110.
21. M. G. Debije and P. P. C. Verbunt, *Adv. Energy Mater.,* 2012, **2**, 12–35.
22. T. Markvart, *J. Opt. A: Pure Appl. Opt.,* 2008, **10**, 015008.
23. T. Markvart, *Appl. Phys. Lett.,* 2011, **98**, 071107.
24. T. J. J. Meyer and T. Markvart, *J. Appl. Phys.,* 2009, **105**, 063110.
25. A. J. Chatten, D. Farrel, C. Jermyn, P. Thomas, B. F. Buxton, A. Buchtemann, R. Danz and K. W. W. J. Barnham, in *Proceedings of the 31st IEEE Photovoltaic. Specialists Conference,* Institute of Electrical and Electronic Engineers, New York, 2005, pp. 82–85.
26. T. J. J. Meyer, J. Hlavaty, L. Smith, E. R. Freniere and T. Markvart, in *Proceedings of SPIE,* ed. M. Osinski, B. Witzigmann, F. Henneberger and Y. Arakawa, SPIE, 2009, Vol. 7211, pp. 72110N–72110N-11.
27. T. J. J. Meyer, *Photon Transport in Fluorescent Solar Collectors,* PhD Thesis, University of Southampton, 2009.
28. P. Kittidachachan, L. Danos, T. J. J. Meyer, N. Alderman and T. Markvart, *Chim. Int. J. Chem.,* 2007, **61**, 780–786.

29. J. Roncali and F. Garnier, *Appl. Opt.*, 1984, **23**, 2809.
30. R. Bose, D. J. Farell, A. J. Chatten and M. Pravettoni, in *Proceedings of the 22nd European Photovoltaic Solar Energy Conference*, 2007, p. 210.
31. L. Fang, T. S. Parel, L. Danos and T. Markvart, *J. Appl. Phys.*, 2012, **111**, 076104.
32. T. Förster, *Ann. Phys.*, 1948, **2**, 55.
33. B. Valeur, *Molecular Fluorescence: Principles and Applications*, Wiley-VCH, Weinheim, Germany, 2001, vol. 8.
34. J. R. Lakowicz, *Principles of Fluorescence Spectroscopy*, Springer, New York, 3rd edn, 2006.
35. A. Zastrow, in *Proceedings of SPIE*, ed. V. Wittwer, C. G. Granqvist and C. M. Lampert, SPIE, 1994, vol. 2255, pp. 534–547.
36. G. Seybold and G. Wagenblast, *Dyes Pigm.*, 1989, **11**, 303–317.
37. BASF, *Lumogen® F. Collector Dyes*, Technical Information TI/P 3201, BASF Aktiengesellschaft, Ludwigshafen, Germany, 1997, available at http://www2.basf.us/additives/pdfs/p3201e.pdf [accessed 27 June, 2014].
38. L. H. Slooff, E. E. Bende, A. R. Burgers, T. Budel, M. Pravettoni, R. P. Kenny, E. D. Dunlop, A. Buchtemann and A. Büchtemann, *Phys. Status Solidi RRL*, 2008, **2**, 257–259.
39. W. G. J. H. M. van Sark, K. W. J. Barnham, L. H. Slooff, A. J. Chatten, A. Buchtemann, A. Meyer, S. J. McCormack, R. Koole, D. J. Farrell, R. Bose, E. E. Bende, A. R. Burgers, T. Budel, J. Quilitz, M. Kennedy, T. Meyer, C. D. M. Donegá, A. Meijerink, D. Vanmaekelbergh and A. Büchtemann, *Opt. Express*, 2008, **16**, 21773–21792.
40. B. C. Rowan, L. R. Wilson and B. S. B. S. Richards, *IEEE J. Sel. Top. Quantum Electron.*, 2008, **14**, 1312–1322.
41. L. R. Wilson and B. S. Richards, *Appl. Opt.*, 2009, **48**, 212–220.
42. J. B. Birks, *Photophysics of Aromatic Molecules*, John Wiley & Sons, London, 1970.
43. S. Akimoto, A. Ohmori and I. Yamazaki, *J. Phys. Chem. B*, 1997, **101**, 3753–3758.
44. R. W. Olson, R. F. Loring and M. D. Fayer, *Appl. Opt.*, 1981, **20**, 2934–2940.
45. H. Kuhn, *Colloids Surf., A*, 2000, **171**, 3–12.
46. R. Koeppe, O. Bossart, G. Calzaferri and N. Sariciftci, *Sol. Energy Mater. Sol. Cells*, 2007, **91**, 986–995.
47. T. Kobayashi, *J-Aggregates*, World Scientific Publishing, Singapore, 1996.
48. U. Rau, F. Einsele and G. C. Glaeser, *Appl. Phys. Lett.*, 2005, **87**, 171101.
49. T. Markvart, *J. Appl. Phys.*, 2006, **99**, 026101.
50. J. M. Drake, M. L. Lesiecki and J. Sansregret, *Appl. Opt.*, 1982, 212945.
51. L. Andrews, B. C. McCollum and A. Lempicki, *J. Lumin.*, 1981, **24–25**, 877–880.
52. J. Bomm, A. Büchtemann, A. J. Chatten, R. Bose, D. J. Farrell, N. L. A. Chan, Y. Xiao, L. H. Slooff, T. Meyer, A. Meyer, W. G. J. H. M. van Sark and R. Koole, *Sol. Energy Mater. Sol. Cells*, 2011, **95**, 2087–2094.

53. R. Reisfeld, *Inorg. Chim. Acta,* 1987, **140**, 345–350.
54. T. S. Parel, L. Danos and T. Markvart, in *Proceedings of the 2013 IEEE 39th Photovoltaic Specialists Conference*, Institute of Electrical and Electronic Engineers, New York, 2013, pp. 1761–1765.
55. A. Devaux and G. Calzaferri, *Int. J. Photoenergy,* 2009, **2009**, 1–10.
56. M. J. Currie, J. K. Mapel, T. D. Heidel, S. Goffri and M. A. Baldo, *Science,* 2008, **321**, 226–228.
57. S. Bailey, G. Lokey, M. Hanes, J. Shearer, J. Mclafferty, G. Beaumont, T. Baseler, J. Layhue, D. Broussard and Y. Zhang, *Sol. Energy Mater. Sol. Cells,* 2007, **91**, 67–75.
58. H. J. Hovel, R. T. Hodgson and J. M. Woodall, *Sol. Energy Mater.,* 1979, **2**, 19–29.
59. L. Danos, T. Parel, T. Markvart, V. Barrioz, W. S. M. Brooks and S. J. C. Irvine, *Sol. Energy Mater. Sol. Cells,* 2012, **98**, 486–490.
60. E. Klampaftis, D. Ross, K. R. McIntosh and B. S. Richards, *Sol. Energy Mater. Sol. Cells,* 2009, **93**, 1182–1194.
61. K. R. McIntosh, G. Lau, J. N. Cotsell, K. Hanton, D. L. Ba, F. Bettiol and B. S. Richards, *Prog. Photovoltaics,* 2009, **17**, 191–197.
62. S. Irvine, V. Barrioz, D. Lamb, E. Jones and R. L. Rowlands-Jones, *J. Cryst. Growth,* 2008, **310**, 5198–5203.
63. J. D. Major, Y. Y. Proskuryakov, K. Durose, G. Zoppi and I. Forbes, *Sol. Energy Mater. Sol. Cells,* 2010, **94**, 1107–1112.
64. L. Danos, T. Parel, T. Markvart, V. Barrioz, W. S. M. Brooks and S. J. C. Irvine, *Sol. Energy Mater. Sol. Cells,* 2012, **98**, 486–490.
65. J. D. Joannopoulos, S. G. Johnson, J. N. Winn and R. D. Meade, *Photonic Crystals: Molding the flow of light*, 2nd Edition, Princeton University Press, Princeton, NJ, 1995.
66. J. C. Goldschmidt, M. Peters, A. Bosch, H. Helmers, F. Dimroth, S. Glunz and G. Willeke, *Sol. Energy Mater. Sol. Cells,* 2009, **93**, 176–182.
67. S. Fischer, J. C. Goldschmidt, P. Löper, G. H. Bauer, R. Brüggemann, K. Krämer, D. Biner, M. Hermle and S. W. Glunz, *J. Appl. Phys.,* 2010, **108**, 044912.
68. S. Knabe, N. Soleimani, T. Markvart and G. H. Bauer, *Phys. Status Solidi RRL,* 2010, **4**, 118–120.
69. M. G. Debije, P. P. C. Verbunt, B. C. Rowan, B. S. Richards and T. L. Hoeks, *Appl. Opt.,* 2008, **47**, 6763–6768.
70. M. G. Debije, M.-P. Van, P. P. C. C. Verbunt, M. J. Kastelijn, R. H. L. van der Blom, D. J. Broer and C. W. M. Bastiaansen, *Appl. Opt.,* 2010, **49**, 745–751.
71. (a) C. L. Mulder, P. D. Reusswig, A. M. Velázquez, H. Kim, C. Rotschild and M. A. Baldo, *Opt. Express,* 2010, **18**, A79.
   (b) C. L. Mulder, P. D. Reusswig, A. P. Beyler, H. Kim, C. Rotschild and M. A. Baldo, *Opt. Express,* 2010, **18**, A91.
72. J. A. Delaire and K. Nakatani, *Chem. Rev.,* 2000, **100**, 1817–1846.
73. T. Markvart and L. Danos, in *Proceedings 5th World Conference on Photovoltaic Energy Conversion*, 2010, pp. 2692–2693.

74. E. Yablonovitch, *J. Opt. Soc. Am.,* 1982, **72**, 899–907.
75. A. Goetzberger, *Sol. Cells,* 1981, **4**, 3–23.
76. M. Sidrach de Cardona, M. Carrascosa, F. Meseguer, F. Cusso and F. Jaque, *Appl. Opt.,* 1985, **24**, 2028–2032.
77. K. Heidler, *Appl. Opt.,* 1981, **20**, 773–777.
78. A. H. Zewail and J. S. Batchelder, in *Polymers in Solar Energy Utilization,* ed. C. G. Gebelein, D. J. Williams and R. D. Deanin, American Chemical Society, 1983, pp. 331–352.
79. L. Desmet, A. J. M. Ras, D. K. G. de Boer and M. G. Debije, *Opt. Lett.,* 2012, **37**, 3087–3089.

Subject Index